JOHN TRESCH

The Reason for the Darkness of the Night

John Tresch teaches the history of science and art at the Warburg Institute in London. He previously taught in Philadelphia at the University of Pennsylvania and has lived and worked in Chicago, Paris, and New York. He is the author of *The Romantic Machine: Utopian Science and Technology After Napoleon*, which won the 2013 Pfizer Award for an outstanding book from the History of Science Society.

ALSO BY JOHN TRESCH

*The Romantic Machine: Utopian Science
and Technology After Napoleon*

The Reason
for the Darkness
of the Night

and the Forging of
American Science

JOHN TRESCH

PICADOR ❖ FARRAR, STRAUS AND GIROUX ❖ NEW YORK

Picador
120 Broadway, New York 10271

Illustration credits can be found on pages 433–434.

The Library of Congress has cataloged the Farrar, Straus and Giroux
hardcover edition as follows:

Names: Tresch, John, 1972– author.
Title: The reason for the darkness of the night : Edgar Allan Poe and the forging of
American science / John Tresch.
Description: First edition. | New York : Farrar, Straus and Giroux, 2021. | Includes
bibliographical references and index.
Identifiers: LCCN 2020058350 | ISBN 9780374247850 (hardcover)
Subjects: LCSH: Poe, Edgar Allan, 1809–1849. | Poe, Edgar Allan, 1809–1849—
Knowledge and learning. | Science—United States—History—19th century. |
Authors, American—19th century—Biography.
Classification: LCC PS2631 .T74 2021 | DDC 818/.309 [B]—dc23
LC record available at https://lccn.loc.gov/2020058350

Paperback ISBN: 978-1-250-84940-3

Designed by Gretchen Achilles

Our books may be purchased in bulk for promotional, educational,
or business use. Please contact your local bookseller or the Macmillan Corporate
and Premium Sales Department at 1-800-221-7945, extension 5442, or
by email at MacmillanSpecialMarkets@macmillan.com.

1 3 5 7 9 10 8 6 4 2

"Tresch is not the first to comment on Poe's many connections with science, but his new biography succeeds in giving the full view, combining an ample tour through the major scientific ideas of the early nineteenth century with deft and frequently profound readings of Poe's large body of work . . . The book is valuable reading for anyone drawn to the strange beauty within and just beyond science, and who wonders, as Poe did, what role a poet has to play."

—James Dinneen, *Undark*

"Richly engaging . . . [Tresch] succeeds in placing Poe's life story and the development of his art in the context of the maturation of science . . . A moving, insightful biography."

—Robert J. Scholnick, *Poe Studies*

"Tresch has produced a steady, clever, engaging literary biography that provides an excellent survey of an overlooked aspect of Poe's writing."

—Bob Blaisdell, *The Christian Science Monitor*

"An admiring, intellectually stimulating portrait . . . Edgar Allan Poe's status as a major American literary figure is long established. But John Tresch's *Reason for the Darkness of the Night* now should encourage the curious and open-minded to devote fresh attention to the artist's contributions to the advancement of scientific thought."

—Harvey Freedenberg, *Shelf Awareness*

"One of the many compelling aspects of this book is its treatment of the first half of the nineteenth century as a distant mirror of our own times . . . It's taken a long time to set things right. This book helps to do that. A reader comes away not only with admiration for Poe but affection, too."

—David Brown, *The American Scholar*

"[Tresch] situates [Poe] in a maelstrom of competing tides, as a new class of engineers and experimentalists splashed up against philosophers, theologians, and cranks . . . Tresch suggests [Poe's writings] can be seen as endorsements of a synthesis of science and Romanticism."

—Daniel Engber, *The Atlantic*

"Tresch sheds light on Edgar Allan Poe's engagement with science in this intriguing biography . . . While [he] addresses the common impression of Poe as a 'morbid dreamer' and a penniless writer, he takes things further by offering a nimble account of the emerging science of Poe's day. Fans of Poe's work—and science enthusiasts—will appreciate Tresch's fresh angle."

—*Publishers Weekly*

"[Balances] insightful discussions of Poe's literary works alongside his intriguing scientific pursuits. A surprising side of Poe splendidly revealed."

—*Kirkus Reviews* (starred review)

"This biography is a masterwork on a master, and one of my favorite reads in years. John Tresch lets the reader see how Poe's imagination was not only wild but also procedural, not only unbounded but also formal. In these pages, we meet the engineer of horror, the trickster of reason, and the mutinous captain of mystery." —Rivka Galchen, author of

Everyone Knows Your Mother Is a Witch and *American Innovations*

"In this original and moving biography, John Tresch brilliantly integrates Poe the man and the writer with his deep engagement with the sciences of his day. He lucidly reveals the content of these fascinations, and convincingly illuminates their powerful influence on Poe's writings and thought. Tresch discerns Poe's privileging of mind and mystery over frigid empiricism and his terrifying perception that only spiritual darkness lay beneath the surface of the material knowledge that was transforming the world. A transfixing and eye-opening portrait."

—Daniel J. Kevles, professor emeritus of history
at Yale University and author of
In the Name of Eugenics and *The Physicists*

"At last, a biography of Poe that places him in the thick of the philosophical and scientific investigations of his time, reclaiming the rigor of his thought, the inventiveness of his writings, and a personality as visionary as it is audacious. John Tresch gives us the Poe whose deepest preoccupations became a threat to the smug morality, highfalutin cant, and 'doggerel aesthetics' of his contemporaries. An astonishing feat of research and beautiful prose, this book gives Poe what he has long deserved."

—Colin Dayan, author of *In the Belly of Her Ghost* and *Animal Quintet*

"This marvelous time machine transports us deep into the nineteenth century and the company of one of its most fascinating citizens. Poetry, mesmerism, the rise of American science—it's all here, as vivid as can be."

—Fred Turner, Harry and Norman Chandler
Professor of Communication at Stanford University
and author of *The Democratic Surround*

"John Tresch's *The Reason for the Darkness of the Night* is a lively and learned investigation of the life of Poe through his literary engagement with science—an accessible and exciting exploration."

—Richard Kopley, author of *Edgar Allan Poe and the Dupin Mysteries*

"John Tresch's engaging biography of an enigmatic genius puts Edgar Allan Poe right where he belongs: present at the creation of American science."

—Richard R. John, author of *Network Nation*

"Poe's devilish analytical wit and radical aesthetic genius shine through John Tresch's marvelously informative biography. The great poet's delirious cosmological artifices, Tresch shows, are in dialogue with his wide-ranging knowledge of technology, empirical method, materialism, and the newly forming scientific institutions—and trickster spectacles—of his time."

—Charles Bernstein, professor emeritus of English
at the University of Pennsylvania
and author of *Near/Miss* and *A Poetics*

"John Tresch juxtaposes Poe's biography with the rise of American science to turn a familiar story into a mesmerizing narrative, cast like a magic lantern show. Almost everything the author wrote takes on stunning new meaning. This is Poe wrestling with the big questions—and Tresch unfolding the history of a momentous cultural revolution."

—J. Gerald Kennedy, author of *Strange Nation*

To my mother, who read me scary stories

Art is the perfection of Nature: Were the world now as it was the sixt day, there were yet a Chaos: Nature hath made one world, and Art another. In briefe, all things are artificiall, for Nature is the Art of God.

—Thomas Browne, *Religio Medici*, 1643

Contents

Contents

The Reason for
the Darkness of the Night

INTRODUCTION

Subject: The Universe

At the start of February 1848, New York newspapers announced a mysterious impending event: "Edgar A. Poe will lecture at the Society Library on Thursday evening . . . Subject, 'The Universe.'" There could be no grander topic. But no one could say just what to expect: a story, a poem, a critical diatribe? It might contain anything and everything.

The Home Journal ventured, "There is but one thing certain about it, that it will be compact of thought, most fresh, startling, and suggestive." As a "born anatomist of thought" Mr. Poe "takes genius and its imitations to pieces with a skill wholly unequalled on either side the water." The announcements fueled speculation about the man himself. Despite his renown, he had been out of the public eye for more than a year.

The venue, newly relocated to Leonard and Broadway, offered few clues. The Society Library's board included social luminaries such as the banker Cornelius Roosevelt. Its fare was more polished than at P. T. Barnum's American Museum, ten blocks south; Ralph Waldo Emerson had delivered his "Lecture on the Times" there, but all manner of entertainments passed through its doors. Recent performances featured Swiss bell ringers, the American Daguerreotype Association, and the stage magician Signor Blitz.

Would Poe's lecture be a story, a poem, a scientific treatise, or some unheralded novelty? The *Weekly Universe* remarked, "Mr. Poe is not merely a man of science—not merely a poet—not merely a man of letters. He is all combined; and perhaps he is something more."

The speech was Poe's return after a fretful absence. He had become famous three years earlier with his poem "The Raven." Its odd, enticing melody and haunting refrain, uttered by a cryptic bird to a scholar racked with grief, had engraved itself on the public mind: "Quoth the raven: *Nevermore*." First published under a pseudonym, it was celebrated, reprinted, and parodied. A Gotham paper raved, "It is written in a Stanza unknown before to gods, men, and booksellers, but it fills and delights the ear strangely with its wild and clashing music. Everyone reads the Poem and praises it."

They also heard it. Poe became a fixture of New York's literary salons, where he magnetized listeners with intense and hushed recitations. The poet Frances Sargent Osgood recalled "his proud beautiful head erect, his dark eyes flashing with the electric light of feeling and of thought." Another author reported his reputation as a practitioner of mesmerism, the new science of invisible fluids and the vibrations that unite minds: "People seem to think there is something uncanny about him, and the strangest stories are told, and, what is more, *believed*, about his mesmeric experiences."

"The Raven" opened other doors. In 1845 he gave a lecture titled "The Poets and Poetry of America" to a crowd of three hundred, denouncing the feeble state of American writing and criticism, its regional cliques and inflated reputations. By the end of 1846, Poe's poems, tales, unflinching opinions, and flair for provocation had brought him near to his dream of running his own magazine. A sketch of him in the first flush of fame suggests a man of poise, wit, and discernment, with good reason to be optimistic (if perhaps a bit anxious).

But his luck changed. For most of 1847 he dropped from salons and lecture halls, to whispers of scandal and tragedy. He moved to Fordham, twelve miles north of the city, with his aunt and his ailing wife, Virginia. He later confessed to a friend, "I became insane, with long intervals of horrible sanity. During these fits of absolute unconsciousness I drank, God only knows how often or how much."

Both allies and enemies speculated about his condition. His friend George Eveleth, a medical student, wrote to the editor Evert Duyckinck, "Where is Mr. Poe—what is he doing—and what is he likely to do? . . . *Does* he continue to drink hard yet, or has he reformed?" His rival Thomas Dunn English, who had written a novel with a vicious caricature of a drunken Poe,

Poe around 1846, miniature sketch by John A. McDougall

ridiculed his lowered state: "We understand that Mr. E. A. POE has been employed to furnish the railing for the new railroad over Broadway. He was seen going up the street a few days ago, apparently laying out the road."

In fact, withdrawn from the squabbles of the New York papers and salons, safe from public scrutiny and petty attacks, Poe was plotting the next phase of his career. While he was nearly isolated, at one of his lowest moments, his imagination soared. Rambling through Fordham's lush meadows and on the stark, stony cliffs along the Hudson, he undertook daring new works: essays articulating the "science of composition"; "Ulalume," an incantatory ballad lit by the "nebulous lustre" of a newly born star; a visionary tale, "The Domain of Arnheim," scrawled on a single, long scroll, in which an artist of unlimited wealth engineers a vast landscape garden resembling a heaven or a hell; and, most audaciously, the lecture titled "The Universe," which the New York papers announced in early 1848.

Poe was pinning his hopes on this work, the first step of his concrete plan "to re-establish myself in the literary world." As a magazine writer, he was lucky to receive more than twenty dollars for an article, no matter how widely it was read. A lecture with a decent audience paying fifty cents each could earn him several months' rent. He planned to follow the New York event with a national tour, reaping ticket revenues and subscriptions for his reborn literary magazine, *The Stylus*, starting with his former classmates from the University of Virginia and the U.S. Military Academy: "I must get a list of, at least, five hundred subscribers to begin with:—nearly two hundred I have already. I propose, however, to go South and West, among my personal and literary friends—old college and West Point acquaintances—and see what I can do."

His lecture's published title would be *Eureka: An Essay on the Material and Spiritual Universe*. "Eureka!—I found it!"—was the exclamation of the ancient philosopher Archimedes upon discovering a method for testing the purity of gold. "Eureka!" was also the joyful shout of prospectors in California. Poe was convinced that the discoveries in his essay would secure his immortal fame, make him rich, and, by plumbing the mysteries of the universe, save his life.

The strategy was less crazy than it might seem. Poe had trained in mathematics and engineering at West Point and spent subsequent decades reviewing fast-breaking developments in electromagnetism, chemistry, natural history, and astronomy; he was positioned as well as nearly any of his contemporaries to speak on cosmology. The origin and makeup of the universe obsessed great minds of the age—Pierre-Simon Laplace, John Herschel, Alexander von Humboldt—and had captivated the American public. The Scottish minister and astronomer Thomas Dick published popular astronomy books that harmonized natural science and Protestant theology, while the eight volumes of the *Bridgewater Treatises* had updated "natural theology" to keep pace with scientific advances.

The bestselling *Vestiges of the Natural History of Creation*, published in Edinburgh in 1844, was still being ferociously debated on both sides of the Atlantic. Scandalously, *Vestiges* retold the origin and development of the solar system, the sun, the earth, and humanity as the result of uniform natural law—without divine intervention. Was the book's anonymous author a radical, a crank, or a respectable man of science? No one could say. Building on the notoriety of *Vestiges*, just days before Poe was to speak at the Society Library in early 1848, John Pringle Nichol, professor of astronomy at Glasgow—one of the prime suspects for having written *Vestiges*—gave a series of well-attended talks in New York.

Nichol had concentrated popular attention on nebulae—remote, shimmering patches of light being closely inspected with powerful new telescopes. According to the "nebular hypothesis" he advanced, our own sun had once been a whirling, luminous cloud of gas before it condensed, leaving the planets orbiting in its wake.

The theological consequences were stark: if true, this hypothesis meant the heavens had gradually evolved into their present state and the universe had a markedly different history than in Genesis. If such processes of evolution occurred as a result of natural laws, without a creator's intervention, they might still be at work, beyond our tiny planet, or even here on earth, in the formation of new species.

In and around the cottage in Fordham, his aunt and vigilant protector, Maria Clemm, accompanied Poe as he worked. "I used to sit up with him, often until four o'clock in the morning, he at his desk, writing, and I dozing in my chair," she recalled. "When he was composing 'Eureka,' we used to walk up and down the garden, his arm around me, mine around him, until I was so tired I could not walk. He would stop every few minutes and explain his ideas to me, and ask if I understood him. I always sat with him when he was writing and gave him a cup of hot coffee every hour or two." Restlessly walking late into the night, Poe gazed up at the undimmed stars arched above the countryside, wondering where they had come from and what hints they might hold for those below.

Since childhood, Poe had been looking to the stars: peering through his British-made telescope from the balcony of his foster father's Richmond mansion; polishing lenses as an artillery engineer in the U.S. Army; setting one of his earliest poems, "Al Aaraaf," on a new star observed by the astronomer Tycho Brahe. Like C. Auguste Dupin, the gentleman investigator Poe introduced in "The Murders in the Rue Morgue"—the first modern detective story—Poe suspected that the constellations he observed on his night walks held clues to the early history of the universe and the laws governing its life and death.

Poe spent January 1848 revising his lecture, like the scholar in "The Raven," "wondering, fearing, doubting, dreaming dreams no mortal ever dared to dream before." He hired the Society Library and called on friends in the press to announce the event. Yet the weather was beyond his control. The chosen night, February 3—a year and four days after his wife's death—a storm struck the city.

When Poe stepped to the lectern, dressed with simple elegance in a black suit with an immaculately clean (if threadbare) collar and neckerchief, and placed before him a stack of manuscript pages covered with his minuscule and regular handwriting, a mere sixty-odd people had braved the cold to attend—a "select, but highly appreciative audience."

Undaunted, Poe unraveled the mystery of the universe. "I have seen no portrait of POE," a listener remarked, "that does justice to his pale, delicate, intellectual face and magnificent eyes. His lecture was a rhapsody of the most intense brilliancy. He appeared inspired, and his inspiration affected the scant audience almost painfully." The core of Poe's lecture was a new creation story, marked with strange and poetic symmetries: a theory of the formation of stars, extending the nebular hypothesis to the universe as a whole.

Everything, in his telling, began with a single, unitary particle that exploded outward to the limits of "the Universe of Stars," its diffused matter gathering in nebular clouds. These clouds then condensed inward, forming suns and planets. Yet the inward force of gravity was countered by a rival, repulsive force that Poe called electricity—the cause of all "phaenomena of vitality, consciousness and *Thought*." For eons, the two principles had been locked in struggle, generating the diverse beings that filled Earth and other planets. Eventually, gravity would take the upper hand, and "with a million-fold electric velocity" all matter would rush back together, hurtling once more into the "absolute unity" of the "primordial Particle."

A reporter for the *Morning Express* described the lecture as "the most elaborate and profound" he had ever heard; it was "greeted with warm applause by the audience, who had listened with enchained attention throughout." Poe saw *Eureka* as the culmination of his writings, his dreams, and his ill-starred life. He told a friend that it was destined to "revolutionize the world of Physical & Metaphysical Science. I say this calmly—but I say it."

When it was published, he wrote to his mother-in-law, "I have no desire to live since I have done *Eureka*. I could accomplish nothing more." He died the next year.

The Poe Effect   Many of Poe's devoted fans—whether they revere him as the mastermind of horror, inventor of the detective story, pioneer of science fiction, high priest of symbolist art, or brooding prince of the goths—have never read *Eureka*, the cosmological theory he propounded that night in 1848. Beyond its length, and the difficulties of its form and argument, *Eureka* seems out of place among Poe's most famous works, whether his outrageous tales of horror and madness or his hymns to ethereal beauty.

This book tells the full story of Edgar Allan Poe's life—but it does so from a new angle. It returns Poe's cosmology to its place at the summit of his life and thought, showing his work as a singular expression of the tumultuous ideas and passions of his age, thoroughly bound up with the emergence of modern science.

Poe explored the exhilirating prospects and treacherous blind spots of science's powerful new ways of assembling the world. Understanding his life and work demands close attention to his multiform engagements with these theories, methods, and discoveries. In return, Poe's life and works are vividly revealing of modern science in a decisive moment. The historian and philosopher Thomas Kuhn labeled the first half of the nineteenth century "the second scientific revolution." With methods of precise measurement and calculation, researchers were consolidating the programs of the seventeenth century—the first "scientific revolution," identified with Bacon, Kepler, Descartes, Galileo, and Newton—while scientific fields diversified and expanded. Poe's case sheds distinctive light on the obsessions and controversies of early nineteenth-century science as it unfolded in America. His work embodies its defining tensions: between popular diffusion and elite control, between empathy and detachment, between inspired enthusiasm and icy materialism.

Poe argued that every word and image in a poem or tale should contribute to a single, deliberately chosen effect. His works deliver a dazzling array of shocks and delights—effects of terror, humor, disgust, sublimity. Yet a second moment often follows the initial blow. Attentive readers might wonder, how did he do it? What combination of words, expectations, and publicity allowed him to strike such a concussion upon the individual and collective mind? Further questions might follow: Were these the techniques of a high and inspired art, or crude tricks to trigger base reactions? In cases where Poe attained an effect of reality, or "verisimilitude," readers might ask, is this a true report of facts or a hoax, a humbug?

The "Poe effect" was *an exclamation point followed by a question mark*: a striking, concentrated impact that left the reader wondering after a chain of causes. These led back to a captivating but elusive source: Poe himself. Poe's fantastic tales, detective stories, and nonfiction writings dramatized the act of inquiry and the struggles, fears, hopes, and delusions of the human being undertaking it. His invention of new sensations and his search for hidden causes place him at the center of the maelstrom of American science in the first half of the nineteenth century.

The senses of Poe and his contemporaries were bombarded with new technical effects: electromagnetic signals, brilliant light shows, musical innovations, clattering city streets, mesmeric emanations, machine-printed words. They were also confronted with new methods and theories for analyzing the universe, curing illness, justifying political decisions, organizing society, and shaping minds. Today's image of science is of laboratories, microscopes, and white coats: a regulated and uniform undertaking, heavily funded by government and recognized as the best—indeed, for some, the only—means of obtaining reliable knowledge about the world. Yet as Poe embarked on his career in the 1830s, this image of science was at best a distant dream.

The term "scientist" was coined only in 1833, to replace the generalist "natural philosopher." In Europe, the fallout from Napoleon's wars and the dawning Industrial Revolution prompted violent conflicts over rights and property, belief and expertise. The scientific situation in the United States was even more chaotic, because national traditions and institutional frameworks did not yet exist. In the popular press and lecture halls, self-appointed experts announced dubious observations and flamboyant theories on every topic. No sufficiently powerful authorities existed—whether scientific associations, national academies, or peer-reviewed journals—to separate reliable claims from errors or outright fraud.

But the current was beginning to shift. In Poe's time a number of aggressive, well-connected reformers—whose paths Poe crossed and re-crossed and whose discoveries and inventions he digested and reviewed—began to refashion science into a set of interwoven claims about how to inhabit the world, a driving force for material improvements, a unified image of knowledge and nature.

As we will see, for Benjamin Franklin's great-grandson Alexander Dallas Bache and his close allies the physicist Joseph Henry and the mathematician-astronomer Benjamin Peirce, fostering science was fundamental to building America as a unified nation. They strove to establish well-organized, feder-ally supported institutions for training and research. They saw these as crucial supports for state power, industrial development, and territorial conquest. Woven through the era's conflicts, science was just beginning to coalesce into the powerful (if at times threatened) institution we recognize today.

Writing and reflecting on the range of his era's sciences, Poe often joined his voice to these calls for scientific institutionalization and expansion. At the same time, he scrutinized and satirized science's rise. He made it clear that science would take root only by persuading people to follow a specific, utopian vision, and that the lines between truth and illusion, reason and irrationality, common sense and madness were shifting and unstable.

At times he explicitly aligned himself with the projects of scientific reform-ers such as Bache and Henry, echoing their calls to put American intellectual life on a new, more solid footing. At other times he mocked the reductive sim-plifications that modern science offered up as the last word on any topic, and tweaked his contemporaries' credulity with hoaxes to rival P. T. Barnum's. In wide-ranging philosophical works, including *Eureka*, he offered an alternative vision of science and the cosmos in which intuition, feeling, and imagination played leading roles. Poe wrote both with and against the emerging scientific consensus; his deep familiarity with science was the fulcrum on which his thought balanced.

Though later thinkers have drawn heavily on his work—including the linguist Roman Jakobson, the philosophers of science Charles Sanders Peirce and Gaston Bachelard, and Freud's student Marie Bonaparte, who promoted psychoanalysis by examining his thought—there was of course much more to Poe than his scientific obsessions. Every junior high student knows him as an exacting artist of atmosphere and anxiety. He was a verbal virtuoso attuned to words' sounds and depths of meaning and a compelling advocate for beauty of an exquisitely irregular sort.

Working in a stunning variety of styles, genres, and tones, he conjured up sublime landscapes, hypnotic interiors, and compellingly disturbed characters, demonstrating to later writers all that the short story could be: an aesthetic experiment, a study of anomalous psychology, a philosophical investigation, a wild ride of beauty and fear. Poe spoke for the tormented, the cursed, the exiled, and the strange—even as he explained with rigorous clarity the abstract and universal principles of art.

Through the writers he profoundly influenced—Baudelaire, Dostoevsky, Verne, Lovecraft, Conan Doyle, Nabokov, Highsmith, and Borges among them—Poe shaped the course of modern literature. He is now one of the most widely read writers of fiction in any language. With the frantic, grotesque juxtapositions of his work, he is also perhaps the most American of authors. Observing the national experiment from high and low, from positions of both extreme privilege and extreme deprivation, Poe was a seismograph registering the volatility of his place and time. Written in dialogue with the distinctive cultural activity of the country's major cities—Richmond, Boston, Baltimore, Philadelphia, New York—his works bring to light the violence, anxiety, feverish idealism, and horror that were indissociable from the building of the United States.

Despite the myths and clichés about him, Poe was not a morbid, melancholy dreamer, prone to slip at the slightest suggestion into an earthshaking alternate reality—or rather, he was that, and something more. Poe experienced great misfortune, much of which he brought upon himself. But as his portrait suggests, he faced up to his life with dignity, kindness, a sense of duty, and a sense of fun. One of the first Americans to earn his living by literature alone, he forced himself through unprecedented contortions to capture readers and fame.

But he aimed higher. Glancing compulsively up to the skies, he was an analyst, a philosopher, and a detective, seeking to crack the code of the universe. With exceptionally keen eyes, Poe followed the dazzling trail of the strange and contradictory nation that produced him and the sublime, terrifying modernity it was bringing down to earth.

PART I

From Allan to Poe

A dark unfathom'd tide
Of interminable pride—
A mystery, and a dream,
Should my early life seem.

—POE, "Imitation"

Magic lantern slide, "Half Moon" and "Crescented Moon," England, ca. 1847

I

The Young Astronomer

I n the swampy heat of the summer of 1825, anyone walking past the mansion on the corner of Fifth and Main could look up and see a slender young man on the balcony making fine adjustments to a telescope. The house stood on Shockoe Hill, sloping down to the James, the tidal river that opened Richmond to the sea. As he peered through the instrument of glass, wood, and brass, the youth alternated between careful motions and patient stillness. By day he could track boats coming in and out on the tide; at night he followed the stars.

The sixteen-year-old astronomer was Edgar Allan—sometimes called Eddie, sometimes Edgar Poe. He knew the constellations as maps, legends, and calculable motions. With bright hazel eyes and sharp, delicate features, he had a quick and supple mind; some compared his animated, dignified bearing to a bird's. At Richmond Academy, the school for young gentlemen he'd attended since 1821, he excelled in languages, rhetoric, and the "Astronomy, Conic Sections, Algebra, Fluxions, Mechanics" its curriculum boasted.

Edgar was an orphan, the only child in his foster family. He and his friends would run through the woods beyond the city, testing each other in boxing matches, pranks, and dares. Once he swam six miles across the widest expanse of the James, a feat he compared to Lord Byron's paddle across the Hellespont. He also followed Byron as a poet. His heart, he wrote, was "a chaos of deep passion" that he poured into musical, brooding verses. He lost himself in novels, adventure stories, history books, and the magazines sold in the store owned by his foster father, John Allan.

In appearance and character, Allan had little in common with Edgar. Large and heavyset, Allan was a man concerned with social standing and the satisfaction of his desires. He traded Virginia's pungent tobacco, the crop grown on plantations to the west and south. Harvested by African slaves, on land seized from Native Americans, its leaves were bundled together and loaded onto carts to Richmond, where it was sold, cured, and packed for shipment to northern states and England. Allan cared about shipping times and arrivals, about weather, prices, and costs.

The mansion, named Moldavia, was steps from Richmond's markets for tobacco and cotton; slaves were bought and sold a few streets away at Shockoe Bottom. When he bought the property in June 1825, John Allan had secured his arrival among Richmond's elite and the self-described southern aristocracy. The white columns and domes of the city's largest buildings echoed the imperial proprieties of Greece and Rome. Standing three stories high with a pillared portico, Moldavia proclaimed Allan's rise from struggling businessman to plantation owner and oligarch. Like the home's portraits, sumptuous drapery, and ornate furnishings, Edgar's telescope demonstrated wealth and

Richmond from the hill above the waterworks, engraved by W. J. Bennett from a painting by G. Cooke, 1843

taste. It was also part of the world of facts dear to Allan—a tool of navigation and cartography, crucial for transatlantic trade.

Decades earlier Allan had sailed to Richmond from Scotland, joining his uncle William Galt, a successful trader. With a partner, Charles Ellis, Allan began importing household goods, fabric, and magazines to a well-heeled clientele. He married Frances Keeling Valentine, a planter's daughter. The House of Ellis and Allan was shaken in the years following the War of 1812, but when his uncle died in March 1825, Allan became one of Virginia's richest men. He inherited three large estates "with the slaves, stocks, and properties of all kinds belonging thereto," as well as Galt's well-placed pew in Richmond's Episcopal Monumental Church.

John and Frances Allan had taken in the two-year-old Edgar Poe in 1811, when his mother, a celebrated actress and "one of the handsomest women in America," died in Richmond. In many ways they treated him as a son and a future member of Virginia's ruling class, giving him hope for a magnificent inheritance of his own one day. Yet Allan refused to adopt the boy. Edgar's childhood became an unrelenting test to earn Allan's love and recognition.

By the summer of 1825, when Poe was sixteen, tensions between the blustery patriarch and his ward had increased. Frances Allan had always doted on Edgar, and he returned her affection. Yet to John Allan's annoyance, Edgar showed little inclination for the life of business; the earliest existing poem in Poe's hand was written on a piece of paper on which Allan worked out the compound interest on a loan.

Allan had grown indifferent to his delicate, demanding wife. Toward Edgar he behaved condescendingly, dismissively, and angrily. He was a stern disciplinarian, whipping the boy for childish infractions. At times he lavished him with spending money and gifts, but he could be inexplicably stingy. Above all, he never let Edgar forget that he was neither a blood relative nor legally adopted and had no right to any expectations. Other disputes—over Allan's infidelities and those of Edgar's birth mother—had begun to fester.

With the telescope, Edgar could travel far. As he crouched before the lens, blurred mist resolved into an absorbing darkness broken by pinpricks of light, revealing a precise, eerie vision of alien places. He also escaped into poetry, plotting a celestial drama in the lines of "Evening Star":

The stars, in their orbits,
Shone pale, thro' the light
Of the brighter, cold moon,
'Mid planets her slaves,
Herself in the Heavens,
Her beam on the waves.

The arrangement flatters the moon, the ruler of the night. But after gazing "on her cold smile; Too cold—too cold for me," he turns instead to Venus, the evening star. He is charmed by "the proud part" it plays

in Heav'n at night,
And more I admire
Thy distant fire,
Than that colder, lowly light.

Though the regal moon draws the world's attention, the poet reaches for a rarer, more distant pleasure.

These precise, otherworldly lines, written as a teenager, anticipate Poe's later poetry. They also echo the world of his anxious childhood: a capital city on the water, a commercial center of contradictory values in which classical ideals, fascination for modern science, violently enforced subservience, and the longings, contrasts, and passions of romantic poetry all played their part.

Son of the Stage

Poe was born in the North—in Boston, on January 19, 1809. His mother, Eliza Arnold, had arrived from England in 1796, age nine. She grew up on the stage, charming audiences in roles from singing maids to Ophelia, praised for her beauty and vivacity. Acting was a morally dubious profession in early America; plays were banned in Boston until 1795. On tour, Eliza caught the eye of David Poe, son of a respected Baltimore family; his father, called General Poe, had served as Lafayette's quartermaster during the Revolutionary War. They married and toured together, though he drank and cut a feeble figure onstage.

Their first son, Henry, was born in 1807, followed by Edgar. Eliza Poe had a third child, Rosalie, though by the time of her birth (and possibly long before) David had abandoned the family. Mrs. Poe's theatrical run in 1811 in Virginia was enormously well received. Yet in Richmond, where she played Juliet and other celebrated roles, she grew deathly ill.

A beautiful, dying actress surrounded by her small children, the youngest still nursing, made for a pathetic scene. Ladies paid their respects, according to a local merchant: "A singular fashion prevails here this season—it is— charity—Mrs. Poe, who you know is a very handsome woman, happens to be very sick, and (having quarreled and parted with her husband) is destitute. The most fashionable place of resort now is—her chamber." The *Richmond Enquirer* noted, "*Mrs. Poe*, lingering on the bed of disease and surrounded by her children, asks your *assistance*, and *asks it perhaps for the last time.*"

She died on December 8. The feckless David Poe died in Norfolk, perhaps only a few days later. Henry was taken in by Poe relatives in Baltimore; baby Rosalie was adopted by William and Jane Mackenzie, close friends of the Allans'. At the urging of his wife, John Allan brought Edgar home.

Another public tragedy followed: during a performance of *The Bleeding Nun* on December 26 at the same Richmond theater in which Eliza Poe had performed, a fire broke out. Seventy-two people died. The city council commissioned a memorial. Guided by the Supreme Court's chief justice, John Marshall, and designed by a pupil of Jefferson's, the neoclassical, octagonal Episcopal Monumental Church rose up on Broad Street. Allan's uncle William Galt purchased a pew, as did several senators, congressmen, and the governor. Accounts of the fire and of his mother's death were joined in Poe's memory.

Richmond was the political center of the South, with a strong hold on national government. Plantation owners depended on traders, many of them Scottish immigrants like Allan and his uncle. Their calculating logic and penny-pinching contrasted with the planters' largesse and long-established bloodlines, but Frances's family connections and Galt's wealth secured the Allans' social position. Each summer, the family left Richmond's humid heat for White Sulphur Springs. There Edgar was known for "charming everyone by his childish grace" and his "frank, affectionate, and generous" disposition;

he was "a lovely little fellow, with dark curls and brilliant eyes, dressed like a young prince."

Virginia society prided itself on traditions of honor and hierarchy. These were inscribed in the rituals of balls and duels, injected with a dreamy urgency by Walter Scott's heroic warrior sagas, *Ivanhoe* and *Waverly*. Virginia's slaveholding gentry sent sons to represent their interests in Congress, the courts, and the White House. They prized coolness and bravado, elevated feminine purity and delicacy, and anxiously policed racial borders. Such ideals were undercut by the sexual liberties frequently taken by male slaveholders over enslaved women—a recurrent theme of sensationalist and abolitionist literature.

Poe moved between the parlor, where the Allans entertained friends and family, and the servants' quarters. In Richmond, more than one-third of the population was enslaved. Races mixed in countless ways, and much of Poe's early childhood was passed in the care of Black servants, including Allan's house manager, Dabney Dandridge. His caretakers told him stories. African tales—passed down generations, modulated and reinvented through the Middle Passage—often depicted souls possessed through witchcraft and dangerous obsessions, dead bodies brought back to life, malevolent spirits tormenting the living. Poe's later tales and poems of hauntings, sepulchres, and zombielike revenants would channel these African and Creole themes.

Some critics have found it peculiar that Poe embraced gothic literature, a European tradition associated with haunted castles and cursed aristocracies, popularized by *Blackwood's Magazine* in Edinburgh and the fantastic tales of the German author E. T. A. Hoffmann. Yet in the preface to his first collection of tales, Poe maintained that "terror is not of Germany, but of the soul." More concretely, in the aristocratic airs of Virginia society and the everyday horror of slavery, Poe found direct inspiration for the atmosphere of morbid decadence in "The Fall of the House of Usher" and "The Masque of the Red Death," the sentimentalized abuse of "The Black Cat," and the torture devices of "The Pit and the Pendulum." The central place of slavery in Virginia society marked Poe's later writings as surely as it defined his childhood. The possibility of revolt was constant; rumors of rebellion were followed by ruthless repression.

As Allan's ward, Poe was steeped in Virginia privilege, inseparably tied to the plantations. Yet he was removed from full membership twice over. Allan belonged to a rising mercantile class only recently arrived among the established gentry. More damningly, Edgar was constantly reminded that he was an orphan and the son of actors, in a culture where ancestry was paramount. The tragedy of his parents' deaths, especially that of his mother—of whom he cherished a vague, perhaps imagined memory—also hung over his childhood. These were ideal conditions for producing a gloomy, nervous, extremely observant child.

Experimental Republic

The white elite of Richmond welcomed displays of wealth, learning, and aesthetic sensitivity. They also embraced the modest rationalism inscribed in the Declaration of Independence and the Constitution. Common sense—a combination of observation and ordinary reckoning—could uncover both the underlying laws of nature and the motive springs of politics. As Virginia's governor during the Revolution, Washington's secretary of state, and the nation's third president, Thomas Jefferson exerted a powerful influence on Virginia society. Along with Benjamin Franklin, he was also one of the early republic's best-known contributors to science.

Science in the colonies had been largely an affair of gentlemen enthusiasts. Jefferson and his planter friends discussed historical and natural matters in Richmond and sent their sons to William & Mary for a classical education. In the North, mechanical arts were valued, but sciences other than medicine were pursued more as pastime than profession; while savants in London and Paris toasted his 1751 *Experiments and Observations on Electricity*, Franklin had earned his living in Philadelphia as a printer. Until his retirement in 1747, he conducted his research on lightning and Leyden jars—working out the principles of conservation of electrical force, the complementarity of negative and positive electricity—in his spare time.

In colonial colleges, ancient languages, history, and theology prevailed; professors were expected to profess, not to collect, classify, or experiment, and resources for instruments and other equipment were scarce. Medicine offered some scientific opportunities; many who pursued botany, natural history, chemistry, and physics were doctors. Much discussion of natural philosophy happened in informal, locally organized associations. Franklin founded the American Philosophical Society (APS) in 1743 in Philadelphia, and John Adams launched the American Academy of Arts and Sciences in 1780 in Cambridge, Massachusetts. Despite their illustrious participants, they existed without national support, surviving on membership fees and state subventions.

In such clubbish settings, reports of astronomical observations, natural historical curiosities, and new agricultural practices were politely exchanged. At the APS, William Bartram discussed plants, trees, and lichens; David Rittenhouse displayed telescopes, clocks, and orreries—mechanical models of the solar system. While Revolutionary-era men of science cherished the austere proportions of Newtonian theory, much of their knowledge reached people through lively exhibitions, curious objects, and sensory displays, at times open to wider audiences including women and children. In electrical performances, Franklin's associates literally shocked audience members, bringing them bodily into the sensations of enlightened natural science.

The Revolution was at times cast as the application of rational laws to society. This "bold, sublime experiment" was for Franklin "a glorious task assigned to us by Providence." In *The Rights of Man*, Thomas Paine averred that the Revolution "presented in politics what was only theory in mechanics." Jefferson owned portraits of Bacon, Newton, and Locke, his "trinity of the three greatest men that had ever lived, without any exception." Madison and Adams referred to Newtonian laws in their arguments for the Constitution's balance of powers. Attempts to present the early republic as a direct outgrowth of science were mainly rhetorical, however; the framers made no provision for national support for research, keeping science as primarily a local affair.

In *Notes on the State of Virginia*, Jefferson wrote as a meteorologist, surveyor, and naturalist; he was fascinated by materialist philosophies, setting his five editions of Lucretius's *De rerum natura* alongside Erasmus Darwin's cosmological poems, *The Temple of Nature* and *The Botanic Garden*. He paid particular attention to the care of plants. "Botany I rank with the most valuable sciences," he said, for its contributions to "subsistence," "adornments," "perfume," and "medicaments," while "to a country family it constitutes a great portion of their social enjoyment." As president, he arranged the purchase of the Louisiana Territory from France in 1803 and sent Meriwether Lewis and William Clark to explore the continent clear to the Pacific coast, convinced that future settlement would extend that far. Lewis and Clark were given a crash course in cartography and natural history by scientists connected to the APS—including instructions to look for evidence of large animals, extinct or alive, to prove (against the French naturalist Buffon) that America's nature was large and vigorous, not "shriveled and diminished."

Jefferson's *Notes* also notoriously proclaimed "the real distinctions which nature has made" between races and presented Africans as inferior in beauty and intelligence to Europeans. Benjamin Banneker, a free Black astronomer and surveyor who had helped plot the District of Columbia, urged Jefferson and others "to wean yourselves from these narrow prejudices." Though Jefferson denounced slavery as a "hideous blot," his actions were unhurried; rather than abolition he preferred schemes of "colonization" for resetting Blacks in Africa.

Jefferson's sense of the pleasures of botany, his interest in using science to claim the continent, and his reluctance to extend "universal" rights to those other than males of European descent made him a fairly typical early American man of science. Strictly speaking, such enthusiasts for science were not "amateurs," because there were few "professional" scientists against whom to contrast them. Jefferson tied the sciences to local practical concerns (he designed the dome and columns of his home at Monticello, filling it with books and mechanical inventions), the conquest of western lands and Native peoples, and international networks. He corresponded frequently with European experts, though he claimed to prefer one evening with David Rittenhouse to a week of parties in Paris.

James Madison replaced Jefferson as president in 1809, the year of Poe's birth. Two years earlier a British warship had fired on an American frigate off Virginia's coast and seized four sailors. While it waged its lengthy war against Napoleon, Britain's navy captured more than ten thousand American sailors and impressed them into service.

Goaded by hawks in Congress—including Andrew Jackson, resentful of his family's bad treatment at the hands of British soldiers during the Revolution—Madison stoked tensions with England. Hungry for distraction from domestic dissent, in 1812 he declared war on Britain, calling for a hundred thousand volunteer troops. John Allan and his comrades in a local militia, the Richmond Riflemen, prepared for action. The city was not directly touched by the war, though its merchants anxiously watched their stocks of tobacco and flour accumulate unsold, with English trade embargoed and all shipping under threat.

The War of 1812, the "Second War for Independence," ended in a stalemate, with a peace treaty negotiated at Ghent in 1814 by a delegation that included John Quincy Adams, Albert Gallatin, and Henry Clay. The country emerged with a strengthened sense of national identity, support for a standing army, and a massive increase in the navy. The war also fueled the political career of Andrew Jackson, the victor of the Battle of New Orleans.

With the embargo lifted, John Allan needed to unload his stores of tobacco. He sought new buyers for Virginia leaf in Scotland and England. In June 1815 he and Frances, along with Edgar and Frances's sister, packed up their necessities and auctioned off the rest of their belongings—selling one slave, Scipio, and hiring the others out.

They set sail from Norfolk to New York and on to Liverpool. After the lurching voyage across the Atlantic, Allan sent a letter home to his business partner, Ellis: "Edgar says Pa say something for me, that I was not afraid coming across the Sea." The storms and heaving waves of the crossing fired the six-year-old Edgar's imagination. So did London, nerve center of a growing empire.

London Years   Poe spent five formative years in Britain, from age six to eleven. The family first traveled to Scotland, visiting Irvine, Kilmarnock, and Greenock, Allan's hometown. There Edgar saw the dramatic landscapes—plunging cliffs, hills and glens, chiaroscuro skies—glorified by Walter Scott in his novels of the time. They passed through Edinburgh, a bustling modern city overlooked by a cramped medieval town, a center of scientific experiments, commonsense philosophy, and publishing.

The city that indelibly marked Poe's memory—where John Allan purchased the telescope Poe would use in Richmond—was London. Allan wrote home of "sitting by a snug fire in a nice little parlour at 47 South-

Russell Square and Bedford Square, by Edward Walford

ampton Row, Russel [*sic*] Square where I have procured lodgings for the present with Frances and Nancy sewing and Edgar reading a little Story Book." The mother of one of his school friends wrote, "I expect Edgar does not know what to make of such a large city."

The gaslit crowds of London were a dazzling novelty, but the Napoleonic Wars had taken a toll. Allan described the situation to a friend: "Taxes heavy, debt large, People discontented & desperate . . . the Prince's Carriage was attacked with stones." During their stay, troops on horseback stampeded a peaceful workers' demonstration in Manchester, perpetrating the Peterloo massacre of 1819. Workers were increasingly desperate and the government was increasingly willing to use violence to keep them down.

Fatigued by travel, Frances Allan suffered failing health; as her husband traveled to Manchester, Scotland, and Liverpool, she was sent to the spa town of Dawlish and the Isle of Wight. Though he frequently invoked "the God of heaven" in his letters, John Allan was a man of the world, an admirer of "natural Beauties." Frances wrote to him with muted resentment, "I only wish my health would admit of my entering into all the gaieties of this place I would soon let you see I could be as happy and contented without you as you appear to be in my absence as I hear of nothing but partyes at home and abroad."

Despite the political malaise, Edgar was present for a remarkable moment in British history. The arts and sciences were in bloom. The press followed every move of Lord Byron, the most visible and scandalous figure of the romantic movement. Explorers, gardeners, anatomists, and geologists collected and classified living things and stones, while chemists tested nature's fundamental building blocks. At London's Royal Institution, Humphry Davy and Michael Faraday demonstrated the possibilities of chemistry, electricity, and magnetism in dynamic, explosive performances. London was a city at the forefront of modernity, confident in its arts and sciences, its manufacturing prowess, its navy, and its military victories.

Allan enrolled "Master Edgar Allan" at a nearby school before sending him to board at the Manor House School, five miles north in Stoke Newington, a village once home to radicals including Richard Price and Mary Wollstonecraft (and her daughter by William Godwin, Mary Shelley). The school's director, the Reverend John Bransby, was an eager if stern teacher of languages, mathematics, and geometry; Allan was billed for *Cambridge Mathematics*, a grammar book, and Cicero's *Orator*. By the time Poe left, "he was able to speak the French language, construe any easy Latin author, and was far better acquainted with history and literature than many boys of a more advanced age."

Years later, Poe would feature Bransby's academy in his story "William Wilson," whose title character lives a life of moral recklessness, haunted by a rival and doppelgänger who shares his birth date and name. The two William Wilsons meet for the first time at a school that is a double of Bransby's, "a palace of enchantment" with labyrinthine corridors: "Our most exact ideas in regard to the whole mansion were not very far different from those with which we pondered upon infinity." The schoolroom is crowded with desks and benches, "black, ancient, and time-worn, piled desperately with much-bethumbed books." Looming over all is "a clock of stupendous dimensions."

Bransby remembered Edgar Allan as a "quick and clever boy" who might have been very good "if he had not been spoilt" by "an extravagant amount of pocket-money, which enabled him to get into all manner of mischief." Yet John Allan wrote home with paternal satisfaction: "Edgar is a fine boy and I have no reason to complain of his progress."

While they were in London, however, the British branch of Allan's tobacco trading firm collapsed. The family packed up to return to Richmond. Memories of this stay in Britain—the monarchical mirror to America's upstart republic—remained a touchstone for the future poet and storyteller.

The Old Dominion Again   When their ship, the *Martha*, docked in New York in July 1820, Frances was "so unwell" that Allan called in a doctor. They returned via Norfolk on a steamboat—newly introduced since their departure, with the first steam-powered transatlantic journey undertaken the previous year.

Allan's firm was saddled by debts. Edgar enrolled at Clarke's Academy; the lines of his personality were becoming visible. The headmaster, Joseph Clarke, found him "remarkable for self-respect, without haughtiness, strictly just and correct in his demeanor with his fellow playmates, which rendered him a favorite"—though "in his difference of opinion with his fellow students, he was very tenacious, and would not yield till his judgement was convinced." Edgar also flexed his "imaginative powers" in "juvenile compositions addressed to his young female friends." He had "a sensitive and tender heart, and would strain every nerve to oblige a friend." Thomas Ellis—the son of Allan's business partner—called Edgar "a leader among boys"; the boundless admiration he inspired led Ellis "to do many a forbidden thing." Beyond teaching him to swim, skate, and shoot, Edgar once kept him out long after dark, shooting fowl that belonged to another landowner—for which they both got whipped.

Poe's origins cast a shadow. He devoured any communication from his brother, Henry, living with Poe relatives in Baltimore and making plans to sail around the world. His younger sister, Rosalie, still cared for by the Mackenzies, had grown physically but not mentally; she would never marry. One classmate, noting Richmond's "aristocratic" attitudes, said, "Of Edgar Poe it was known that his parents had been players, and that he was dependent upon the bounty that is bestowed upon an adopted son." This reputation gave him "a fierceness he would otherwise not have had."

Though encouraged to see himself as a member of an aristocracy, Poe was constantly reminded of his lack of any claim to an estate. As John Allan's unadopted ward, he acquired a lifelong sense of resentment for affection expected but denied. His sense of high status and entitlement stood alongside his awareness of a complete lack of security. He had grown competitive, even combative, convinced that he was exceptional, unique—which seemed to him both a high destiny and a curse.

By 1825, gazing at the stars from Moldavia's porch, he was a moody teen—fighting with his foster father, dreaming of running off to sea.

2

In Jefferson's Experiment

I n the summer of his sixteenth year, Edgar dragged his feet through Allan's office, doing odd jobs for the firm as clerk and messenger. Allan wrote to Poe's brother, "He does nothing & seems quite miserable, sulky & ill-tempered to all the Family. How we have acted to produce this is beyond my conception." He thought Edgar was without "a particle of gratitude" despite "a much superior Education than ever I received myself." Poe's friends in Richmond, Allan thought, were leading him to "a line of thinking & acting very contrary to what he possessed when in England."

One cause for friction was a slight Allan made against Poe's mother: that Rosalie was the offspring of an affair. The accusation, though possibly true, infuriated Edgar. It was an open secret that Allan was paying for the schooling of an illegitimate son; he would eventually father a set of twins outside his marriage. Allan's moralizing was hurtful and hypocritical, while his philandering was an injury against Poe's foster mother, Frances.

Poe briefly reconnected with his brother that summer. Henry passed through Richmond on his way out to sea on the USS *Macedonian*, bound for South America; he would carry on to Greece and Russia. Edgar would later embellish his own biography with his brother's adventures.

British magazines nourished Poe's teenage obsessions: history, travel, science, literary celebrity, poetry. He kept abreast of European developments with the publications Allan stocked in his shop, including *The Spectator, The Edinburgh Review*, and most impressively *Blackwood's Magazine*—an Edinburgh imprint full of racy first-person "tales of sensation," literary gossip, scientific discoveries, and philosophical debates. In its pages, Coleridge, Carlyle, De Quincey, and other essayists introduced readers to the German idealists—Kant, Fichte, Schelling, and the Schlegel brothers—and to romantic poetry.

Poetry—what a beautiful idea! To enter into a private world as vivid and real as this one; to find the images and sounds to bring others into that more perfect, intimate space and return them to their own world, transformed. In his rambles around the Richmond countryside, Poe fixated on one isolated spot "lovely" for its "loneliness": a "wild lake, with black rock bound." It gave him a terror that "was not fright,"

> But a tremulous delight,
> And a feeling undefin'd,
> Springing from a darken'd mind.

He imagined the pool as a "fitting grave" for him—or for any creature strange enough to feel reassured by such an eerie vision, anyone whose wild thoughts "could even make / An Eden of that dim lake." In the uncultivated woods beyond Richmond's facades, and in the empathic dream of poetry, he discovered a mixture of pleasure and terror, where death promised comfort and hope.

Though Poe was following Byron's example, posing as a world-weary exile, his gloom had real causes. He later wrote a sonnet, "To Helen," in memory of Jane Stith Stanard, the mother of a friend whom he visited when he "was unhappy at home (which was often the case)." She went mad and died in 1824—adding to a growing line of beautiful dead women in Poe's life. His affections turned to Sarah Elmira Royster, the fifteen-year-old girl with dark curls who lived across the street from the Allans. The two were secretly engaged in the summer of 1825.

The death of Allan's uncle William Galt earlier that year made Edgar's position no more secure. Frances Allan, perpetually ill, could do little to protect him from Allan's outbursts, which were heightened by bouts of drinking. As Sarah Elmira Royster later wrote, Poe "was pleasant but his manner was sad—He was devoted to the first Mrs. Allan and she to him."

Poe placed his hopes in changes ahead. Early in 1826 he would start advanced studies in Charlottesville, where Thomas Jefferson's new university had just opened. If Allan would recognize him as a son, he could pursue a life of erudite leisure; if not, he could follow his brother to sea and a life of adventure.

Lafayette, Humboldt, and the Lighthouses of the Skies

In 1824, the United States had seen the return of the Marquis de Lafayette, the Frenchman who had fought in the American Revolution. He now toured the nation. The old soldier and statesman was welcomed to Virginia's capital by the Richmond Junior Riflemen. They were led by Lieutenant Edgar Allan Poe, who presented the troops and exchanged words about his grandfather, Lafayette's quartermaster in Baltimore, whose grave, it turned out, the general had just visited.

Lafayette's tour of the United States, nearly half a century after 1776, marked a moment of national optimism. In the decade following the War of 1812, hostility had faded between Jefferson's farm-loving Democratic-Republican Party and the urban, centralizing Federalist Party originally led by Alexander Hamilton. The war convinced President Madison, Jefferson's ally, of the need for stronger national coordination in finance and industry: with the peace, he approved the Federalist-sponsored Second National Bank and a protective tariff. James Monroe, president from 1817 to 1825, faced no opponent in the election of 1820. Political antagonisms revived with the hard-fought election of 1824, with John Quincy Adams squeaking past Andrew Jackson, winning the Electoral College via a bargain while losing the popular vote.

During his tour, Lafayette was awed by "the immense improvements, the admirable communications, the prodigious creations" of the new republic. On his inauguration, John Quincy Adams announced his plans to push the nation's improvements further. He promised massive investment in communications—roads, canals, railways, and scientific resources, building on Henry Clay's "American System" of internal improvements. Adams proposed a network of astronomical observatories to follow "the phenomena of the heavens" as a matter of national honor. Europe boasted of "130 of these light-houses of the skies," while every year Americans must learn "at second hand" of "some new astronomical discovery." Adams, a former professor of rhetoric, lamented the fact that Americans were deprived of "the means of returning light for light, while we have neither observatory nor observer upon our half of the globe." Meanwhile, "the earth revolves in perpetual darkness to our unsearching eyes."

Adams's opponents jeered at the phrase "light-houses of the skies." Jackson had won the popular vote but was defeated by the electors, and his followers plagued Adams, stonewalling his plans for national improvement. In particular, Jacksonians dismissed scientific institutions as an aristocratic luxury. Yet Adams's plans for organizing national science were channeling a widespread thirst for knowledge.

After 1812, public lectures and experimental performances were on the rise. Lecturers traveled from town to town; they rhapsodized on the pleasures of collecting and classifying plants, rocks, and fossils, observing the birds and other animals native to America, studying the processes behind the formation of soils and stones. Some brought machines with weights on pulleys to demonstrate Newton's laws of motion, glass disks spun to generate static electricity to shock audiences, chemical experiments producing foul and fragrant odors, sparks and explosions, and optical devices such as magic lanterns, colored slides projected by candle or oil lamp. Some showed off inventions, including automatons, elaborate music boxes, and the kaleidoscope—invented in 1816 by the Edinburgh optician David Brewster—which offered the eye a fantastic, symmetrical play of colored lights.

"Lyceum" halls sprang up across the nation, dedicated to informing and uplifting local citizens. Boston and other New England towns were especially active in the lyceum movement; an early entry was New York's Lyceum of Natural History, opened in 1819. Women played important roles in lyceums—as listeners, organizers, and speakers. The lyceum movement showed "that science is confined to no favored spot under heaven; that intellects and affections are coextensive with the race of man, and that science is as boundless as the earth and the heavens." Direct sensory experience of the sciences—through specimens, instruments, lights, and sounds—emboldened men and women with little formal education to step onto the stage of knowledge. One theologian saw "the land of our birth and our affections rising as rapidly in scientific fame as in wealth and political power."

America's scientific projects had received a forceful stimulus in 1803, thanks to the visit of another representative of the European Enlightenment: the exuberant Prussian polymath Alexander von Humboldt. As a last stop after his years-long expedition through the Spanish Americas, Humboldt docked in Philadelphia and continued on to Washington to meet Jefferson. Having explored the Orinoco River farther than any (European) predecessor, climbed to the dizzying peak of Mount Chimborazo (or nearly), braved storms, illness, hostile Spanish officials, and understandably wary Native Americans, he now paid his respects to "the country of the future."

Accompanied by a staggering arsenal of precision instruments (some lost at sea) and a fabulous trove of plants, animals, maps, and notebooks, Humboldt poured out his thoughts and observations at the APS and in the president's offices. His monologues whizzed between the formation of clouds, the variety of soils, the habits of lizards, the magnetic energy of the earth, the indignities of Spanish treatment of Native and African slaves, and his hopes for revolution in Europe. Though an aristocrat, Humboldt opposed slavery; an ardent republican, he would later encourage Simón Bolívar to lead revolutions in New Spain.

Humboldt's approach to natural environments was both local and global. He analyzed the variations of flora, fauna, and atmospheric phenomena—temperature, air pressure, its chemical composition, the blueness of the sky—according to latitude, longitude, and altitude. He depicted these milieus in their variations, identifying global patterns in the distribution of plants, geological formations, animals, and weather, making unprecedented use of diagrams and maps. One of his innovations was "isothermic bands," regions across the planet that shared similar temperatures—such as northern Europe and the northern United States, a commonality later used to justify the expansion of white settlement across North America. Jefferson was particularly keen to hear Humboldt's reports about Spanish holdings and whether these might yield to influence (or invasion) from the United States.

During his 1803 visit, Humboldt dazzled both scientists and statesmen, including Albert Gallatin, Charles Pickering, and the young John Quincy Adams. Twenty-two years later, President Adams's plans for a network of "light-houses of the skies" and a unified national scientific infrastructure dovetailed with Humboldt's vision for the future of international science. Globally distributed networks of instruments and observers would form a dynamic, living whole, weaving together nature and knowledge. With his whirlwind visit, Humboldt became American science's godfather. Four decades later, Poe would dedicate his cosmology, *Eureka*, to him.

Humboldt was to American science what Lafayette was to American politics. Each helped galvanize an enlightened, republican vision that was distinctive to the former colonies yet tied to elite-led progressive movements in Europe. Jefferson welcomed the elder statesman as warmly as he hosted the youthful explorer. He saw both pointing the way to the country's future.

A Most Liberal Curriculum   When Jefferson retired to his plantation at Monticello in 1809, he focused on improving the state of learning in the United States. The crowning institution in his plan for public education would be the University of Virginia at Charlottesville, where Poe would enroll in 1826.

UVA was a bold experiment: a nationally oriented university with a modern curriculum and no theological oversight. In Jefferson's plan for the state of Virginia, elementary schools fed into colleges to teach navigation, surveying, languages, and "higher branches of numerical arithmetic." At the top of this "ladder of learning" was the university. Writing to offer a professorship to the Massachusetts astronomer Nathaniel Bowditch, Jefferson explained his design: an "Academical Village" of buildings, each in a distinct style, on an open quadrangle forming a miniature society "independent, hospitable, correct and neighbourly." On the advice of the architect Benjamin Latrobe, Jefferson placed at the center a library: a miniature Pantheon echoing the shape of Monticello, his nearby plantation house. The rotunda's red brick and white wooden frames echoed colonial styles, while Jefferson and Latrobe's devotion to classical rationalism was visible in its columns, dome, and proportions.

The Declaration of Independence's axiom that "all men are created equal" clashed with Jefferson's tolerance for slavery and his belief in a "natural aristocracy" based on inherited differences in talent and virtue. Although Jefferson saw the aristocracies of Europe as the soil of corruption, his university would favor those "destined for learned professions as a means of livelihood" and the "wealthy, who, possessing independent fortunes, may aspire to share in conducting the affairs of the nation"—the sons of planters and slave owners.

In contrast to colonial colleges, Virginia's curriculum gave new emphasis to natural sciences, though languages and rhetoric were still encouraged. Jefferson's university would allow its students "uncontrolled choice in the lectures they shall choose to attend." As at the University of Berlin, recently founded by Wilhelm von Humboldt (Alexander's brother), the guiding principle in Charlottesville was to let "everyone come and listen to whatever he thinks may improve the condition of his mind."

On the West Range   A young man with Poe's social standing, training, and talents could expect a bright future from such an education. After graduating, he could marry Sarah Elmira Royster and take up some small employment—perhaps in law or politics—and with luck might inherit his foster father's fortune and do as he pleased.

But his relations with Allan had been strained for months when he made his way to Charlottesville in February 1826, in a carriage driven by Allan's enslaved servant James Hill. It was the university's second year. At Charlottesville, Poe joined rich young white men acquiring a final gentlemanly polish before returning to manage their inherited estates. They clattered onto the lawn with horses and carriages, accompanied by African servants, fine clothing, dueling pistols, and large allowances.

Poe witnessed the grand ambition of Jefferson's experiment in freedom and virtue and its early failures. He was taught by professors recruited from Europe, studying ancient languages with George Long of Cambridge and modern languages with the German George Blaettermann. From the library he checked out works on ancient history, volumes of Voltaire, and Dufief's *Nature Displayed in Her Mode of Teaching Language to Man*, a French textbook with examples from astronomy, natural history, and mathematics. He joined the Jefferson Literary and Debating Society. He was likely invited to dine with

Jefferson, who regularly welcomed students to Monticello, and witnessed the great man's funeral in 1826.

The university's students gambled, cursed, dueled, horsewhipped each other, and broke out into drunken riots. Poe's early tale "Mystification"— part of a genre of literary hoaxes and deceptions aiming to lead readers to reevaluate preconceptions—drew upon his university experience. Though set in the German university town of Göttingen, its events could have taken place in Charlottesville in 1826. Nothing "was done beyond eating and drinking and making merry"; duels are all the rage, and the mere hint that one's "opinions are not the opinions to be expected from a gentleman" is cause for a challenge.

As he settled in, Poe wrote to Allan,

> The pillars of the Portico are completed and it greatly improves the appearance of the whole—The books are removed into the library— and we have a very fine collection.
>
> We have had a great many fights up here lately—The faculty expelled Wickliffe last night for general bad conduct—but more especially for biting one of the student's arms with whom he was fighting—I saw the whole affair—it took place before my door— Wickliffe was much the stronger but not content with that—after getting the other completely in his power, he began to bite—I saw the arm afterwards—and it was really a serious matter.

A refined setting of classical aesthetics and learning is shattered by an act of depraved brutality: this would be Poe's signature style. It was also a faithful transcription of early American reality.

Poe was seen by a fellow student as "very excitable & restless, at times wayward, melancholic & morose, but again in his better moods frolicksome, full of fun & a most attractive & agreeable companion." He became known for "quoting poetic authors and reading poetic productions of his own, with which his friends were delighted & entertained," until "suddenly a change would come over him" and he would sketch on his dormitory walls "whimsical, fanciful, & grotesque figures." Among the sketches on his walls were illustrations from works of Byron, the bestselling author in Charlottesville at the time.

Byron's concise, melodic poetry channeled wit, longing, and morbid sorrow. His fame was made with *Childe Harold's Pilgrimage*, a long poem recounting the journey of a bold and sensitive hero embroiled in scandalous affairs and much misunderstood. Subsequent works, including *Manfred* and *Don Juan*, worked in tragic and satirical modes. Byron toyed with his audiences, stoking the rumors around him, including that of an incestuous affair with his half sister. He eventually fled to the Continent. In Geneva, a ghost-story session with Mary and Percy Shelley led to *Frankenstein*. After Percy drowned, Byron died from a fever contracted while fighting in the Greek revolution.

At the university, Poe grew into the Byronic role of the dissolute, haunted poet. To look the part, he bought on credit three yards of "'Super Blue Cloth,' a set of 'Best gilt Buttons,' and a velvet vest." He also accepted the frequent offer of a drink. As a friend put it, "To calm & quiet the excessive nervous excitability under which he labored, he would too often put himself under the influence of that 'Invisible Spirit of Wine.'"

Alcohol, particularly the peach brandy favored by students, had a devastating effect. According to a classmate, "It was not the *taste* of the beverage that influenced him; without a sip or smack of the mouth he would seize a full glass, without water or sugar, and send it home at a single gulp. This frequently used him up." Small quantities of liquor made him deranged. In his first months, he also gambled—a habit more forgivable in those with a fortune to lose. At one point, discovering a book of Hogarth prints he longed to own, he made a bet—and wound up having to buy the book for another young man.

In Poe's later tale "William Wilson," the title character's descent into debauchery is confirmed while he is a student at Oxford:

The uncalculating vanity of my parents furnished me with an outfit, and annual establishment, which would enable me to indulge at will in the luxury already so dear to my heart—to vie in profuseness of expenditure with the haughtiest heirs of the wealthiest earldoms in Great Britain. Excited by such appliances to vice, my constitutional temperament broke forth with redoubled ardour, and I spurned even the common restraints of decency in the mad infatuation of my revels . . . I added no brief appendix to the long catalogue of vices then usual in the most dissolute university of Europe.

The narrator's doppelgänger—the embodiment of his guilty conscience—exposes him cheating at cards, driving him from the university in shame and on to further perfidy.

Poe thought the root of his own troubles in Charlottesville was not Allan's excessive generosity but rather his "mistaken parsimony." Poe made little effort to be thrifty, but the allowance Allan gave him covered only two courses and board, leaving him with little for books, bed, and furniture. His request for further funds was answered with "terms of the utmost abuse—if I had been the vilest wretch on earth you could not have been more abusive than you were because I could not contrive to pay $150 with $110." This dire situation, he admitted to Allan, led to a fall: "Books must be had, if I intended to remain at the institution—and they were bought accordingly *upon credit* . . . I then became desperate, and gambled—until I finally involved myself irretrievably."

After several months, Poe buckled down and aced his exams, but Allan's stinginess decided his fate. He gambled to pay his debts and lost. Poe was near the top of his class while in a bottomless financial pit, with debts estimated at twenty-five hundred dollars—nearly sixty thousand today.

Too Little, Too Late   Finally grasping the extent of Poe's troubles after eight months, Allan came to Charlottesville. Allan "inquired into his ways, paid every debt that he thought ought to be paid." His refusal to pay gambling debts put Poe into dishonor with classmates and university officials and left a number of angry creditors.

Allan took Poe back to Richmond and forbade him to return. He put him to work in the counting room of Ellis and Allan, "attempting to give him some knowledge of bookkeeping, accounts, and commercial correspondence." He soon withdrew even this opportunity, leaving Poe in disgrace with nothing to do. Poe learned that the father of his secret fiancée, Sarah

Elmira Royster, had hidden Edgar's letters to her; she married someone else, prompting a rueful lyric from Poe, "I saw thee on thy bridal day."

Hostilities between Allan and his ward exploded in early 1827 with a fierce quarrel. In a huff, Poe detailed his charges against Allan, including humiliations both public and private, "exposing me before those whom you think likely to advance my interest in this world" and subjecting him "to the whims & caprice, not only of your white family, but the complete authority of the blacks." Like his fellow squires, Poe was sharply conscious of rank; within Virginia's deep-seated ideology of white supremacy, he found submitting to the command of Allan's family and servants an unbearable humiliation: "These grievances I could not submit to; and I am gone."

He took a room in an inn, registering with the pseudonym Henry le Rennet—a nod to his adventurous brother. He wrote to Allan, "My determination is at length taken—to leave your house and indeavor [*sic*] to find some place in this wide world, where I will be treated—not as *you* have treated me."

Poe saw Allan's cruelest injury as removing him from the university. "Since I have been able to think on any subject, my thoughts have aspired, and they have been taught by *you* to aspire, to eminence in public life—this cannot be attained without a good Education, such a one I cannot obtain at a Primary school—A collegiate Education therefore was what I most ardently desired, and I had been led to expect that it would at some future time be granted—but in a moment of caprice—you have blasted my hope."

Other cuts went as deep: "I have heard you say (when you little thought I was listening, and therefore must have said it in earnest) that you had no affection for me—You have moreover ordered me to quit your house, and are continually upbraiding me with eating the bread of Idleness, when you yourself were the only person to remedy the evil by placing me to some business." He asked for his trunk and some money so he could travel "to some of the Northern cities" and support himself for a month until he could "obtain some livelihood."

In reply, Allan chastised him for wasting his time on literature: "I taught you to aspire, even to eminence in Public Life, but I never expected that Don Quixote, Gil Blas, Jo. Miller & such works were calculated to promote [that] end . . . The charge of eating the Bread of idleness, was to urge you to perseverance & industry in receiving the classics, in perfecting yourself in mathematics, mastering the French."

Homeless, Poe roamed the streets. He again wrote to Allan for a small sum for travel and on March 24 made his way to Norfolk. He found passage on a coal ship heading north.

Three days later, John Allan remarked with lethal casualness in a letter to his sister, "I'm thinking Edgar has gone to sea to seek his own fortunes." At the age of eighteen, Edgar Poe was on his own.

3

Exile, Artificer, Cadet

Poe then caught a boat to Greece. There, like Byron, he joined a ragged band of revolutionaries fighting for independence from the Ottomans. After a series of heroic battles and thrilling misadventures he resurfaced in St. Petersburg, rescued from deep waters by the U.S. ambassador to Russia.

That's what he later said. The truth was grimmer and more mundane. It was shockingly easy to starve.

He went to Boston, carrying a collection of manuscript poems that included his epic, "Tamerlane," about a proud Turkish warlord of the fourteenth century who abandons his first love—named Ada, like Byron's daughter—for a life of conquest.

Poe's mother had written on the back of a watercolor painting she left him, along with a locket portrait, that Boston had always been kind to her. He found otherwise. The city's discriminating social and intellectual circles were closed to him. A former classmate, Peter Pease, ran into him near the harbor, "a rather stoop-shouldered clerk, emerging from a mercantile house . . . very shabbily appareled." Edgar hurried him into an alleyway, begging him not to speak his name: "he had left home to seek his fortune, and until he had hit it hard"—found incontrovertible success—"he preferred to remain incognito." Poe had toiled for two months at a waterfront warehouse "at a very small salary, the most of which he had been too proud to ask for"; he then worked as market reporter for "an obscure paper," but "the proprietor being a man of shady reputation, the office soon got into debt."

This was not the last time Poe would be laboring for pennies, nor his last streak of hard luck. But in Boston he realized a dream: he published his first poetry collection. The run of *Tamerlane and Other Poems* was between forty and at most two hundred; it was printed on coarse paper, with the title page rendered in a commercial type. Despite the title poem's theme of conquest and glory, the publication sank into obscurity.

With no other options, Poe followed in the footsteps of his grandfather, Lafayette's supply officer. He joined the U.S. Army with a five-year contract. Only eighteen, he lied about his age and his name and signed on as Edgar A. Perry, clerk from Boston, age twenty-one.

He moved into the barracks of the First Artillery Regiment at Fort Independence, Massachusetts. Army life was no easy business. Basic training lasted weeks, followed by drilling, hard labor, and boredom. Probably half of the army's soldiers were illiterate. In October 1827, a malaria outbreak forced his battalion to relocate to South Carolina, to Fort Moultrie on Sullivan's Island: a large, irregular-shaped brick fort, surrounded by beaches and trees. Leaving the South was turning out to be more difficult than he thought.

For those who know only Poe's poems and tales of "mystery and imagination," his time in the army may be a surprise; some biographers pass over the episode as insignificant for his life as a writer. Yet Poe spent nearly four years as a soldier at an extremely impressionable time of his life. He worked in conditions of rigorous discipline and hierarchy, where questions of detail and procedure could be matters of life and death. Furthermore, the U.S. Army was the nation's most reliable body for technical knowledge, through the Corps of Engineers, formed during the Revolution and rechartered by Jefferson in 1802. Its members mapped the Louisiana Territory, built forts, laid down roads, canals, and railroads, and were engaged in exploring western lands for further settlement. They were the nation's chief experts in building up resources for industry, defense, and conquest.

Poe's technical abilities served him well. At the end of a year he was promoted to "artificer," ranked above four hundred others in his battalion. A position of responsibility and expertise, the artificer oversaw ammunition supplies. The work required minute attention, with constant awareness that a slip could bring a deadly explosion. He inventoried, maintained, repaired, and at times constructed the equipment used by his artillery regiment, from bullets, rifles, cannons, and carriages to surveying equipment—including field telescopes.

In his spare time Poe worked on another long poem, "Al Aaraaf," inspired by astronomy and Islamic mythology and its view of the afterlife; the poem grew into a colorful and fragrant reverie of otherworldly beauty and passion. He also gathered future materials: Sullivan's Island's sandy shores, with specimens for shell collecting and amateur entomology, would be the setting for "The Gold-Bug."

At the end of 1828, he swallowed his pride and wrote to John Allan. A year and a half of army life had changed him, he said; he was "no longer a boy tossing about on the world without aim or consistency." But the experiment had run its course: "I have been in the American army as long as suits my ends or my inclination, and it is now time that I should leave." All that was required was Allan's assent. He sent "dearest love to Ma," hoping "she will not let my wayward disposition wear away the love she used to have for me."

If Allan sent a reply, it was lost, as Poe's regiment relocated once again, this time to Fort Monroe, off the southern tip of Virginia. There he befriended Colonel House, a literary man who had known Poe's grandfather. House promoted him to sergeant major—the highest rank for a noncommissioned officer, making him "complete master of all exercises of the battalion from the first drill to the movements in line of battle." This was a meteoric ascent—smoothed by his pedigree—after less than two years of service.

Poe now saw his shortest route to regaining the class privileges of his childhood in a commission to the corps at West Point Military Academy. Most slots went to the sons of officers, politicians, and other grandees. He tried Allan once more, asking only two favors: permission to leave the army, and a word of support from Allan's well-placed friends.

Death brought a temporary reconciliation. Edgar's beloved foster mother, Frances Allan, died on February 28, 1829. He was granted leave and rushed to Richmond but arrived the night after she was buried in Shockoe Hill Cemetery. Allan welcomed Edgar with his fitful generosity, buying him a wardrobe suitable for an officer and gentleman: "a suit of Black clothes, 3 pair Sock of Half Hose," a "London Hat," a knife, suspenders, and a pair of gloves.

Poe returned to Fort Monroe. While Colonel House was in Washington congratulating the newly elected president, Andrew Jackson—John Quincy Adams's replacement—Poe resolved to pursue an appointment at West Point. He found a substitute, "an experienced soldier & approved Sergeant" ominously named Bully Graves. He paid him a few dollars, with the promise of fifty more, to finish out Poe's term in the army.

On April 15, "Edgar Perry" ceased to exist. In his place was Edgar A. Poe—aspiring officer, engineer, and poet.

Two Routes to Reputation

Poe headed to Baltimore to live with his relatives while he waged his campaign for a West Point commission.

He was warmly welcomed by his father's sister Maria Clemm, now a widow with an eight-year-old daughter, Virginia. Clemm was well educated, practically minded, sensitive, and emotionally resilient. She ran a tight household and was not above wheedling for necessities to keep her family fed. She welcomed him as a son; he called her Muddy.

His brother, Henry, had returned from sea and regaled him with tales of exotic lands, hardship, and daring. The two composed poems together, printed under Henry's name in Baltimore and Philadelphia newspapers. Poe made his sense of a singular destiny clear in a poem he penned in the autograph book of Lucy Holmes, the daughter of a Baltimore physician:

From childhood's hour I have not been
As others were—I have not seen
As others saw—I could not bring
My passions from a common spring—
From the same source I have not taken
My sorrow—I could not awaken
My heart to joy at the same tone—
And all I lov'd—I lov'd alone.

In the spring of 1829, at the age of twenty, he set out with dogged tenacity in pursuit of two career paths. Both were routes to the "eminence in public life" he had been raised to expect. One began at West Point and would make him an army officer and engineer. The other was poetry.

The composition of a poem was a different matter from the construction of a bridge. Yet both poet (or fiction writer) and army engineer (or scientist) were novel professions in America. Until Poe's generation, writers of fiction and verse had to be born rich or supported by aristocratic patrons; otherwise they pursued their art as a sideline. None of Poe's major contemporaries earned their living only through writing: Hawthorne worked at the Boston Custom House, Longfellow was a lawyer and a professor, Irving and Cooper were diplomats. Similarly, most of those who worked in scientific fields still did so in their spare time, many of them employed as physicians, with only a handful of universities offering courses in natural philosophy. The Army Corps of Engineers offered one of the very few routes to employment that depended on formal scientific training.

Whether poet or engineer, Poe would have to secure influential supporters. A West Point commission required the direct approval of the secretary of war, John Eaton. Poe's Fort Monroe commanders testified that his "habits are good, and intirely [sic] free from drinking," his education of "a very high order." More consequential letters came from Allan's circle of powerful Virginians. Colonel James Preston wrote of Poe "having been born under circumstances of great adversity" and noted "undoubted proofs that he is a young man of genius and talents" who would "remunerate the Government at some future day, by his services." The Speaker of the House of Representatives, Andrew Stevenson, confirmed testimonials "from the highest authority"; Major John Campbell averred that Poe was "a gentleman in whose word you may place every confidence." Meeting Poe in person, Secretary Eaton was impressed—not least because Poe hand delivered his sheaf of reference letters after traveling from Baltimore to Washington on foot. There were no openings, however. He would have to wait.

Back in Baltimore, he introduced himself to the Richmond-based politician and lawyer William Wirt, hoping for both patronage for West Point and literary support. Though Wirt found that the footnotes for "Al Aaraaf" contained "a good deal of curious and useful information," he wondered "whether the poem will take with old-fashioned readers." Nevertheless, he put Poe in touch with publishers in Philadelphia; Poe sent another copy of "Al Aaraaf" to Isaac Lea, who ran a distinguished publishing house with his father-in-law, Mathew Carey, and was a respected geologist and expert on shells. Poe avowed, "If the poem is published, succeed or not, I am 'irrecoverably a poet.'"

Emboldened by his progress, Poe asked Allan to lend his name as a guarantee against any losses from publishing "Al Aaraaf." At his young age, he explained, "there is much in being *before the eye of the world*—if once noticed I can easily cut out a path to reputation—It can certainly be of no disadvantage as it will not, even for a moment, interfere with other objects which I have in view."

Allan did not share Poe's vision of a bright literary future. He wrote back furiously: "men of genius ought not to apply" to his aid.

The poetic campaign met modest success. The Maine novelist John Neal commented on an excerpt from "Al Aaraaf" in the *Boston Literary Gazette*, saying its author "might make a beautiful and perhaps a magnificent poem." Poe sent him sections of "Tamerlane" and gratefully confessed, "I am young—not yet twenty—*am* a poet—if deep worship of all beauty can make me one." Excerpts also appeared in a Baltimore daily, while in the December *Yankee* Neal wrote, "If the remainder of Al Aaraaf and Tamerlane are as good as the body of the extracts here given . . . he will deserve to stand high—very high—in the estimation of the shining brotherhood."

On December 10, Poe published his second poetry collection with a small Baltimore press: *Al Aaraaf, Tamerlane, and Minor Poems*. It was noted in the Boston "Catalogue of American Poetry" and cautiously applauded by Sarah Josepha Hale's *American Ladies' Magazine*: "A part are exceedingly boyish, feeble, and altogether deficient in the common characteristics of poetry; but then we have parts . . . which remind us of no less a poet than Shelley."

Poe's name subsequently appeared in an anonymous satire of contemporary poets in Baltimore, the *Musiad, or, Ninead, a Poem, by Diabolus*: "Next Poe who smil'd at reason, laugh'd at law." What better encouragement for a young poet than to be recognized as worthy of satire—and better yet, in the role of romantic jester, scoffing at reason and law?

Ode to a New Goddess   But Poe's perspective on reason and science was already complicated, as he revealed in his collection's very first poem. A classic sonnet—fourteen lines of iambic pentameter in alternating rhymes with a final couplet—the poem would be revised several times over his lifetime. Later called "Sonnet—To Science," in 1830 it read,

> SCIENCE! meet daughter of old Time thou art
>> Who alterest all things with thy peering eyes!
> Why prey'st thou thus upon the poet's heart,
>> Vulture! whose wings are dull realities!
> How should he love thee—or how deem thee wise
>> Who wouldst not leave him, in his wandering,
> To seek for treasure in the jewell'd skies
>> Albeit, he soar with an undaunted wing?
> Hast thou not dragg'd Diana from her car,
>> And driv'n the Hamadryad from the wood
> To seek a shelter in some happier star?
>> The gentle Naiad from her fountain-flood?
>> The elfin from the green grass? and from me
>> The summer dream beneath the shrubbery?

The poem was an ode not to an eternal deity but to science, a product of history—a "daughter of old Time." At the turn of the seventeenth century, Francis Bacon had written, "Rightly is truth called the daughter of time, not of authority." In other words, truth is not etched in eternal tablets, but gradually emerges through human actions, such as observation, experiment, and discussion—the program of research Bacon set out in his *Great Instauration*.

Yet where Bacon held that science revealed hidden realities, Poe sounded a different note: the "peering eyes" of science *alter* things. Science has "dragg'd Diana from her car": when moderns look at the moon they see a cratered satellite, not an Olympian on her nightly journey. Science has chased away nature deities, sprites, and muses, "the Hamadryad from the wood," elves from the meadow. This predator picks at the poet's heart, a vulture with "dull realities" for wings—replacing meaning and myth with facts. To a poet, such a deity seems neither wise nor lovable.

Many critics have read this sonnet as a poet's "attack upon Science." Poe insisted he *was* a poet, with a "deep worship of all beauty." But by the time he wrote this sonnet, he was also a soldier, experienced in technical and scientific matters. With one foot on each path, he sought a way to move forward on both. Consider the sonnet's punctuation: after four *exclamation points* in the first four lines, the rest of the poem has only *question marks*. In the fifth line, Poe asks Science the poem's central question: "How should he love thee—or how deem thee wise[?]"

The question might be rhetorical, inviting a simple answer: a true poet cannot love science at all. This would lead to a familiar romantic interpretation: that the poem is a polemic against science, bewailing its hostility to dreams, myths, and imagination. Read this way, it recalls the lament of John Keats's *Lamia*: "Philosophy will clip an Angel's wings, / Conquer all mysteries by rule and line, / Empty the haunted air, and gnomed mine— / Unweave a rainbow."

But what if Poe's question was genuine? What if Poe was sincerely asking how poetry might confront, engage, challenge, reimagine, or align with science? How might poetry stay true to science's specific form of wisdom, without being destroyed by it?

If this was a genuine question, we might expect an answer. Poe provided one with "Al Aaraaf," the long poem that immediately followed "Sonnet—To Science" in his collection. Its title came from Islamic mythology: "Al Aaraaf" was a place where souls went after death, "a medium between Heaven & Hell where men suffer no punishment, but yet do not attain that tranquil & even happiness" of heaven. As he explained to Isaac Lea, Poe imagined Al Aaraaf as a real place, locating it on "the celebrated star discovered in 1572 by Tycho Brahe which appeared & dissapeared [*sic*] so suddenly."

The key image of "Al Aaraaf"—the planet of its title—was thus based on a decisive event in modern science. Tycho Brahe was Europe's best-funded stargazer. Though astronomers were not yet using telescopes, the king of Denmark gave Tycho an island for his observatory, whose giant quadrants and sextants allowed for the most exact observations anywhere in Europe. His announcement of the new star, "Stella Nova," came as a shock. For Aristotle and medieval philosophers, the moon and the domains above it traveled on crystalline spheres, perfect and unchanging; only on earth was there imperfection, transformation, and death. Brahe's discovery, a suddenly appearing and disappearing star, was a novelty taking place *above the moon*. It suggested, against all received wisdom, that the heavens might be part of history.

In his letter to Isaac Lea, Poe described Al Aaraaf as a "messenger star," recalling the epithet for the Prophet Muhammad, as well as another primal event in modern science. Galileo's *Starry Messenger* of 1610 reported observations he made with the "perspicillum," or telescope. Like Brahe's new star, Galileo's observations—pockmarked irregularities on our moon, four moons revolving around Jupiter—suggested imperfection and change in the "superlunary" realm.

Time, change, and decay were part of the heavens: this was the cosmos-shattering message of Bacon, Tycho, and Galileo. In "Al Aaraaf," Poe aligned this new star—exploding into existence, fading away soon after—with a wavering limbo "between Heaven & Hell," in which strange new forms of beauty and desire flower into life.

With the seeming hostility between science's "dull realities" and poetry's "summer dream," how might a poet love science or find it wise? "Al Aaraaf" answered: if science has "altered" human experiences of the night sky and the natural world, poetry can return the gesture. The imagination can seize the facts of science—observations of new stars, the discoveries of history and comparative mythology, strange plants and insects, the grandeur of space—and weave out of them new experiences of beauty, new emotions and images, new myths.

"Sonnet—To Science" laid out a program for Poe's life's work. Even while exploring the outer limits of imagination and irrationality, he would continue to ask how a poet might love science and might deem it wise. He would arrive at a panoply of answers, always thoughtful, frequently contradictory, often sublime.

In the spring of 1830, just after *Al Aaraaf*'s publication, Poe's future was upon him. After a last nepotistic push—a recommendation letter from Senator Powhatan Ellis of Mississippi, the brother of John Allan's business partner—Poe was granted an appointment at West Point.

At the Blackboard   Perhaps happy to be rid of him at last, Allan received Poe in Richmond at the end of May; he bought him clothing and blankets and saw him off at the steamboat dock. Poe once more addressed his letters "Dear Pa," desperate to maintain Allan's affection. Though his debts from Charlottesville were paid or forgotten, he still owed fifty dollars to his army replacement, Bully Graves. Allan refused to cover it, which Poe explained frankly in a letter to Graves: "Mr. A is not very often sober."

In June, Poe arrived at West Point after a day's journey north from New York City. He moved into a tent on the plain, on a cliff with stunning views of the Hudson bending below, wooded mountains receding across the swell. As in Charlottesville, Poe again joined a select group of young white men from well-connected families. But now they had to demonstrate "a disposition and capacity to be of future usefulness to the country," with an entrance exam in writing, reading, and "ground rules of arithmetic": "reduction," "simple and compound proportion," and "vulgar and decimal fractions." Poe passed with ease, but "a great many cadets of good family" were "rejected as

The Plain at West Point, *1828, engraved from a painting by George Catlin*

deficient," including the son of Virginia's governor. "One fellow, on being sent up to . . . the black board to divide 5 by 2/3 began blubbering most tremendously." Out of 130 cadets appointed each year, "only 30 or 35 ever graduate."

Poe and the other survivors of the exams spent the summer in drills and military training, including construction and munitions—for which his experience as an artificer prepared him. On August 30 they broke camp and moved into the barracks to begin classes, maintaining an hour and a half of drills in the afternoon followed by evening parade.

After Jefferson's stumbling experiment to adapt the German university model to Virginia, at West Point Poe entered a quite different (though also European-inspired) experiment for creating a republican elite. A national school for officers had first been advocated by Hamilton and the Federalists, while Jefferson thought national defense should depend on local militias. The academy opened in 1802; it proved its value by supplying engineers to the War of 1812. A new superintendent, Colonel Sylvanus Thayer, was appointed in 1817 and sent by the Department of War to France to survey the École Polytechnique—an engineering school founded during the Revolution. Napoleon turned it into a military academy. The École Polytechnique became the model, imitated worldwide, for training mathematicians and mechanics to serve modern states and empires.

Thayer brought back from France the largest collection of scientific textbooks in the United States and a plan for reform. As at the École Polytechnique, his students were called "cadets," trained as both scientists and soldiers. They all learned French, for Thayer "the only means of opening to the student the scientific works of Europe." Lessons in grammar and mathematics were taught with the same rigor as combat training.

Thayer's reforms sought to erase from cadets' memory "an unauspicious period of the institution when they were allowed to act as though they had rights to defend." The spartan barracks lacked running water and heat; the food was unappetizing. Thayer filled every minute of cadets' fifteen-hour day with drills and work. His "General Merit Roll" depended on nearly omniscient surveillance, with each professor preparing weekly reports on each student.

Poe found the regulations "rigid in the extreme." He and his classmates followed a curriculum of French mathematics, geometry, and engineering, featuring Lacroix, Legendre, Lagrange, and Gaspard Monge, the École Polytechnique's revered founder. The library also contained volumes by the engineer and science popularizer Charles Dupin—including a recent translation, *Mathematics Practically Applied to the Useful and Fine Arts*—a predecessor for Poe's detective, C. Auguste Dupin, who would be both a mathematician and a poet.

Considerable shelf space went to works by Pierre-Simon Laplace, one of the École Polytechnique's directors, who embodied the rationality and cosmological scope of French mathematics. In his *Essai philosophique sur les probabilités*, Laplace evoked an all-knowing intelligence who could predict any future state of the universe if only it knew its starting conditions—a touchstone for the idea of a determinist, mechanical universe. His *Traité de Mécanique céleste* perfected Newton's "clockwork universe," depicting a self-adjusting solar system whose parts interacted according to uniform laws. The book would be translated by the Yankee Nathaniel Bowditch, better known for writing *The American Practical Navigator*.

Another work of Laplace's, *Exposition du système du monde*, advanced the theory that our solar system had formed through the gradual condensation of a whirling cloud of gas, or nebula. This controversial idea, also proposed by the astronomer William Herschel, was later called the nebular hypothesis. Laplace's mechanical theory of the formation of the sun, earth, and planets was seen as a revolutionary—and theologically dangerous—replacement for the creation story of Genesis. Napoleon asked him why, in such a long book on the universe, he never mentioned God. Laplace is said to have replied, "I had no need for that hypothesis."

West Point was one of the first American institutions to use blackboards—another import from the École Polytechnique. Cadets worked out problems and proofs at the board, often simultaneously, making each step of their reasoning, and its speed, visible to inspection and correction. One student who had studied for three years at the University of Pennsylvania wrote to his parents that after three weeks at West Point "we have already gone through as much Algebra as I have looked at." Each recitation demanded mastery of a dozen textbook pages, "which it is absolutely necessary not only to be able to put on the board but to explain every particular."

Laplace fixed the image of a universe run by mechanical laws, perfectly knowable through reason and mathematics. The cadets at West Point were disciplined to align themselves, body and mind, with this cosmic machine.

Franklin's Heir One of the first students to imbibe Thayer's curriculum was Alexander Dallas Bache, who graduated first in his class without a single demerit—a feat unheard of, given the high standards of conduct and constant surveillance. Great things were expected of him. Bache's great-grandfather was Benjamin Franklin. His father was the son of Franklin's only daughter, while his maternal grandfather, Alexander Dallas, had been secretary of the Treasury.

Born in 1806, Bache had large eyes, a soft chin he later covered with a gigantic beard, and a confident, knowing, at times impishly skeptical expression. According to a friend, his "superiority in scholarship was freely acknowledged by every member of his class, while his unassuming manner, friendly demeanor, and fidelity to duty secured him the affection as well as the respect" of both students and officers. When he needed to be, Bache was charming, but he did not suffer fools gladly; when he was opposed, his resolve was unbending, and his temper was fierce.

On his graduation in 1825, Secretary of War James Barbour wrote to Bache's mother, Sophia, to tell her "how greatly I was gratified in the evidences given by your son in his examination of the excellence of his attainments . . . I knew and loved your Father [Dallas]—his great paternal ancestor [Franklin] I knew only by his works. I persuade myself to believe that I saw the excellence of both branches about to be united in your son." Bache's fellow pupils—who included his friend Jefferson Davis—formed a pact to keep him away from any foolishness that might interfere with the glorious achievements to which he seemed destined.

At West Point, Bache acquired the military perspective that he would apply in lifelong projects of scientific research and administration. After graduation, he stayed on as instructor for three years, teaching alongside the influential New York chemist and botanist John Torrey and the astronomer and mathematician Ormsby Mitchel, who would found the Cincinnati

Observatory. Just before Poe's arrival, Bache returned to Philadelphia, where both of them would live from 1838 to 1842—decisive years for both.

Encouraged from his earliest youth to fulfill the expectations of his ancestors, Bache would carve out an exemplary career for a West Point cadet. As an orphan with precarious family ties, Poe's initial conditions were as different from Bache's as his eventual trajectory. But he, too, devoured the curriculum at West Point and applied its lessons in his future career. Despite their eventual divergence, Poe's life and Bache's frequently ran in parallel.

Poe wrote to Allan of his "excellent standing," though "the study requisite is incessant, and the discipline exceedingly rigid." He spoke with General Winfield Scott, a Virginia hero of the War of 1812 he had met in Richmond. A translator of Napoleon's training manuals, Scott was nicknamed Old Fuss and Feathers for his love of pomp and ceremony. "I am very much pleased with Colonel Thayer," Poe wrote home, "and indeed with everything at the institution."

Having earned his own way for more than two years, Poe was older and more experienced than his classmates. One cadet observed his "worn, weary, discontented look, not easily forgotten," though his tentmate thought him "the greatest fellow on earth"—a seasoned adventurer whose "ambition seemed to be to lead the class in all studies." He had "a wonderful aptitude for mathematics, so that he had no difficulty in preparing his recitations in his class and in obtaining the highest marks."

Poe earned a reputation for his "forceful and vicious doggerel," often at the expense of instructors, passed around the barracks; such "poems and squibs of local interest" gave him "a high reputation for genius." He took part in the school's "Lyceum," which offered philosophical and literary lectures, "as our course of education here is almost purely scientific," one student noted, "which does not fit one very well to palaver in the world." Other publications available to cadets included Silliman's *American Journal of Science and Arts* and *The North American Review*.

The periodic revelries of West Point—a riotous celebration on July 4, a masquerade at the end of the summer encampment—gave cadets explosive relief from the school's rigors. One classmate believed Poe "had already acquired the more dangerous habit of constant drinking." A joke went about that "he had procured a cadet's appointment for his son, and the boy having died, the father had substituted himself in his place."

On arriving, Poe believed that his earlier education and army service would allow him to complete the two-year course of study in six months. He was wrong. He was in it for the long haul.

That autumn, a deadly blow to his soldierly resolve came from Richmond: John Allan was getting married again, to Louisa Patterson, daughter of a wealthy New Jersey family. A new Allan household was forming, with no place for a full-grown foster child.

Burning Bridges   The wedding took place in October in New York. Poe was not invited. "Dear Sir," he wrote to Allan (no longer "Pa"), "I was greatly in hopes you would have come to W. Point while you were in N. York, and was very much disappointed when I heard you had gone on home without letting me hear from you." He closed with formal "respects to M^{rs} A"—the new wife he had yet to meet.

Soon after, Poe's army substitute, Bully Graves, confronted Allan in Richmond for repayment of his fifty-dollar debt. He showed the letter in which Poe called Allan a drunkard. Allan wrote to Poe in a rage, denying any further obligation toward his former ward.

Poe replied defensively: "Did I, when an infant, solicit your charity and protection, or was it of your own free will, that you volunteered your services in my behalf?" He might well have remained in Baltimore with his grandfather General Poe, but this "natural protector" believed Allan's promises to adopt and educate Poe. "Under such circumstances, can it be said that I have no *right* to expect any thing at your hands?"

As for the "liberal education" Allan did provide, eight underfunded months at the University of Virginia hardly counted. "Had you let me return, my reformation had been sure—as my conduct the last 3 months gave every reason to believe." Poe only regretted returning too late to Richmond to see Frances before she died. "*Your* love I never valued—but she I believed loved me as her own child." After her death, Allan had promised "to forgive all— but you soon forgot your promise. You sent me to W. Point like a beggar." He admitted to writing the letter about Allan's drunkenness: "As to the truth of its contents, I leave it to God, and your own conscience."

Poe declared that his "future life (which thank God will not endure long) must be passed in indigence and sickness." West Point, and the lack of "necessaries" he endured there, had worn him out; he only wanted Allan's written permission to resign.

The same day he wrote to Allan, he began his midyear exams. Out of eighty-seven students, he placed seventeenth in mathematics and third in French. If he had pursued it, Poe could have had an outstanding career as an officer and engineer. Yet Allan's indifference, and his own wounded feelings, pushed him over the edge. From January 7 he stopped attending lectures, drills, and church services. At the month's close he was summoned to a court-martial for dereliction of duty. Secretary Eaton approved the judgment: "Cadet E. A. Poe will be dismissed the service of the United States."

Days later, a fellow cadet wrote to his mother in Philadelphia—the literary editor Sarah Josepha Hale, who wrote "Mary Had a Little Lamb" and had published an appreciation of "Al Aaraaf"—that Poe was considered a "fellow of talent here but he is too mad a poet to like Mathematics." Another soldier later wrote, "Edgar Poe, at West Point, was the wrong man in the wrong place—although, from an intellectual point of view, he stood high there." According to one biographer, "Poe's stay at West Point must be looked upon as an interruption of his real career."

Such observations are superficially true: from the time he left West Point, Poe earned his livelihood as a writer and editor, not as a member of the Army Corps of Engineers. Yet even without the family support and unshakable social status of a cadet like Alexander Dallas Bache, Poe excelled at West Point. In the midst of a complete dissolution of his hopes, he finished his first term near the top of his class, at the school producing the best-trained mathematicians and engineers of his country.

West Point's analytic, standardized, and systematic training prepared cadets to play crucial roles in the intellectual and industrial development of the United States. In the Army Corps of Engineers and private firms they built roads and canals and plotted maps of western territories. Some, like Alexander Dallas Bache and Ormsby Mitchel, became researchers and scientific administrators. Others systematized the production of cloth, iron, and guns. West Point engineers went on in great numbers to work in the public and private partnerships that were building the railroad. Their training prepared them to evaluate the efficiency of routes and solve logistical puzzles posed by long-distance supply chains and schedules. In these corporations, they replicated the management and disciplinary structures of West Point; as railroad employees and directors, many made fortunes.

Though Poe took quite a different path, his West Point training decisively shaped his career as poet, critic, and author. He would make constant use of his knowledge of mathematics, geometry, and astronomy—as well as the techniques of analysis and careful reconstruction he learned at the school, with a drive to excel, strategize, and command.

West Point was for Poe a turning point, not only because it marked the end of his hope for significant support from Allan. At West Point he was immersed in a thoroughly modern, mechanical way of thinking and living. This heady social experiment—a system of highly controlled violence—drilled its way into him, becoming an unshakable part of his poetic and intellectual equipment. The pride and bearing of a soldier were his. He had acquired tools that would serve him in literary campaigns in the years ahead.

Setting out now on a literary career, he would continue to ask, as a poet, story writer, and magazinist, how should he love science? How should he deem it wise?

PART II

Setting Sail

*I was ambitious—have you known
The passion, father? You have not:
A cottager, I mark'd a throne
Of half the world as all my own.*

—POE, "Tamerlane"

Magic lantern slide with simulated motion mechanism, nineteenth century

4

A Baltimore Apprenticeship

L odged in a miserable room in New York City after leaving West Point, Poe wrote to Allan in a shaky hand: "I have no money—no friends—I have written to my brother—but he cannot help me—I shall never rise from my bed—besides a most violent cold on my lungs my *ear* discharges blood and matter continually." Allan did not reply.

By spring, he was making plans. He wrote to Colonel Thayer asking for a "certificate of standing" to use "with the view of obtaining, thro' the interest of the Marquis de La Fayette, an appointment (if possible) in the Polish Army." Nothing came of the request, but 131 West Point cadets (out of 232) contributed $1.25 each from their pay to put Poe's poems into print. In April his third collection, *Poems*, was published.

He dedicated it to "The U.S. Corps of Cadets." His classmates were disappointed: it was "a miserable production mechanically, bound in green boards and printed on inferior paper, evidently gotten up on the cheapest scale." Nowhere was the sarcastic doggerel Poe recited in the barracks. Instead, they found the medieval Mongolian epic "Tamerlane," the dense astronomical fantasy "Al Aaraaf," and briefer, equally obscure pieces such as "The Valley of Nis" and "The Doomed City," describing mournful, cursed landscapes, and love poems, including the classically tinged "To Helen."

Literary critics greeted the volume somewhat more favorably. It was noticed in Philadelphia's *Saturday Evening Post* and *Casket* and received a tepid review in *The New-York Mirror*: "Every thing in the language betokens poetic inspiration, but it rather resembles the leaves of the sybil when scattered by the wind"; a few lines from "The Doomed City" were "less incomprehensible than most in the book." Another reviewer called Poe "a fellow of fine genius" and, despite the book's passages of "sheer nonsense," judged that its author "has the *gift*, and betrays the *presence*," as demonstrated in the opening lines of "To Helen":

> Helen, thy beauty is to me
> Like those Nicéan barks of yore,
> That gently, o'er a perfumed sea,
> The weary, way-worn wanderer bore
> To his own native shore.

Yet such passages of "pure poetry" stood in contrast to what the critic considered the "pure absurdity" of the "Sonnet—To Science."

Poe's preface announced his serious intent. The "great barrier in the path of an American writer," he wrote, was to be negatively compared with European poets: "It is with literature as with law or empire—an established name is an estate in tenure." In America, "fops glance from the binding to the bottom of the title-page, where the mystic characters which spell London, Paris, or Genoa, are precisely so many letters of recommendation."

Yet English poets had gone astray: Wordsworth, Poe said, mistakenly used poetry to teach morals, while Coleridge's verse was embroiled in metaphysics. Poetry, Poe argued, requires a particular form of attention; if examined too closely, both beauty and truth disappear. "Poetry is a beautiful painting whose tints, to minute inspection, are confusion worse confounded"; by contrast, the "cursory glance of the connoisseur" learns to elide the details and see the beautiful unity. The point held in astronomy as well: "He who regards it directly and intensely sees, it is true, the star, but it is the star without a ray—while he who surveys it less inquisitively is conscious of all for which the star is useful to us below—its brilliancy and its beauty," a description of "indirect" observation, a phenomenon described by the astronomer John Herschel and emphasized by the physicist and optical inventor David Brewster.

The preface announced Poe's intention to establish himself among the great poets. Yet it touched upon all three of the fields in which he sought to make a mark. While *poetry* aims at "indefinite pleasure," *romance*, or fiction, aims at "presenting perceptible images with definite" sensations, and *science* aims at truth. Poe's first published works were poems; he would soon turn his hand to stories. His fascination for science never waned.

On Mechanics Row

Poe left New York for Baltimore in the spring of 1831, to join his ailing grandmother, General Poe's widow, his aunt Maria Clemm, her daughter, Virginia, and his brother, Henry. Poe wrote to an editor there looking for work: he was "anxious to remain and settle."

With a population of just over eighty thousand, Baltimore was the nation's second-largest city in 1830, a few hundred ahead of Philadelphia and well behind New York's two hundred thousand. The city prided itself on industry. Beyond its shipyards on the Chesapeake Bay and its shot towers (large chimneys down which molten lead was dropped, turning into buckshot pellets as it cooled), it was the starting point for the country's first interstate railroad, opened just before Poe's arrival in 1830: the Baltimore and Ohio line.

The city also boasted literary and scientific societies, with more than seventy journals and magazines starting in the 1830s. It was home to a spinoff of Charles Willson Peale's American Museum, directed by Peale's son Rembrandt. Like other U.S. cities, around 1830 Baltimore witnessed several hot-air balloon launches. Balloon flights were taken both for scientific

View of Baltimore (detail), by William H. Bartlett, ca. 1830

purposes—as in the ascent of Joseph Gay-Lussac and Alexander von Humboldt, to measure the chemical composition of the atmosphere at unprecedented heights—and for thrills.

Poe's family lived near the harbor in Fells Point on Wilks Street, known as Mechanics Row. The small, crowded house stood a few blocks from a house whose residents included Frederick Bailey—later known as Frederick Douglass. Douglass was enslaved, his labor lent to his owner's brother, who hired him out to a shipbuilding firm.

The conflict over slavery was becoming the preeminent national issue. Since his election in 1828, Andrew Jackson had reshaped the presidency with the "spoils system"—replacing government officeholders with loyal followers. His policy of Indian removal by military force made cheap land available to settlers, while his belligerent appeals to white supremacy and nativism heightened a sense of lawlessness, often erupting in mob violence. Jackson presented himself as the autocratic defender of democracy and common people—which meant white males and slaveholders, particularly his friends.

In 1831, inspired by prophetic visions and a solar eclipse, Nat Turner led a slave uprising through Southampton County, Virginia, in which sixty whites and twice as many Blacks were killed. Southern planters, along with northern weavers and traders dependent on cotton, supported increasingly harsh treatment on plantations and sponsored legislative campaigns to entrench and expand slavery's claims.

The movement to abolish slavery also grew with religious intensity. The 1829 *Appeal to the Coloured Citizens of the World* by David Walker, a clothier in Boston connected to the African Methodist Episcopal Church, critiqued Jefferson's racial theories and called for active resistance to establish conditions of equality. When Walker died suddenly in 1830—perhaps poisoned—his campaign was continued by the white publicists William Lloyd Garrison and Arthur Tappan in New York, who injected the abolitionist movement with the millenarian energies of the Second Great Awakening. Driven by the expectation of Christ's imminent return, this wave of religious enthusiasm had begun in upstate New York. In revival meetings, the Holy Spirit brought convulsions, cries, and ululations. In the West and the South, Baptists caught the fever, although many southern evangelicals tamped down the gospels' implications of equality and reform.

Maryland was the northernmost of the southern states. For Frederick Douglass, Baltimore meant a temporary reprieve from the terrors of plantation slavery. "A city slave is almost a freeman, compared with a slave on the plantation," he recalled in his autobiography. "He is much better fed and clothed, and enjoys privileges altogether unknown to the slave on the plantation." In unassigned hours he studied, spurred by his observations as a laborer: "The idea as to how I might learn to write was suggested to me by being in Durgin and Bailey's ship-yard, and frequently seeing the ship carpenters, after hewing, and getting a piece of timber ready for use, write on the timber the name of that part of the ship for which it was intended." After his escape north, he would become a writer, publisher, and leading figure of the abolitionist movement—an advocate of intellectual and technical improvement, and a fiercely informed opponent to the American sciences of race.

For Poe, Baltimore meant living among caring relatives—and poverty. His brother, Henry, remembered their dead mother and could describe her; he had picked up contacts in the newspaper business, along with a severe drinking habit. He worked sporadically at best. Their grandmother received a small pension in recognition of the General's wartime service. Other Baltimore branches of the Poe family were well-off, but the fortunes of Maria Clemm declined with her husband's death; she took up irregular work as a teacher. The poor had no social net beyond the aid of friends and neighbors. Worse yet, a cholera epidemic appeared on the East Coast starting in 1832, taking thousands of lives. Every touch and breath was imbued with the fear of contagion.

Poe sought work as a teacher and an editorial assistant. A West Point graduate spotted him working as a day laborer at a kiln, making bricks. He was also writing short stories—encouraged by a competition announced in the Philadelphia *Saturday Courier* in June 1831, with a prize of a hundred dollars. To his friend Lambert Wilmer, a journalist at the *Baltimore Saturday Visiter*, Poe seemed "one of the most hardworking men in the world. I called to see him at all hours, and always found him employed."

While crafting his tales on Mechanics Row, Poe ruminated on the general conditions of literary production. What made for a successful journal, a prizewinning story? His chief literary models came from Edinburgh's

Blackwood's Magazine, which featured first-person accounts of life-threatening scrapes—Poe called these "tales of effect" or "sensation"—as well as comic tales mocking literary fashions and exaggerating the foibles of high society, which Poe called "grotesques."

To the *Saturday Courier*'s contest Poe submitted not one but five richly varied tales, each drawing on his wide reading. They included a parody of the stories of the American novelist Nathaniel Willis, showing a monarch in hell who saves himself by cheating the devil at cards, and "A Tale of Jerusalem," a fable written in a biblical style.

On August 1, 1831, he faced a new tragedy. Battered by alcohol and hard living, his brother, Henry, died, aged twenty-four. Henry's death drew him closer to his aunt and his young cousin Virginia, whom he affectionately called Sissy. Wilmer visited Poe and found him "giving Virginia lessons in Algebra"; passing a funeral on a walk, Virginia "became affected and shed more tears than the chief mourner. Her emotion communicated itself to Poe."

After Henry's death, Poe wrote to John Allan in a mood of regret and reminiscence: "When I think of the long twenty one years that I have called you father, and you have called me son, I could cry like a child to think that it should all end in this." He was briefly out of debt and wanted nothing: "It is only at such a time as the present when I can write to you with the consciousness of making no application for assistance, that I dare to open my heart." Yet he was "truly conscious that all these better feelings *have come too late*."

By November, Edgar was eighty dollars in debt, and creditors were knocking. He again wrote to Allan, his earlier tranquility gone: "I am in the greatest distress and have no other friend on earth to apply to except yourself if you refuse to help me I know not what I shall do. I was arrested eleven days ago for a debt which I never expected to have to pay"—because it was Henry's. There is no record of Poe's imprisonment, but half of the inmates of Baltimore's jail were there for unpaid debts. In December he begged Allan, "Do not let me perish for a sum of money you would never miss." Allan started a letter to Poe on December 7, enclosing a hundred dollars, but neglected to post it until January 12.

In December, Poe lost the short story contest, his stories defeated by a sentimental tale by Delia S. Bacon (the leading advocate of the theory that Shakespeare's works had been written by her ancestor Francis Bacon). The *Courier* nevertheless published Poe's tales over the next several months, starting with "Metzengerstein"—a story of gothic revenge, "in Imitation of the German," about a demonically possessed stallion—with no payment beyond publicity. This, too, was a valuable prize, though hard to eat. The *Baltimore Saturday Visiter* observed that "few American authors in our opinion have produced anything superior." Starvation, illness, and prison were chasing him, but his work was being published and read.

The Folio *Sampler*   Emboldened by this response, he sent a tale to *The New-England Magazine*, describing it as part of a larger collection "supposed to be read at table by the eleven members of a literary club." Each member was modeled on a well-known literary figure, and their comments on the tales would make for "a burlesque upon criticism." He offered the entire set: "If you like the specimen which I have sent I will forward the rest."

He called the collection *Tales of the Folio Club*. Each was written in a distinct style, exaggerating the conventions and clichés of established genres and authors, often uproariously. The starting point for "A Decided Loss" was *Blackwood's* breathless "tales of effect." Taking the word "breathless" literally, it tells of a man who "loses his breath" while berating his wife. Searching for the elusive object, he gets smothered by a coach passenger, taken for dead, prepared for burial, nibbled by cats, put in place of a condemned criminal, and hanged. Still alive—still breathless—he finds himself on a dissecting table, yet even his kicks are ignored, "attributed to the effect of the new Galvanic Battery" that an experimentally inclined doctor applies to what he takes for a corpse. The tale's manic play upon the relation between body and spirit masks its grim autobiographical truth: it is the confession of an anxious, overeducated man whose life-threatening indignities are met by indifference.

Poe's other proposed *Folio Club* tales were modeled on German fantastic tales, biblical parables, and fashionable "silver fork" novels, best represented by Edward Bulwer-Lytton's novels and *Vivian Grey* by Benjamin Disraeli. First published anonymously, *Vivian Grey* followed an ambitious young man's rise from poverty to literary fame, satirizing along the way London's "lions" and "bluestockings," male celebrities and female authors and readers.

In Poe's "Lionizing," the hero builds a career on the admiration prompted by his remarkable nose; he pens a treatise titled "Nosology" that sets the literary world atwitter:

> "Fine writer!" said the Edinburgh . . .
> "One of us!" said Blackwood.
> "Who can he be?" said Mrs. Bas-Bleu [Poe's quasi-French for "bluestocking"].

In this burlesque of Disraeli's satire about fame's fickleness, success means being "lionized," crowned literary king for a day.

The location in which Poe composed these tales was far from the genteel drawing rooms of *Vivian Grey*. In 1833 he and his family moved from Mechanics Row to a narrow row house on Amity Street, five blocks north of the Baltimore and Ohio train yard.

Yet the industrial setting was in peculiar harmony with the *Folio Club*'s literary experiments. Poe treated literary "genres" as a form of mass production. Applying the habits of his engineering training to the writing of fiction, Poe surveyed the field, analyzed the construction of earlier products, and applied these formulas in a series of works of his own; the result, the *Folio Club* tales, was a display case of "specimens"—samples of his wares. Poe's first batch of stories was grounded in a logic akin to the system of standardized manufacture at Harpers Ferry, which Eli Whitney borrowed from French precedents, just as Poe drew on European literary models.

Identifying and replicating the principles of a genre or style did not have to mean dull repetition. Poe optimized his formulas, magnifying them into "grotesques," or rarefying them into more concentrated forms. While Congress debated Henry Clay's American System—the project to use the federal government to fund and coordinate a unified infrastructure of roads, canals, railroads, and communications—Poe, steps from the first successful railway to western hinterlands, was tinkering with the springs and motive powers of the American system of literature.

The Author Unbottled   Another prize competition, announced in the *Baltimore Saturday Visiter* in June 1833, gave Poe a chance to test his formulas. One judge was John Latrobe—son of Benjamin Latrobe, "the father of American architecture," who helped design the city of Washington, D.C., the Capitol, the White House, Philadelphia's National Bank, and extensive works in Baltimore and New Orleans. John Latrobe had attended West Point and would make technical improvements to steam engines; he worked as a lawyer and adviser for the B&O Railroad. Another judge was J. H. Miller, a physician and founder of Baltimore's Washington Medical College. The third was John Pendleton Kennedy, a thirty-eight-year-old former navy officer and novelist whose *Swallow Barn*, published in 1832, offered a rosy view of life on a southern plantation—though hearing Frederick Douglass speak would later convert him to abolitionism. He would become a congressman and secretary of the navy and was considered a possible running mate for Abraham Lincoln.

All three judges were concerned with national improvement and rational reform. Associated with the Whig Party, they were continuing the Federalist vision of strong national government led by an intellectual and moral elite. The Whigs rose in the 1830s in reaction to Jackson's "democratic republicanism," with its localist tendencies and strong support of slavery. Latrobe and Kennedy were involved with the B&O line; both would help Samuel Morse secure a federal patent for his telegraph.

These proponents of American innovation were perfectly prepared to recognize Poe's combination of scientific acuity, classical learning, wit, and daring imagination. According to Latrobe, the collection of tales Poe submitted "was so far, so very far superior to anything before us, that we had no difficulty in awarding the first prize to the author." They gave his classical poem "The Coliseum" second place for poetry (simply to avoid awarding him twice, Poe thought). Their only difficulty was selecting just one of his tales.

The judges summoned him to their offices. Latrobe was intrigued by this fellow West Point cadet "who carried himself erect and well, as one who had been trained." Poe was "dressed in black, and his frock coat was buttoned to the throat, where it met the black stock, then almost universally worn. Not a particle of white was visible." Though his clothes had seen better days, "there was something about this man that prevented one from criticizing his garments." His dignity shone through: "Gentleman was written all over him."

Latrobe was entranced by Poe's powers of imagination: "To judge by the outward man, the world was then going hard with him . . . yet his appearance was forgotten, as he seemed to forget the world around him, as wild fancy, logical truth, mathematical analysis, and wonderful combinations of fact flowed, in strange commingling." Latrobe was struck with "the power that he seemed to possess of identifying himself with whatever he was describing. He related to me all the facts of a voyage to the moon, I think, which he proposed to put upon paper, with an accuracy of minute detail and a truthfulness as regarded physical phenomena, which impressed you with the idea, almost, that he had himself just returned from the journey."

First column of "MS. Found in a Bottle," prize announcement in Baltimore Saturday Visiter, October 19, 1833

The judges bestowed the prize on "MS. Found in a Bottle," a grippingly realistic tale of a shipwreck and an encounter with the Flying Dutchman, the ghost ship of legend. A skeptical man of science sailing from Sumatra believes only his senses. After a wreck, he boards a ship piloted by superhuman sailors on a trancelike mission, guided by "singular-looking instruments, and decayed charts of navigation." As the boat approaches a terrifying whirlpool, he finds himself "hurrying onward to some exciting knowledge—some never to be imparted secret, whose attainment is destruction."

The story's power arose from the contrast between resolute factuality and wild, seemingly supernatural events—much like the polar explorations and alpine chases of Mary Shelley's *Frankenstein*, which also shifted the gothic tale out of the mansion into natural extremities. In Poe's skillful hands, scientific language and concepts heightened the dreadful sense of a catastrophic revelation just ahead.

Thanks to this last-ditch communication of a dying man, Poe now had fifty dollars to placate his creditors, and powerful new allies. But he was still in danger. He wrote to tell John Allan that he was "perishing—absolutely perishing for want of aid."

Allan himself was ailing, confined to his bedroom, "greatly distressed by dropsy." Poe went to Richmond for a last visit. At first refused entry at Moldavia, he rushed in past the second Mrs. Allan and up to the bedroom, where, according to local rumors, "Mr. Allan raised his cane, & threatened to strike him if he came within his reach, ordered him out; upon which Poe withdrew, & that was the last time they ever met."

Allan died on March 27, 1834, in possession of eight houses, plantations, dozens of slaves, and business concerns worth upwards of $750,000, over fifteen million today. His will left substantial provisions for his sons by the second Mrs. Allan, less for his three illegitimate children, and nothing for Poe. The orphan was utterly disowned.

Poe wrote to John Pendleton Kennedy to ask for help finding a teaching job. When Kennedy invited him to dinner, Poe admitted that the invitation "has wounded me to the quick. I cannot come—and for reasons of the most humiliating nature in my personal appearance." He lacked the minimum kit for bourgeois decency. Kennedy "found him in a state of starvation. I gave him clothing, free access to my table, and the use of a horse for exercise whenever he chose—in fact brought him up from the very verge of despair."

The "Berenice" Experiment Kennedy's paternal support opened another door. Thomas W. White, a printer from Richmond, was launching a new monthly periodical, the *Southern Literary Messenger*. With a good word from Kennedy, White commissioned Poe to write reviews and publicity.

When White agreed to consider one of his short stories, Poe sent his shocking *Folio Club* tale "Berenice"—another "tale of sensation." Poe compared it to Thomas De Quincey's *Confessions of an English Opium-Eater* and the "MS. Found in a Madhouse" published in London's popular *New Monthly Magazine.*

"Berenice" brought the reader into the consciousness of a nervous scholar suffering from "monomania"—a diagnosis recently announced by the French alienist Étienne Esquirol. He grew morbidly obsessed with his fiancée's teeth: "Would to God that I had never beheld them." His betrothed, Berenice, would agree; she met a gruesome fate through unholy late-night dentistry. The story conveyed the murky states of an unmoored intelligence feeding on its own fixations. This was one of the first, most disturbing of Poe's unreliable narrators; through their voices, Poe told readers exactly what it felt like to be driven to acts of horrific extremity, detailing the bizarre perceptions and self-justifications of the deranged.

Poe admitted to White that the subject of "Berenice" was "far too horrible"; he had hesitated before sending it as "a specimen of my capability." Yet he was offering the tale not for its content but as a test of his theories of popular literature.

"The history of all Magazines," he confidently explained, "shows plainly that those which have attained celebrity were indebted for it to articles *similar in nature to Berenice.*" Such articles consist of "the ludicrous heightened into the grotesque: the fearful coloured into the horrible: the witty exaggerated into the burlesque: the singular wrought out into the strange and mystical." (These could have been the tags attached to the specimens from the *Folio Club.*) "Berenice," he said, was the result of sound market research: "Whether the articles of which I speak are, or are not in bad taste is little to the purpose. To be appreciated you must be *read*, and these things are invariably sought after with avidity. They are, if you will take notice, the articles which find their way into other periodicals, and into the papers," building the reputation of the magazine that first published them.

Poe was describing the conditions of success—for magazines, for authors—of his place and time. The way to raise a magazine's profile was to publish striking, memorable work (even if it was frightening or in bad taste) and to depend on other magazines and newspapers to discuss, debate, and reprint it. In a logic familiar from today's social media economy of clicks, likes, and retweets, the value of a text lay in how many other texts talked about it, excerpted it, or reprinted it.

Not just any work would do, Poe claimed: "Originality is an essential in these things—great attention must be paid to style, and much labour spent in their composition, or they will degenerate into the turgid and the absurd." Thus "Berenice" was a skillfully crafted, rare—and therefore valuable—commodity. Poe claimed to possess its general recipe and was willing to use it on White's behalf.

He offered to furnish the *Messenger* with a new tale each month: "No two of these Tales will have the slightest resemblance one to the other either in matter or manner." The results would speak for themselves: "The effect—if any—will be estimated better by the circulation of the magazine than by any comments upon its contents."

Poe took an anticipated criticism of his tale, turned it into an argument about how to build a magazine's reputation, and closed with a bet. He was offering his wares as a brag, a wager, a business proposition, and as a well-prepared experiment.

In June 1835, White raised the possibility of a job for Poe in Richmond. Poe lunged. "Nothing would give me greater pleasure," he declared. "Indeed I am anxious to settle myself in that city."

He left behind his aunt and cousin, with plans to bring them soon. With John Allan dead, Edgar Poe, twenty-six years old, might now claim a position in the city of his childhood on his own terms.

5

Richmond: The Palpable Obscure

The Prodigal Returns   Thomas White, a mild, round-faced, curly-headed man, possessed the means and energy to launch the *Southern Literary Messenger* but doubted his own judgment. He had to rely on the counsel of his contributors, many of them prominent figures in southern intellectual and political circles. When Poe arrived, White needed advice and assistance, though he was ready to override Poe's every decision.

White saw the *Southern Literary Messenger* as part of the lyceum movement: a high-minded but popular vehicle to elevate the intellectual life of his city, state, and region. Literary progress was akin to economic development, he thought: a journal "imparts the same energy, and exercises the same influence upon mental improvement, that a rail road does upon agricultural labor." But "south of Washington," he complained, "there are but two Literary periodicals."

White eagerly published reports from the newly founded Virginia Historical and Philosophical Society, hoping to spur a movement "to establish something like a literary and scientific character for the Old Dominion. Is there not something, besides politics, worth living for?" He wanted the *Messenger* to bring his region into the national march of progress, "to see the North and South bound endearingly together forever, in the silken bands of mutual kindness and affection."

This national mission meant White and his editors had to step carefully to avoid "controversial theology" and "politics"—especially slavery. Abolitionism was gathering steam, as was the movement to entrench slavery further. In New York, Philadelphia, Boston, and Baltimore, white workers who feared competition were rioting and attacking free Blacks and abolitionists, encouraged (and sometimes paid) by industrialists who foresaw higher prices for southern goods. The fight over slavery was becoming the polarizing, life-or-death conflict that would erupt as the Civil War.

White did not want his journal to be dismissed as a vehicle for pro-slavery views. The *Messenger* only rarely approached the topic. Occasionally it published works in support of African colonization (to send liberated Blacks to Africa—a mission seen by abolitionists as an unacceptable halfway measure). In 1836, unusually, it ran an unsigned review of two books defending the "peculiar institution": *The South Vindicated from the Treason and Fanaticism of the Northern Abolitionists*, later attributed to William Drayton, and *Slavery in the United States*, by the New York novelist James Kirke Paulding. The reviewer, most likely Nathaniel Beverley Tucker, a novelist, judge, and law professor at William & Mary, claimed that slavery improved Africans' conditions and character; as John Calhoun declared, slavery was not a necessary evil but a "positive good."

Following the Vesey rebellion—a foiled plot to liberate thousands of slaves in Charleston in 1822—elite southern opinion had hardened. Arguments in favor of slavery grew more aggressive. Meanwhile, the omnipresent ideology of white supremacy went largely unquestioned, even by many of slavery's opponents. At the *Messenger* and afterward Poe largely avoided offering an opinion on slavery, keeping within the bounds of what the literary critic Terence Whalen has called "average racism": the baseline of pre–Civil War attitudes among whites from both the North and the South. Like Thomas White—and like all white Americans—Poe benefited from the system of chattel slavery and racial exclusion in myriad ways. In 1940 a journalist found evidence in the Baltimore Court House that in 1829 Poe, aged twenty, acted as an "agent for Maria Clemm" and sold a twenty-one-year-old enslaved man named Edwin to a "labourer" named Henry Ridgway, listed by a city directory as a "person of colour." The bill of the sale—for forty dollars—testifies to slavery's inhuman complicities.

The polarization between "white and Black" that so rigidly framed the U.S. social order was deeply engraved in Poe's experience and imagination.

Some of his tales contain crude stereotypes of African Americans, and his use at times of light and dark imagery to figure oppositions between reason and ignorance, between spirit and matter, resonated with a sense of the "African presence" as a fearsome other above which to prop up a fragile white American identity. As Toni Morrison wrote, "No early American author is more important to the concept of American Africanism than Poe."

Yet if Poe's writings often echoed his era's hierarchies of race—and sex, and class—he was also sharply aware of the violence, threat, and fear behind "polite" divisions and the cruelty with which they were enforced. A perpetual outsider, he repeatedly staged subversions and reversals of hypocritical and unjust hierarchies in his writings. Though Poe largely avoided overt political statements, he sympathized with the downtrodden and the mad, recognizing the "perverse" and destructive impulses that motivate even supposed paragons of reason and virtue. Far from a lifelong partisan of the South, or a nationalist blind to the faults of the United States, or even an advocate of what passed for modern civilization, Poe would take a frightful pleasure in imagining the downfall—or apocalyptic destruction—of the world in its present state.

Lifting the Level

As he started work in the offices of the *Messenger*, just a few streets from his childhood home on Main Street, Poe's chief concern was the state of American letters and learning. Dutifully taking up White's project of enlightenment, his reviews urged educational reform and government-funded schooling.

He promoted accessible works of general information, lauding, for example, John W. Draper's *Introductory Lecture to a Course of Chemistry and Natural Philosophy*. Draper, an energetic young physics professor relocated from England to Virginia's Hampden-Sydney College, whirled readers through new discoveries in chemistry and optics, Laplace's astronomy, mechanics, and physiology. Practically hopping behind his lectern, Draper declared that "the last half century has added more to human acquirements than the preceding thousand years"; he and his contemporaries "tremble upon the brink of discovering the elementary constitution of the material world. We can feel as it were the molecules of light itself, that most subtle of all fluids. We can

Office of the Southern Literary Messenger, *Richmond*

almost perceive their sides and their ends, and can actually control, regulate and arrange the constituent parts of a *sunbeam!*" Draper would soon move to New York University, to direct its teaching in natural sciences and harness sunbeams in research on photography.

Poe lavished praise on *The American Almanac and Repository of Useful Knowledge* and on the *Cabinet Cyclopaedia* of Dionysius Lardner, a popular Anglo-Irish lecturer on science and mechanics. He found *The Western Journal of the Medical and Physical Sciences* "highly creditable to Cincinnati" and recommended *The American Journal of Science and Arts*, the only national science periodical, published in several cities simultaneously and edited by the Yale chemist Benjamin Silliman, for its "well written articles of pure science" of "universal and practical interest."

In the *Messenger*, Poe compiled his own informative series, "Pinakidia," offering it as a storehouse of knowledge to be "sprinkled into any work" to give it an air of profundity—such as the fact that Erasmus Darwin's "Great Poem" on "the loves of plants" borrowed much of its content from a French predecessor, *Connubia florum*. He drew his facts about ancient and rare texts from other reference works, including the philological tables of the Columbia professor Charles Anthon and the Baron Bielfeld's *Elements of Universal Erudition*—his go-to search source for intellectual arcana.

Each month, Poe also printed a new tale. One of his first, "The Unparalleled Adventure of One Hans Pfaall," showed a new way to "love science" and "deem it wise"; it used scientific facts to heighten the realism of a lighthearted tall tale. Hans Pfaall, a natural philosopher, flees from his creditors in Rotterdam in a hot-air balloon. Free from the earth's gravitation, his vehicle flips and is drawn *downward* to the moon and its strange flora and intelligent quadrupeds. On his return, the flying Dutchman's exploits are greeted by Rotterdam's astronomers as a hoax, "but hoax, with these sort of people, is, I believe, *a general term for all matters above their comprehension.*"

The *Richmond Whig* cheered the tale: "We have never perused one which caused such a dizziness of sensation"; its author's speculations "assume a true philosophical character, exhibit genius and invention." Poe drew from various scientific and literary sources and specifically noted John Herschel's 1833 *Treatise on Astronomy*. Herschel—son of William Herschel, the astronomer who discovered Uranus and inventoried hundreds of nebulae—had recently moved to the British colony of South Africa to direct one of the world's largest observatories. "Hans Pfaall" extended science's imperial reach to the moon.

At first Poe seemed to think of this story as a satire; it was introduced as "a capital burlesque upon balloonings, which have *recently* been carried to a ridiculous extent." The tale's two essential characteristics—its impossible absurdity, and its exact and plausible details—pulled in opposite directions: it was either a spoof, mocking credulous enthusiasm for science, or a hoax, making use of that very credulity to make readers believe the unbelievable. This jarring instability made it a story to be admired, rather than loved—though, as Poe would soon learn, it was also a story to be imitated.

Family Ties   Within months, Poe's "Berenice" experiment—printing a strikingly original tale each month—had the effect he predicted. The *Messenger's* reputation and subscriber list grew. Poe and White had favorable reviews of the *Messenger* printed on the green paper that wrapped each issue, making readers aware of the value of the object in their hands. The writer James Kirke Paulding testified that the *Messenger* "is decidedly superior to any Periodical in the United States, and Mr. Poe is decidedly the best of all our going writers." Another reviewer described the artist at work: "He first touches so beautifully on what is sure to interest the reader . . . just the outline, with all left to the imagination that deserves to be filled by itself." For a Virginia critic, "Mr. Poe possesses an extraordinary faculty. He paints the *palpable obscure*, with strange power."

The novelist Philip Pendleton Cooke wrote to White, praising Poe as "the first genius, in his line in Virginia," and urging him to value Poe "according to his merits, which are exceeding great." In late 1835, White promoted Poe to editor, with a sixty-dollar monthly salary.

Poe still felt untethered. In Baltimore he had grown dependent on the comforts provided by his aunt and cousin. Virginia was now thirteen, exactly half Poe's age, cheerful, loving, and clever. They planned to marry, out of convenience and family feeling and out of what Poe proclaimed as a devoted passion—a passion at odds with today's sensibilities and viewed with some suspicion at the time.

Though marriage between cousins, sometimes even with great age disparities, was fairly common, Virginia was "small for her age" and appeared childlike. Even Poe's family saw the match as less than ideal. When the cousins' grandmother died, Maria Clemm announced a change of plan: Virginia would not be coming to Richmond, but would move in with another cousin, Neilson Poe. A prosperous lawyer in Baltimore, Neilson could provide Virginia with an education and entrance to society and steer her to a more suitable marriage.

Poe reacted in a letter drenched with raw emotion. "My dearest Aunty," he wrote, "I am blinded with tears while writing . . . I have no desire to live and *will not*. But let my duty be done. I love, *you know* I love Virginia passionately devotedly. I cannot express in words the fervent devotion I feel towards my dear little cousin—my own darling." He described the "sweet little house in a retired situation on Church hill" he had found and "the pride I would take in making you both comfort[able] & in calling her my wife." His salary would give them "comparative comfort & happiness." He could not promise wealth, but asked, "Do you think any one could love her more dearly than I? She will have far—very far better opportunity of entering into society here than with N.P. Every one here receives me with open arms."

Portrait of a girl (Virginia Poe), by Thomas Sully, ca. 1836

He left the decision to Virginia: "Let me have, under her own hand, a letter, bidding me *good bye*—forever—and I may die—my heart will break—but I will say no more."

Poe's melodramatic response—and its attempt at emotional blackmail—might have owed something to *Politian*, the tragedy of love, betrayal, and murder he was then writing, named for the Renaissance scholar and poet Poliziano and based on a famous scandal of passions in Kentucky. Its first installments, published in the *Messenger*, were criticized so vigorously that he abandoned it, though he would return to a similar setting in his plague tale, "The Masque of the Red Death." He might also have been partaking in the local custom of a constitutional julep (brandy with mint and sugar over ice) in the morning. It still took very little to make him not merely drunk but dangerously volatile. According to a printer in White's office, "Mr. Poe was a fine gentleman when he was sober. He was ever kind and courtly, and at such times every one liked him. But when he was drinking he was about one of the most disagreeable men I have ever met." White, who never drank, found Poe to be "unfortunately rather dissipated," and in September 1835, after Poe made a nuisance of himself, White told him to leave.

From Baltimore, Poe asked for a second chance. White believed his promises to reform were sincere, but feared that Poe's resolve would falter and he would "again sip the juice." He set firm conditions: "It must be expressly understood by us that all engagements on my part would be dissolved, the moment you get drunk. No man is safe who drinks before breakfast!"

Poe returned to the *Messenger*. Maria Clemm and Virginia soon joined him in Richmond, taking rooms in a boardinghouse on Bank Street near Capitol Square. Personally and professionally, things were looking up. In January 1836 he wrote to John Pendleton Kennedy, "My health is better than for years past, my mind is fully occupied, my pecuniary difficulties have vanished, I have a fair prospect of future success—in a word all is right."

On May 16, 1836, Edgar and Virginia were married in a Presbyterian ceremony, with White and his family as witnesses. Their marriage certificate was accompanied by an affidavit declaring that Virginia was "of the full age of twenty-one"—rounding up by eight years. Poe told Kennedy she was fifteen.

The role played by Virginia's own wishes in the decision to marry her much older cousin is unknown. Yet whatever physical form the marriage took at its start or over the years (the historical record is as mute as speculations have been extravagant), by all contemporary accounts the love between Edgar and Virginia Poe was patient and affectionate, emotional and intellectual, generous and mutual. After the violent clashes of the Allan home and years of isolation, the bonds of family life—however odd—had become indispensable to him.

Puncturing the Puff   At the *Messenger*, Poe immersed himself in the informal economy of journalism, counted out in reviews, notices, allusions, and excerpts. Each magazine would review others, trading praise and occasional insults; blocks of text, phrases, or entire essays and stories were lifted from one journal and reprinted in another, though often without attribution. A crucial element in these exchanges was "puffing": building up excitement for works by one's friends and allies through favorable advance notices and frothy reviews—the "hype" of the nineteenth century.

Inspired by the fierce literary battles of British journals such as *Blackwood's* and *The Spectator*, Poe would occasionally spice up the *Messenger's* monthly serving of judicious reviews with a flamboyant hatchet job. Though he was a beneficiary and contributor to puffing, he positioned himself as an independent voice standing outside the system.

One of his first targets was *Norman Leslie: A Tale of the Present Times*. Its author, Theodore Fay, was a well-established editor of *The New-York Mirror*—which praised and excerpted Fay's book for four consecutive issues before it was published. The editor of New York's *Knickerbocker*, Lewis Gaylord Clark, raved about the novel's "scenes of great power" and "quiet ease of style and purity of diction" while admitting that he had not actually laid eyes on it.

"WELL!—here we have it! This is *the* book—*the* book *par excellence*—the book bepuffed, beplastered, and be-*Mirrored*," Poe's review began. "For the sake of every thing puffed, puffing, and puffable, let us take a peep at its contents!" He mercilessly ridiculed its improbable, convoluted plot—a love story played out in America and Europe—and deemed Fay's style "unworthy of a schoolboy." He concluded, "Thus ends the Tale of the Present Times, and thus ends the most inestimable piece of balderdash with which the common sense of the good people of America was ever so openly or so villainously insulted."

Journals nationwide applauded Poe for daring to attack the New York literary set. Fay replied with a comedy sketch, "The Successful Novel," which mocked Poe by portraying him as the bilious "Bulldog" of "the *Passenger*." Poe's takedowns brought scandalous success to the *Messenger*. He targeted popular writers with inflated reputations who could afford to be brought down a notch. In an era when public criticism might lead to a duel or a physical attack, Poe's harshest reviews were depicted as gladiatorial entertainment and lawless violence: his writing was likened to "flaying," "cutting," and "dissecting," using a "scimitar" or "iron club." Eventually, as Jackson's program of Indian removal continued, fueled by scare stories of atrocities, Poe was called "the Tomahawk."

As Poe spelled out in his own defense, however, most of his reviews were careful and evenhanded, ruled by explicit critical principles. His guiding notion was the *totality* or *unity of effect*: every element of a poem or tale should contribute to a single, deliberate impact. For instance, while he praised a novel about a voyage to Niagara Falls by Nathaniel Willis (a New York author who had torched Poe's first book of poems), he also noted its drifts between dry comedy, romance, and sublime nature writing. Any writer "neglectful of the *totality of effect*," he cautioned, would soon be "discharged from the memory of man."

He also applied the criterion of unity of effect to verse. He defined poetry by its anticipation of a more-than-earthly ideal: "the sentiment of Intellectual Happiness here, and the hope of a higher Intellectual Happiness hereafter." In another trope of romantic aesthetics—indebted to German idealism, Coleridge, and Victor Cousin, sources also vital for the transcendentalist philosophy then forming in Massachusetts—Poe claimed that poetry in its "intangible and purely spiritual nature" aimed beyond the "tumultuous chaos of human intelligence," toward the "evergreen and radiant Paradise which the true poet knows." He gave the faculty of the imagination a supreme role: "Imagination is [poetry's] soul. With the *passions* of mankind . . . it has no inevitable, and indeed no necessary co-existence."

Despite this lofty idealism, Poe also sounded earthier, even mechanical notes, taking a stand against Coleridge's influential view of the imagination as a quasi-divine faculty. In a review of James Dalton's *Peter Snook*, he downplayed the notion that the artistic imagination created ex nihilo. "To originate," he argued, "is carefully, patiently, and understandingly to combine." Mary Shelley had advanced a similarly materialist view of imagination in the introduction to her 1831 edition of *Frankenstein*, declaring that invention "does not consist in creating out of a void, but out of chaos; the materials must, in the first place, be afforded: it can give form to dark, shapeless substances, but cannot bring into being substance itself."

Poe was working out a paradox. Through sound and sense, imaginative writing led the listener toward an intangible, spiritual ideal. Yet the poet labored not like a spirit or a God but as an *artificer*, a craftsman, perhaps a mechanic: assembling concrete materials into novel arrangements.

At times he would heighten this paradox, swerving between a visceral realism and disembodied flights through spirit realms. Eventually he would set out a unique philosophical vision in which matter and spirit could be rapidly juxtaposed or fused—two sides of a single metaphysical coin.

Bat Men in Gotham   The contradictions in Poe's poetic theories and his love for "the palpable obscure" were of a piece with the philosophical, technical, and scientific tensions of his time. In the age's new sciences—hastily communicated in new journals, lyceums, and lecture halls, carried by rail and steamboat—ideas about the relationship between spirit and matter and between God and humans were being scrutinized, challenged, and overthrown. New discoveries and sciences, along with the frantic pace and relentless surprises of the press, left readers in serious doubt about who and what to believe.

For example, the new and controversial science of phrenology claimed to explain human character by empirical observation of the skull. Launched in the late eighteenth century in Vienna by Franz Gall and promoted in Paris, it gained support among the rising middle classes in Britain and the United States. Phrenology held that people's characters and mental capacities— amativeness (or love), industry, and ideality (or imagination)—were located in different organs of the brain. Because the skull followed the shape of the brain, protuberances in different parts of the head indicated larger or smaller organs for each trait.

Phrenology was an engaging, hands-on science. Sizing up friends and enemies by their appearance and watching others have their character read from the bumps on their skulls brought endless amusement. Dealing in topics of universal interest, it could be mastered with individual study. It also courted controversy. The Scottish phrenologist George Combe's popular *Constitution of Man* flatly declared that "mental qualities are determined by the size, form and constitution of the brain"; despite his arguments for the science's compatibility with Christianity, such statements seemed to deny faith in an eternal soul independent of the body.

Reviewing the fourth edition of a phrenology textbook, Poe said grudgingly, "We might as well make up our minds to listen." A few months later, he scolded another writer for attacking phrenology without studying it. By March 1836, in a review of *Phrenology* by Mrs. L. Miles—sold with printed cards and a ceramic head—phrenology had, Poe decided, "assumed the majesty of a science; and, as a science, ranks among the most important." Phrenology's methods and concepts would run through his criticism and fiction.

In the summer of 1835 another popular scientific sensation came into view: the return of Halley's Comet, last spotted in 1759. Entrepreneurs set up a telescope in City Hall Park in New York, charging six cents for a peek. In the throes of this "peculiar mania" P. T. Barnum observed that "the whole community at last were literally occupied with but little else than star-gazing."

An anonymous newspaper series in the New York *Sun*, the first of the penny dailies sold by the "newsboy system" on street corners, tapped the summer's astronomical fever. In August 1835, the *Sun* ran a front-page exclusive: "Discoveries in the Moon," said to be reprinted directly from *The Edinburgh Journal of Science*. It detailed what John Herschel at the astronomical observatory in South Africa saw with a gigantic telescope; its images were thrown onto a wall using a "hydro-oxygen" lamp. This new form of illumination—applying the chemical principle of theatrical limelight, developed by Robert Hare and Michael Faraday—was powering new and dazzling magic lantern shows in

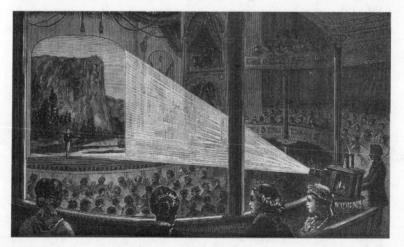

Magic lantern lecture with oxyhydrogen illumination

lyceum halls by projecting natural wonders and brightly colored fantasias. Attached to a microscope, oxyhydrogen magic lanterns could expose hidden worlds of insects, tissues, and tiny natural structures.

In the *Sun* report, this optical technology, attached to a telescope, revealed the world on the moon. On the walls of the Cape Town observatory Herschel and his colleagues observed caverns colored like rubies, vast lakes, soaring mountains, lush forests, and, on closer inspection, horned bears, zebras, and blue unicorns. Most staggeringly, the fifth installment reported, they saw humanlike creatures standing on two legs and flying with wings, along with evidence of the civilization of these "man-bats," including perfect pyramids crowned by reflecting globes.

With the moon story the newspaper's circulation rose to more than seventeen thousand, ten times its competitors'. The series was excerpted and discussed across America and the Atlantic. Visitors bombarded Herschel with questions about lunar life. According to one legend, Yale University's astronomy professors Elias Loomis and Denison Olmsted came to *The Sun*'s office demanding to see the original *Edinburgh* report; the editor sent them on a wild-goose chase from printer to printer. The series ended in catastrophe: the final installment reported that Herschel's telescope, directed at the sun, created a beam of light so intense that the observatory caught fire and burned. At the end of August, the *Herald* debunked the story by documenting its contradictions. Its anonymous author, the abolitionist and reformer Richard Adams Locke, confessed his identity in print—four years later.

"Discoveries in the Moon" appeared two months after Poe's "Hans Pfaall"—also a realistic tale of lunar exploration. Poe publicly accepted Locke's claim not to have seen the earlier tale, but whether or not Locke had been inspired by Poe, the "Moon Hoax" taught Poe memorable lessons. Surprising facts offered a grip for readers' imaginations, especially when expressed in the language of technical proof and precise observation. "Not one person in ten discredited" the account, Poe noted. Even those who did not believe were eager to buy and debate it.

Like Daniel Defoe's *Robinson Crusoe*, which Poe reviewed in a new illustrated edition from Harper & Brothers, Locke's "Discoveries in the Moon" possessed the rare quality Poe named "the potent magic of verisimilitude." Defoe wove a literary spell through his "great power of abstraction" and strong "faculty of identification." Overpowering the reader's imagination through force of will, his profound technical art became invisible. The story became indistinguishable from life itself.

Locke's "Discoveries in the Moon" suggested that both scientists and hoaxers drew from the same tool kit to persuade their audiences and forge conviction. Truth and belief were, at least in part, questions of style. They were effects achieved by a controlled unfolding of information, the language of facts and observation, vivid imagery, a wide distribution network, favorable publicity, word of mouth, good timing, and good luck.

Thinking About Thinking Machines   In 1836, Poe took on another controversial novelty. This time he adopted the role not of a would-be hoaxer but of a rational debunker. His target was Johann Nepomuk Maelzel, a Bavarian showman and instrument maker, and his celebrated partner, "the Turk"—an automaton who played chess. This man-shaped contraption wore a turban and smoked a long pipe, pressed against a cabinet with a chessboard on top of it.

In his performances—including the one Poe described in Richmond—Maelzel would open the cabinet's drawers one by one, to show there was nothing inside but whirring gears. After Maelzel wound a key on its side, the android's eyes, arms, and hands sprang into motion, seemingly under their own power ("automaton" means "self-moving"). This fantastical menace from the Orient moved its own pieces, stopped play if an opponent tried an illegal move, and defeated nearly every contender.

Built in 1770 by a Hungarian craftsman, Wolfgang von Kempelen, the chess player enjoyed success at the Austrian court; in Paris it faced off against Benjamin Franklin, serving as U.S. representative to France. The Turk was then purchased by Maelzel; in one imperial performance he forced Napoleon to a draw. In 1835, Maelzel toured the United States. The press was enchanted, with critics declaring the Turk an astounding mechanical invention, while some suspected it was a fraud.

Poe's pen was ablaze in the *Messenger*. He wrote as an indignant opponent—not in chess, but in the head game Maelzel was playing on his audience. He

Maelzel's chess player as shown to the audience

took issue with those "who make no scruple in pronouncing the Automaton a *pure machine*, unconnected with human agency in its movements." Striking a pose of offended rationality, he set about proving that "the operations of the automaton are regulated by *mind*, and by nothing else." Poe contended that a human must be hidden inside the machine.

His essay culminated in a "train of suggestive reasoning," offering seventeen observations on the machine's construction and the logic of the confidence game—the relation between the impresario and the easily impressed. The fact that "the interior of the trunk is seen to be crowded with machinery" must necessarily "have reference to the spectator." While the motions of Maelzel's other automatons were smooth and lifelike, "free from the semblance of artificiality," the Turk's motions were staccato and unnatural—deliberately, Poe thought, in order to heighten the impression of mechanism.

While Maelzel claimed that the chess player was a machine performing a human function, Poe argued that it was really a human—presumably Maelzel's assistant, hidden inside—pretending to be a machine. Unraveling a tangle of bluff and counter-bluff that would have made Alan Turing dizzy, Poe used machinelike logic to prove that a machine could not think. Later reconstructions of the Turk have shown Poe's argument to be largely correct. The dense gear work seen inside the cabinet was an illusion created by mirrors; behind sliding panels, a short man (who was, not incidentally, extremely good at chess) would hide during the performances, moving the automaton's arms and head and using a system of magnets beneath the chessboard to follow and direct the pieces.

However hostile the essay was toward Maelzel's deceptions, Poe made use of misdirections of his own. Many of his "observations" were plagiarized from David Brewster's *Letters on Natural Magic, Addressed to Sir Walter Scott* from 1832, a bestseller that aimed to enlighten readers about the ways the powerful have used artificial wonders to manipulate the masses. In the book's chapter on automata, Brewster explained the secret of Maelzel's chess player. Brewster also contrasted the Turk with a newer thinking machine, this one genuine: the calculating engine designed by the British mathematician Charles Babbage—one of the first working computers, which could perform complex calculations much faster than a human. While the chess player was a fraud, Brewster saw Babbage's machine as proof of the "one vast miracle" of modern science.

Poe appropriated David Brewster's debunking of the Turk as his own.

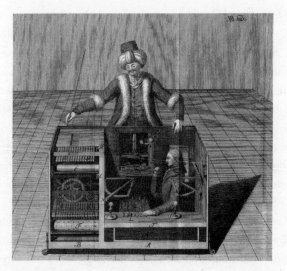

Maelzel's chess player exposed

He also referred to Babbage and his calculating engine but subtly diverted Brewster's argument. Poe compared the logic involved in doing mathematical calculations to the logic required for playing chess. Calculation, he said, is predictable; it requires only a "succession of unerring steps liable to no change." In contrast, a game of chess proceeds unpredictably: "A few moves having been made, *no* step is certain." Compared with what the chess-playing Turk was claiming to do, Poe reasoned, the calculations performed by Babbage's machine were child's play. The chess-playing automaton, if purely mechanical, would be "beyond all comparison, the most astonishing of the inventions of humankind."

Of course the chess player actually *didn't* play chess by mechanics alone; plausible artificial intelligence (repeatedly demonstrated through computerized chess) would have to wait more than a century. Still, a hasty reader of Poe's essay could be forgiven for thinking that by outsmarting Maelzel and his chess automaton, Poe had also surpassed Babbage and his mere "calculating engine." Poe subtly rewrote Brewster's story of progress and enlightenment, blurring the line between a hoaxer and an inventor and placing himself, the infallible reasoner and rhetorician, above them both.

Reading Divine Design   Poe's attack on the specter raised by Maelzel's and Babbage's inventions— the replacement of human thought by a machine—was in line with a more general hostility among romantic authors. Arguments against machines and mathematics and in favor of nature, organisms, and inspiration had been advanced by poets including Coleridge, Wordsworth, Keats, Goethe, and Schiller. In 1829, Thomas Carlyle, a Scottish critic and translator of German literature, raised a rallying cry against the replacement of humans by industrial machinery with his essay "Signs of the Times." Carlyle warned against the deadening effect of "the Mechanical Age" on "the mysterious springs of Love, and Fear, and Wonder, of Enthusiasm, Poetry, Religion."

Poe's insistence that the chess-playing automaton's moves were controlled "by *mind*, and by nothing else" was also reminiscent of arguments from a more traditional source: natural theology. This influential genre— advanced in books, sermons, and proverbs—held up impressive natural phenomena such as eyes, flies, polar bears, and the solar system as evidence of the sublime intelligence that designed them. As Bishop William Paley had argued in 1802, finding a rock on a heath would be no cause for reflection, but finding a watch would suggest the existence of "an artificer or artificers, who formed it for the purpose which we find it actually to answer." Such an intricate device had to be the result of mind; for Paley, every "indication of contrivance" in nature pointed to a wise and benevolent creator.

Recent scientific claims, however, posed a threat to Christian doctrines. Some physiologists and biologists attempted to explain organisms in purely material terms, with no need of a divine breath to set life into motion. Geology's strata and fossils provided evidence for an age of the earth far greater than biblical chronology. Laplace's self-correcting celestial mechanics seemed to dispense with a God seen as "Creator, Governor, and Preserver of the World."

In Britain, radical scientific theories chimed with dissenting religions and popular unrest in the 1820s and 1830s. As a countermeasure, Anglican natural philosophers published the *Bridgewater Treatises*, a series of books to show that science was perfectly compatible with conservative Protestantism. The series was paid for by a legacy from the Earl of Bridgewater (perhaps to clear his conscience after a life as a bon vivant; in his portrait, now in Durham Cathedral, the glass of wine he once held has been painted over). The series was dedicated to demonstrating "the Power, Wisdom, and Goodness of God, as manifested in the Creation."

The eight *Bridgewater Treatises* were embraced in the United States. The most celebrated, *Astronomy and General Physics* by William Whewell, a Cambridge mathematician, astronomer, philosopher, and Anglican priest, focused on the harmonious interactions among the physical laws of the earth and solar system. Their seamless cohesion demonstrated God's infinite foresight. Whewell devoted a chapter to Laplace's account of the formation of the solar system, coining the phrase "the nebular hypothesis." Though the theory had been associated with godless French materialism, Whewell found it compatible with divine creation.

While Whewell took great pains to argue for the beauty and power of God's laws, he insisted that mathematics and physics were useless beyond a certain point, particularly in trying to understand "the first cause" of the universe. If "we establish by physical proofs, that the first fact which can be traced in the history of the world, is that 'there was light,' we shall still be led, even by our natural reason, to suppose that before this could occur, 'God said, let there be light.'" Creation was a miracle before and beyond the system of laws.

For the *Bridgewater* authors, God could suspend these laws. In his contribution, *Geology and Mineralogy Considered with Reference to Natural Theology*, William Buckland explained the discontinuities in the fossil record—extinctions and new species—as successive moments of creation by divine miracle, beyond predictable laws. But even the regular order of nature was due to divine action: without God's constant governance, the delicate equilibrium among all things would run down. Even matter—inert, passive, and dead on its own—would lose coherence.

While they gave a revitalizing jolt to natural theology, the *Bridgewater Treatises* neither dampened the age's controversies nor silenced the opposition. Nor did they bring a final reign of Christian common sense. One curious consequence of *Bridgewater*-style natural theology was the belief in intelligent life on other planets, or "the plurality of worlds"; if God is truly omnipotent, his power and foresight must be expressed through creations on other planets as intricate as on ours. Whewell argued this position (before later changing his mind), as did John Herschel, who claimed that the "magnificent bodies scattered through the abyss of space" were created not "to illuminate our nights" but to provide for "other races of animated beings." In fact, Richard Locke's "Moon Hoax," featuring Herschel's alleged observations of a lunar civilization, was originally intended as a satire of this Protestant orthodoxy.

While the habits of natural theology—seeking a divine reason behind every natural artifice—might seem simplistic and laborious now, its arguments were taken as credible and edifying by most English-speaking readers in the early nineteenth century. Though he would do more than nearly anyone to undermine the movement, Charles Darwin—born, like Poe, in 1809—was steeped in the arguments of design made by Paley and the *Bridgewater* authors. The imprint of natural theology is legible throughout *The Origin of Species* in Darwin's admiration for the snug "adaptations" of organic contrivances to their surroundings, even if these are no longer divinely ordained.

Poe also was fascinated—if troubled and at times incensed—by natural theology. The final volumes of the *Bridgewater Treatises* appeared while he worked at the *Messenger*. He noted the seventh, by the Reverend William Kirby, and called the *London Quarterly*'s review of the series "one of the most admirable essays ever penned." He reviewed the fifth, on physiology, by Peter Mark Roget, now known for his thesaurus. Though Roget demonstrated that many aspects of living things suggested Almighty design, Poe found him too selective; if some trait's providential purpose was unclear, Roget simply skipped it, neglecting "all description of those structures, of which the relation to final causes cannot be distinctly traced."

Poe also furiously marked up his copy of the *Bridgewater Treatise* authored by the Scottish political economist the Reverend Thomas Chalmers, focused on psychology and moral science. He took issue with Chalmers's claim that habitually resisting greed, lust, and anger would eventually lead to "a final deliverance from their intrusion." Poe scrawled in pencil around the page of his copy, "*Mistake: because the antagonist motive is always present—which is not so in the case in which bad habits prevail[,] for conscience subsides.*" In other words, Poe thought that resisting harmful inclinations leads at best to a temporary victory; evil impulses or "antagonist motives" remain. In contrast, yielding to sinful inclinations could lead to their permanent victory: Chalmers himself pointed out that in "the hardihood of crime," the conscience is "cradled into a state of stupefaction," which lets the criminal "prosecute his secret depravities" without hesitating. By Chalmers's own puritanical reasoning, evil held a stronger hand than good.

Poe would constantly return to the central questions of the *Bridgewater* authors. Faced with some set of facts—or trail of clues—he sought the design, plan, reason, or intention behind them and asked what these facts might reveal about the perpetrator, plotter, or creator who left the trace; he borrowed their language of scientifically verified sublimity and their fascination for the intricacy of natural constructions.

But to Poe many of natural theology's answers rang hollow. Where was the divine wisdom or goodness in the universe's tendency toward chaos and decay, as seen in earthquakes and volcanoes, in shipwrecks and fires, in animals' violent instincts, or in humans' willingness to injure others and, at times, themselves? Evidence against harmonious design was especially hard to ignore in his rough-and-tumble republic, built on slavery and the theft of Native lands, constantly threatened by economic turmoil and lawlessness—despite his countrymen's regular invocations of Providence.

Like the vast majority of his contemporaries, Poe had little doubt that nature showed evidence of a divine hand at work. But he dared to wonder whether this creator was as wise and benevolent, or his creation as harmonious and orderly, as mainstream Christianity would have it. Poe's own misfortunes—and self-destructive streak—even made him wonder if at times God was working against him.

At the *Messenger*, Poe's erratic moods and behavior were shaking Thomas White's trust. White had expressed doubts about the effect of Poe's aggressive criticism and wild tales on the journal's reputation and suspected him of continuing to drink. In late 1836, White returned from travel to find the office "in great confusion." Two issues had been delayed, and the printers were on strike.

"Highly as I really think of Mr. Poe's talents," he wrote to a friend, "I shall be forced to give him notice, in a week or so at farthest, that I can no longer recognize him as editor of my Messenger." Despite palpable successes, Poe's first editorship came to an ignominious close. He had to leave Richmond again.

After packing up their few belongings in the boardinghouse, Poe headed to New York with Virginia and Muddy, eager to make an even bigger splash.

6

Delirious Design

Onward to the Pole   When Poe wrote to editors in hopes of publishing *Tales of the Folio Club*, he was warned that there was little public appetite for story collections. "People want something larger and longer," Carey and Lea said. James Kirke Paulding, a reviewer for Harper & Brothers, said Americans preferred works "in which a single and connected story occupies the whole volume."

He took their advice. In late 1836, still in Richmond, he began to write a seafaring novel inspired by *Robinson Crusoe*, with a hero whose name echoed his own: Arthur Gordon Pym. Two installments appeared in the *Messenger*, but the series was cut short when White fired him. He took the incomplete manuscript to New York.

Poe's novel would draw on popular excitement for a national scientific venture he had promoted in the *Messenger*: a government-sponsored expedition to the South Seas. The project was sparked by the lecturer J. N. Reynolds, who had been seized by the "hollow earth" theory of John Cleves Symmes, the "Newton of the West."

Symmes, a former army officer who moved between Kentucky, Missouri, and Ohio, believed that the surface of the earth was the outermost of five concentric spheres; its poles were flat and open, and one might travel smoothly from its extreme north or south into the globe's interior. Lit and heated by reflected light, the inner surface of the outer sphere (and the four smaller spheres it contained) was, Symmes contended, a "warm and rich land, stocked with thrifty vegetables and animals." Declaring

Humphry Davy and Alexander von Humboldt his "protectors," he called for "one hundred brave companions" to depart with him "with Reindeer and slays" from Siberia across "the ice of the frozen sea" and into the earth. Reynolds, a captivating speaker, joined Symmes on a lecture tour and argued that the U.S. government should sponsor an expedition to test the theory.

Reynolds then signed on to a commercial voyage, wound up in Chile, and was recruited to keep records for a U.S. frigate sailing around the globe; Poe admiringly reviewed Reynolds's six-hundred-page account of his adventures, *Voyage of the United States Frigate* Potomac. On his return, Reynolds shored up his political and press contacts, passing through the editorial offices of the *Messenger*.

Though he stopped promoting Symmes's theory, an expedition to the South Pole seemed to Reynolds more urgent than ever. When he spoke on the topic to Congress, Poe took up the cause in the *Messenger*. Nothing less than "national dignity and honor" were at stake, he wrote. The United States was called to increase the world's store of knowledge: "As long as there is mind to act upon matter, the realms of science must be enlarged; and nature and her laws be better understood, and more understandingly applied." An expedition would boost U.S. trade in whale oil, sealskins, sandalwood, and feathers. It should include a "corps of scientific men, imbued with the love of science," to correct navigational charts and "collect, preserve, and arrange every thing valuable" in natural history and anthropology. They would document "man in his physical and mental powers, in his manners, habits, disposition, and social and political relations," studying languages to trace human origins "from the early families of the old world."

Reynolds's campaign succeeded, though without him. President Jackson approved an expedition, but his secretary of the navy, Mahlon Dickerson, botched the preparations. In 1836, Jackson's chosen successor, Martin Van Buren, was elected president. Van Buren's secretary of war, Joel Poinsett of South Carolina (for whom the pointy red flower is named), was interested in both natural history and American expansion. Poinsett steered the United States Exploring Expedition through Washington's bureaucracy. He placed its ships under the command of a recklessly imperious naval officer, Charles Wilkes of New York—who denied Reynolds a place on board. Poe looked forward to the expedition with what he called "an intensity of eager expectation, which we cannot think we have ever experienced before."

Before leaving the *Messenger*, Poe also raved about a new nautical handbook by the Virginian Matthew Fontaine Maury, who frequently passed through Richmond. Maury had briefly hoped to helm the Exploring Expedition but instead became director of the Naval Observatory in Washington. Poe saluted Maury's *New Theoretical and Practical Treatise on Navigation*: "The spirit of literary improvement has been awakened among the officers of our gallant navy. We are pleased to see that science also is gaining votaries from its ranks."

The novel Poe was writing as he left Richmond for New York would also be a literary and nautical work, steeped in current science and imperial ambition. Like the Exploring Expedition itself, it would take bizarre detours from beginning to end.

Chasing Shadows   The Poes rented rooms on Sixth Avenue and Waverly Place. New York's population was continuously growing thanks to the Erie Canal, completed in 1825, connecting the Hudson to the West. Manhattan had become a publishing center with scores of newspapers, magazines, and printing houses. They moved to Carmine Street, where Maria Clemm became the manager of a boardinghouse.

In May the economy screeched to a halt; the panic of 1837 struck. Martin Van Buren had inherited an impending disaster. Interest rates in England had recently risen and cotton prices plunged. Jackson sent gold and silver west to prop up banks, and after his war against the National Bank there was no credit supplier to intervene. The nation plummeted into seven years of stagnation.

Poe worked with a focus sharpened by hunger. He impressed a neighbor, William Gowans, a bookseller, as "one of the most courteous, gentlemanly, and intelligent companions I have met," while Virginia showed a "disposition of surpassing sweetness." Gowans brought him to the Booksellers' Banquet at the City Hotel on Broadway, where he rubbed elbows with Washington Irving and the poet William Cullen Bryant and raised a toast to "The Monthlies of Gotham—Their distinguished Editors, and their vigorous Collaborateurs," hoping to become one himself.

The few pieces Poe managed to publish during this brief stay in New York grew from his deepening interest in ancient scholarship. While reviewing John Stephens's *Incidents of Travels in Egypt, Arabia Petraea, and the Holy Land*, he turned to Charles Anthon, professor of ancient languages at Columbia for help with etymologies and Hebrew terms. In June, he published "Von Jung, the Mystific" in *The American Monthly Magazine*, about the pranks of an expert in ancient languages. His story "Siope," published in a Baltimore annual, channeled his readings in ancient prophecy into a new fable: he personified a deathly plague as a speaking shadow. Poe survived the cholera pandemic, winding down in North America that year. But in New York he and his family again faced dire poverty. His income seems to have been nearly nonexistent as he worked on his novel of shipwreck and starvation.

In June 1837, before the full consequences of the economic crash were realized, Harper & Brothers registered a copyright for Poe's novel. This "single and connected story" wove Poe's excitement about the Exploring Expedition together with his investigations into the decipherment of ancient languages. Packed with shocking passages and ominous imagery, it teased readers with revelations while throwing mystifying obstacles in their way.

Strange Trip   *The Narrative of Arthur Gordon Pym of Nantucket* was published in 1838. Its title page was taken up by an outrageous 107-word subtitle, promising the "DETAILS OF A MUTINY AND ATROCIOUS BUTCHERY ON BOARD THE AMERICAN BRIG GRAMPUS, ON HER WAY TO THE SOUTH SEAS, IN THE MONTH OF JUNE, 1827," followed by a "SHIPWRECK AND SUBSEQUENT HORRIBLE SUFFERINGS," "DELIVERANCE," "THE MASSACRE OF HER CREW," a visit to islands in "THE EIGHTY-FOURTH PARALLEL OF SOUTHERN LATITUDE," and finally, "INCREDIBLE ADVENTURES AND DISCOVERIES STILL FARTHER SOUTH."

One reviewer asked, "What say you, reader, to that for a title page?" The page didn't mention Poe, or that the book was a work of fiction—suggesting that Poe intended the book to be taken, at least at first glance, as a genuine travel account.

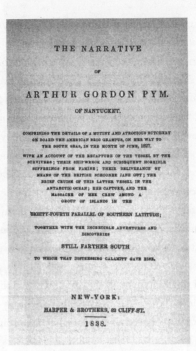

THE NARRATIVE

OF

ARTHUR GORDON PYM.

OF NANTUCKET,

COMPRISING THE DETAILS OF A MUTINY AND ATROCIOUS BUTCHERY
ON BOARD THE AMERICAN BRIG GRAMPUS, ON HER WAY TO
THE SOUTH SEAS, IN THE MONTH OF JUNE, 1827.

WITH AN ACCOUNT OF THE RECAPTURE OF THE VESSEL BY THE
SURVIVERS; THEIR SHIPWRECK AND SUBSEQUENT HORRIBLE
SUFFERINGS FROM FAMINE; THEIR DELIVERANCE BY
MEANS OF THE BRITISH SCHOONER JANE GUY; THE
BRIEF CRUISE OF THIS LATTER VESSEL IN THE
ANTARCTIC OCEAN; HER CAPTURE, AND THE
MASSACRE OF HER CREW AMONG A
GROUP OF ISLANDS IN THE

EIGHTY-FOURTH PARALLEL OF SOUTHERN LATITUDE;

TOGETHER WITH THE INCREDIBLE ADVENTURES AND
DISCOVERIES

STILL FARTHER SOUTH

TO WHICH THAT DISTRESSING CALAMITY GAVE RISE.

NEW-YORK:

HARPER & BROTHERS, 82 CLIFF-ST.

1838.

Title page, Pym, first edition, 1838

Adding to the *Narrative*'s verisimilitude were its precise details about currents, weather, and creatures of the sea and air. It closely resembled first-person voyage accounts—an extremely popular genre. It drew on Reynolds's *Potomac* voyage and its details on the whaling trade (Reynolds's *Mocha Dick* would later catch Herman Melville's attention). *Pym*'s publication was timed to capitalize on excitement about the South Seas Exploring Expedition setting sail in August, which the narrator hoped would "verify some of the most important and most improbable of my statements." The first edition also included notices of other Harper & Brothers books—travel accounts, histories, and biographies—encouraging readers to see the book in their hands as a truthful account of facts and actual experiences.

In that case, its author would be "Arthur Gordon Pym." Yet Poe had published the first chapters the previous year in the *Messenger* as fiction, signed "Edgar A. Poe."

To explain the contradiction, the preface (signed by "A. G. Pym") claimed that after an "extraordinary series of adventures in the South Sea," "Pym" met "several gentlemen in Richmond" who urged him to publish. "Pym" refused, thinking that the events of his journey were "so positively marvelous" that readers would take them as "an impudent and ingenious fiction."

But "Mr. Poe, lately editor of the Southern Literary Messenger," persuaded him that even if the narrative were rough, "its very uncouthness, if there were any, would give it all the better chance of being received as truth." "Pym" agreed to tell his story, on the condition that "Poe" would transcribe and publish it "*under the garb of fiction*"—hence its appearance in the *Messenger*. Yet despite the "air of fable" that "Poe" gave the account, many readers believed it. "Pym" grew convinced that the facts of his journey, if plainly reported, "would prove of such a nature as to carry with them sufficient evidence of their own authenticity." He would tell his tale as it happened, in his "own name."

After this mad squabble between "Pym" and "Poe" about the best means of convincing readers of the truth, the story began calmly enough: "My name is Arthur Gordon Pym. My father was a respectable trader in sea-stores at Nantucket, where I was born." Pym, aged seventeen, sets out one night after a party with his close friend Augustus for a "spree" in a tiny sailboat, the *Ariel*. They are nearly crushed by a large brig, the *Penguin*, which returns to save them.

Pym lets Augustus talk him into another voyage. He stows away below deck on Augustus's father's whaler, the *Grampus*, with a copy of the account of Lewis and Clark's expedition to keep him occupied. He nearly suffocates in the "dismal and disgusting labyrinths of the hold," while above board is a mutiny. Helped by the half-Indian, half-European Dirk Peters and another sailor, Richard Parker, Arthur and Augustus overtake the mutineers by playing on their superstitions. A storm ravages the ship; starving, they resort to cannibalism, drawing lots in a "fearsome speculation" that leaves Parker as the feast. Augustus dies; only Pym and Peters remain.

Rescued by a passing schooner from Liverpool, the *Jane Guy*, they sail farther south than any previous Europeans. They land on the island of Tsalal, whose natives are entirely black—clothing, skin, hair, and teeth—and are fascinated and horrified by the white skins and sails of the Europeans, at which they cry out, "Tekeli-li!" Seeing an opportunity for "profitable specu-

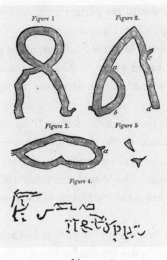

Figure 1.   Figure 2.

Figure 3.   Figure 5

Figure 4.

Images of chasms in Pym

lation," Captain Guy sets up a market, trading European trinkets for edible sea creatures which abound on the island. All goes well for the would-be colonizers until the Tsalalians lure the sailors into a trap, burying them in a deadly avalanche.

Once again, Pym and Peters are their ship's only survivors, hiding in the hills. Hunger forces them down through the black granite chasms of the island, which trace a strange path, like letters, which Pym records (figures 1–3). On one wall of a cavern they also find engraved "indentures" that resemble a pointing human (figure 4). They escape the island in a small canoe, taking a Tsalalian with them. As they paddle furiously away, the vessel is pulled "still farther south." The air grows warm and the sea turns milky; white birds fill the sky, crying, "Tekeli-li!" The current increases and white ash falls on their boat. Before them appears a great white waterfall that they approach with "hideous velocity."

The Tsalalian dies of fear as the darkness of the sky "materially increased, relieved only by the glare of the water thrown back from the white curtain before us." As they rush toward the waterfall, "a chasm threw itself open to receive us. But there arose in our pathway a shrouded human figure, very far larger in its proportions than any dweller among men. And the hue of the skin of the figure was of the perfect whiteness of the snow." There—suddenly, bewilderingly—Pym's narrative ends.

A mischievous "Note" closes the book, just as the preface opened it, explaining that Pym returned to the United States, and died, and that "Mr. Poe" "has declined the task" of reconstructing the final chapters of Pym's voyage.

The author of this final "Note"—neither "Pym" nor "Poe"—tentatively suggests an interpretation of the carved markings on Tsalal. In Egyptian, Arabic, and Ethiopian letters they appear to spell out "shady"; "white"; and "the region of the South." The "Note" concludes with a mysterious, quasi-biblical utterance: "*I have graven it within the hills, and my vengeance upon the dust within the rock.*"

Pym's ending—the "hieroglyphs" in the black chasms, the white figure in the "chasm" of spray and mist, the sudden break in the action, and the note announcing Pym's return and death—provides more questions than answers. It was Pym who urged Captain Guy to push toward the South Pole: "So tempting an opportunity of solving the great problem in regard to an Antarctic continent had never yet been afforded." Though he regretted the "unfortunate and bloody events" that resulted from this advice—the massacre of dozens of natives and the *Jane Guy*'s entire crew—he was pleased to have aided in "opening to the eye of science one of the most intensely exciting secrets which has ever engrossed its attention." Riddled with ambiguities, Pym's tale was about the quest for discovery and its costs.

An Inventory of Altered States

Even though detective fiction didn't yet exist—Poe would invent the genre three years later—*Pym*'s bizarre events gave readers endless puzzles to solve. The book's last paragraph, on the writing in Tsalal's chasms, explicitly invited a variety of interpretations. "Conclusions such as these," it read, "open a wide field for speculation and exciting conjecture." Its call for a "minute philological scrutiny" of the ancient words "written in the windings" of the chasms suggested that the entire book could be studied just as closely.

For example, readers might seek a natural cause for the "whiteout" of the ending: perhaps the sailors are funneled into the hole predicted by Symmes's "hollow earth." Perhaps the "white figure" is an optical illusion, the distorted image of an approaching ship—perhaps the very same ship, the *Penguin*, that saves Pym and Augustus at the book's beginning.

Or perhaps Poe meant readers to see the white figure as an encounter with divine truth, as in the book of Revelation's "vision of the seven candlesticks" with its figure with "hair of white wool." The story might have held a political commentary: some critics have seen in the extreme polarization of black and white in "the region of the South" an allegory of a natural basis for slavery or a reference to the biblical curse of Noah against the descendents of Ham; others read the Tsalalians' deadly rebellion as a warning of slavery's likely consequence.

The book explicitly addressed the slipperiness of interpretation: "In no affairs of mere prejudice, pro or con, do we deduce inferences with entire certainty, even from the most simple data." For his descriptions of optical illusions, Poe drew again on David Brewster's *Letters on Natural Magic*, which had supplied him with an analysis of the chess automaton. Pym experiences mirages, the visual distortions of twilight, and possibly, with the voyage's closing image, "the Specter of the Brocken"—the vision of one's own shadow as a giant when projected against a distant surface. Pym also confirms Brewster's overall message, highlighting the power of optical tricks to manipulate naive believers. Dressing up as a corpse to play on the "superstitious terrors and guilty conscience" of the mutineers, Pym himself is "seized with a violent tremor" when he looks in a mirror; the first mate dies at the sight of what he takes for a ghost.

The book underlined the unreliability of the senses by taking readers through an inventory of altered states of mind. As Pym suffocates below deck, he dreams of serpents, demons, and deserts; starving on the wrecked ship, he drifts into "a state of partial insensibility" with visions of "green trees, waving meadows of ripe grain, processions of dancing girls, troops of cavalry, and other phantasies." His first adventure on board the small boat *Ariel* (the name of the magician Prospero's familiar in *The Tempest*) establishes a narcoleptic rhythm in which Pym drops into a trance or visionary state, then staggers back into consciousness.

Repeatedly taking readers from false appearances to an underlying reality, Poe showed how material conditions—intoxication, hunger, expectation—

affect states of mind. This psychological emphasis added a probing, philosophical dimension to the "explained gothic" novels of Ann Radcliffe and Horace Walpole. Yet much as in De Quincey's *Confessions*, in *Pym* truth was a moving target. "It is utterly useless to form conjectures," he noted, "where all is involved, and will, no doubt, remain for ever involved, in the most appalling and unfathomable mystery." Every appearance might hide a contrasting underlying reality, while that reality's causes remained shrouded in doubt. Illusions and unreliable revelations pull Pym and the reader along, through a fever dream of signs and wonders, collapses, burials, and recoveries.

Writing Backward   Poe always took great care with his writings' typography and physical layout—their visible "composition." Just as he wrote his manuscripts in a precise, minute, and regular hand that resembled type, he worked closely with printers and typesetters. The eye-catching typographical layout of *Pym*'s title page seems to call out for decipherment, suggesting some meaning to its visual appearance. A copy of the French translation of *Pym* appears reflected in a mirror in a 1937 painting by René Magritte—an artist obsessed with the relations between images, words, and things; the suggestive symmetries of Poe's original title page invite a closer look.

The eight words of the main title float above the denser, smaller type of the subtitle. If you look with eyes slightly unfocused—or askance—you can see the title forming a half circle, mirrored by the tapering, slightly rounded cluster of text below. The title and the first part of the subtitle appear to form the two hemispheres of a globe: the upper mostly white, the lower mostly black. The eye is pulled downward, "STILL FARTHER SOUTH," funneling with some bumps down to the publisher and date—the record of the book's birth. This brief visual voyage anticipates the route the story will trace toward the bottom of the earth and, perhaps, to a receding point of origin—right off the page.

Now look again. Can you see the four lines of the title forming two rows of sails, with the subtitle clustered below as the hull of a boat? Imagine a straight line drawn parallel to the line formed by the words "EIGHTY-FOURTH PARALLEL OF SOUTHERN LATITUDE": you can then see the next clus-

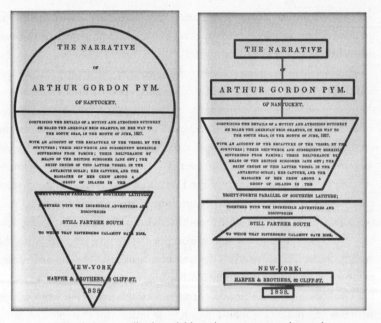

What do you see? *(a) A bisected globe, with a journey tapering downward;*
(b) a ship sailing above its ghostly double

ters of words repeat, on a smaller scale, and upside down, the shape of the blocks of text above. Now we see a boat and its reflection, along with its sails, as if from a distance across a shimmering sea: an apt illustration for the maritime adventures about to unfold, as well as their doublings, inversions, and illusions.

Symmetry and inversion were deeply engraved in *Pym*. As Poe knew from experience, setting pages for print required a typesetter to line up

The compositor's reversal

letters and words in a composing stick—in reverse order. This meant writing and reading *backward*—a mirror effect that could easily go wrong, through misrecognizing or transposing a letter.

Poe built this symmetry and reversal into *Pym*'s structure. Its twenty-five chapters divide neatly in half, folding back upon themselves. Events in the first twelve chapters mirror those at the same distance from the center in the last twelve. In the middle paragraph of chapter 13—the center of the book's central chapter—the *Grampus* crosses the equator, Pym's best friend, Augustus, dies, and the vessel flips over. The cannibalistic feast of the previous chapter—a horrific parody of the Last Supper—is echoed in the chapter that follows, with the ship's departure from Christmas Harbor and Pym's symbolic rebirth. Where before they drifted above the equator, starving, now they drift below the equator among islands with plentiful food. Likewise, the mutiny on the *Grampus* parallels the revolt on Tsalal, and the doomed voyage in the small *Ariel* at the beginning is echoed in the canoe voyage at the end.

The book as a whole embodies the rhetorical figure of chiasmus, where elements of a phrase are repeated in reverse order—for example, "say what you mean and mean what you say." The editor's "Note" suggested a meaning for the shapes traced by Tsalal's chasms—*images of a journey that may form words*, while the title page contained *words that may form images of a journey*. The first and last pages enwrap the verbal voyage between them.

The book's ominous pairings hint at hidden truths about the malleable nature of reality. At the start, Pym speaks of the perverse wishes that drive him to sea, visions of "shipwreck and famine; of death or captivity among barbarian hordes." In its parallel, final chapter, as he hangs from a cliff and imagines himself letting go, he "found these fancies creating their own realities, and all imagined horrors crowding upon me in fact." By that point, his grisly visions have indeed come true; his "fancies" have created "their own realities." It is as if in the second half of the book Pym were walking through the exaggerated projections of his mind. He meets his own thoughts and fantasies, but magnified, turned upside down, fused with the landscape—as if passed through a warped mirror, a kaleidoscope, a camera obscura, or a magic lantern.

Like a natural theologian, Pym seeks evidence of a divine design or providential plan behind his experiences. He doubts, for instance, that the "chain of apparent miracles" on Tsalal could be "altogether the work of nature," hinting that they might be divinely wrought. Yet no unambiguous revelation is at hand. In the central chapter, exhausted and starving but rescued from shipwreck, he reflects on the horrors from which he has "so lately and so providentially been delivered." In comparison, his current pains appear "little more than an ordinary evil—so strictly comparative," he reflects, "is either good or ill."

In other words, any entity, and our judgment of it, depends on the other entities with which it is compared and with which it stands in relation. This theme was echoed in the mirroring between the *Jane Guy*'s sailors and the Tsalalians. Pym and other "civilized" men have become cannibals, while the natives turn out to be no more credulous or savage than the white speculators. If the book implied a racial allegory, it might have been one of a shared damnation.

Pym's final line, "*I have graven it within the hills, and my vengeance upon the dust within the rock*," suggests that Pym's tribulations could be read as evidence that God created not out of generosity and benevolence but from some incomprehensible divine desire for revenge. After all, engraving matter with the originating Word, breathing spirit into dust, has been the cause of boundless human suffering. Perhaps, even more cruelly, the "vengeance" of the creator, whether God or Poe, was that despite the enticing hints of significance at every turn of the journey, there was no ultimate plan or redemptive design to be found. Herman Melville would develop *Pym*'s lessons in the ambiguities of revelation in *Moby-Dick*.

Poe's seafaring novel used remarkable literary precision to raise a set of questions it refused to answer; its meaning was a *definite mystery*.

Grave Particularity   American reviewers found in *Pym* both "horrors" and "quite a treasure." The *New York Gazette* guessed that the book's real author was Richard Adams Locke, "the very ingenious author of the Moon Marvel." Locke wrote to thank the editor but gave the credit to Poe. Several other papers noted *Pym*'s similarities to Crusoe, Gulliver, Sinbad, and the fabulist Baron Munchausen.

The British response was more enthusiastic and more credulous. According to George Putnam, "The grave particularity of the title and of the narrative misled many of the critics as well as ourselves, and whole columns of these new 'discoveries,' including the hieroglyphics found on the rocks," were reprinted "as sober historical truth." *The Spectator* praised its "nautical knowledge, and

Defoe-like appearance of reality," while *The Gentleman's Magazine* classed the book, ambiguously enough, under the rubric "Literary and Scientific Intelligence."

For Poe, one audience now counted most: readers in Philadelphia. Before the book was published, Poe abandoned New York, traveling south in hopes of better opportunities. In Philadelphia, *Alexander's Weekly Messenger* called the book a "very clever extravaganza," while the *Family Magazine* declared that if a reader wanted to "go on an exploring expedition to the south pole, he has only to take up Arthur Gordon Pym's narrative and if he is not led off to the pole scientifically, he will at least find himself, when he gets there, in a situation where science is no longer useful or necessary."

In contrast, William Burton, the owner of *Burton's Gentleman's Magazine*, was appalled. "A more impudent attempt at humbugging the public has never been exercised," he declared. Pym's ultraviolent, mystical tall tale was an insult to the right-minded: "We regret to find Mr. Poe's name in connexion with such a mass of ignorance and effrontery."

But Burton needed an assistant, and Poe needed a job. Soon Poe's name would be connected to Burton's in a city where science was both "useful" and "necessary." In Philadelphia, Poe would craft his most distinctive tales, establishing a unique position as fiction writer, poet, critic, and expert on scientific matters—crossing paths with and learning from those who were producing the "apparent miracles" of modern science.

PART III

Philadelphia

What age can boast improvements like our own,
When men to gods, and idiots bards have grown? . . .
Dreams, clouds, or gas-light, all are made
At cheapest rate by Espy or a blade!
Oh wondrous age! whose glories far excel
All which romancers dream or fictions tell!
When monster banks can raise a monstrous panic,
And infants gain their growth by means galvanic!
Thus population, like the mania, speeds
O'er western wilds and noxious prairie meads.
New states are born, new stars our banner bless,
And struggling realms are caught like men at chess!

—*The Poets and Poetry of America: A Satire*, dubiously attributed to Poe

*Daguereotype of chemist Martin Hans Boyè,
by Robert Cornelius, 1843, Philadelphia*

7

The Athens of America

P hiladelphia is a city to be happy in," wrote the New York wit Nathaniel Willis in 1831. "Everything is well conditioned and cared for. If any fault could be found it would be that of *too much regularity and too nice precision.*" Its very streets, formed in a legible grid, embodied the enlightened values of balance, proportion, and rationality expressed in the Declaration of Independence and the Constitution, signed at Independence Hall. Other civic buildings, including the National Bank, adopted a neoclassical style with friezes and pillars. Well-swept sidewalks and tidy shops lined Walnut, Chestnut, and Market Streets, opening to the Delaware's docks.

As the poet James Russell Lowell observed, American cultural life in the first half of the nineteenth century was a galaxy "divided into many systems, each revolving round its several sun[s]." Philadelphia had been the nation's capital city until 1800. In the 1830s it remained preeminent in publishing, banking, the arts, and science, earning it the nickname—shared with its rival, Boston—"the Athens of America." Painting and sculpture flourished at the Pennsylvania Academy of the Fine Arts; it was home to the engraver John Sartain and the portraitist Thomas Sully, as well as Peale's American Museum, which featured Charles Willson Peale's paintings of heroes of the Revolution alongside natural history specimens, including the bones of a mastodon he had excavated.

From the 1820s, Philadelphia's economy was shifting from trade to manufactures. Eager to replicate the success of New York's Erie Canal, Philadelphians invested in canals to connect the western expanse to the Chesapeake

Bird's Eye View of Philadelphia, *by J. Bachmann, ca. 1843*

Bay, passing through the Schuylkill, Susquehanna, and Delaware, a liquid network soon laden with the hard anthracite coal peculiar to eastern Pennsylvania. By 1838 there were more steam engines there than in any other state, puffing smoke and powering breweries, flour mills, carpet factories, and the iron industry.

The city took pride in its egalitarian attitudes, going back to its founding by the Quaker William Penn. With the rise of manufactures, however, economic inequality grew, while Jacksonian appeals to nativist resentment against Irish immigrants and free Africans erupted in destructive riots in the 1830s and 1840s. Yet the city continued to foster regular contacts among mechanics, scientists, and political leaders. Home to various Working Men's parties, Philadelphia was also the American base for the movement of "Mechanics' Institutes," which offered artisans instruction in the mathematical and physical principles of their trades. At the Franklin Institute, named for its most famous citizen, workers collaborated with scientists and investors on mechanical improvements.

In the 1830s and 1840s, Philadelphia was the nation's most active center for scientific research; Boston was its only real competitor. It was also the city in which Edgar Allan Poe would spend the greatest share of his adult life. The easy exchanges it allowed among journalists, artisans, scientists, and entrepreneurs formed the backdrop for some of his most stunning works.

The concerns of prominent scientists who spent time in Philadelphia in the same years as Poe—including the physicists Joseph Henry and Alexander Dallas Bache, the phrenologist George Combe, the skull collector Samuel Morton, and the geologist and publisher Isaac Lea—channeled crucial forces at work in the sciences of antebellum America. They confronted the formative tensions—between elite concentration and popular appeal, between national pride and dependence on European models, between hardheaded empiricism and controversial speculation—that fueled the pursuit of knowledge and fed Poe's imagination.

Arriving in Philadelphia, Poe would have to find his way in a city shaped by the distinctive combinations of science and invention his contemporaries were forging.

Joseph Henry's
Big Day Out

Joseph Henry, professor of natural philosophy at Princeton, much preferred the scientific culture of Philadelphia to that of New York. "I almost always return from new York dispirited in the way of science," he confided to a friend. "I am there thrown as it appears to me among all the Quacks and Jimcrackers of the land—I am disgusted with their pretentions and anoyed by their communications. How different is my feelings on a return from the [city] of Brotherly love!!! There is there jealousy and rivalry but also science and intelligence, and speculation and money are not the only things which occupy the mind."

In November 1834, while Poe was tinkering with his first short stories in Baltimore, Henry, a newly appointed professor, paid a long visit to Philadelphia. He arrived from Trenton, a cautious, curious man in a homely suit, eager to take in all that the city's industrial shops and scientific institutions had to offer.

Henry was born in Albany. He possessed sturdy, forthright features and was admired as a lucid lecturer. Still, his spelling was atrocious—one of many indications, of which he was painfully aware, of his humble background and

Portrait of Joseph Henry as a young man, ca. 1829

spotty education. Before starting his job at Princeton, he was immobilized with "doubt and self-distrust"; as he admitted much later, his father was a drinker who died young. That family secret might have had some connection with his uncertainty over his birth year—either 1797 or 1799—and with his drive to bring order to his surroundings and work tirelessly to prove his worth.

After only a few years of elementary school, Henry was apprenticed to a silversmith. The newly formed Albany Academy for young men offered him a scholarship. In addition to courses in natural philosophy and chemistry he studied Silliman's *American Journal of Science and Arts* to master the current state of the sciences. His dedication earned him the support of Congressman Stephen Van Rensselaer, who hired him as a tutor. He also gave private lessons to the young Henry James, a rich kid who would become a follower of Swedenborg and raise a clever brood including a novelist and a psychologist, Henry and William James.

Albany Academy hired Joseph Henry upon graduation. To prepare he spent two weeks in June 1826 at nearby West Point, soaking up its scientific culture and befriending Robert Hare, a chemist at the University of Pennsylvania, and John Torrey, a chemist and geologist who would go on to work at Princeton and lead the New York Lyceum. Henry was transfixed by their demonstrations of the recently discovered relationship between electricity and magnetism.

Back in Albany, Henry launched a series of experiments on electromagnetism. His apprenticeship in metalwork gave him the skill to build and modify experimental apparatus and follow out various lines of investigation. Eventually he made stunning discoveries, including the phenomenon of magnetic induction: that a magnet's movement near a conductor generates an electric force strong enough to move a needle and eventually a motor. (Unbeknownst to him, Michael Faraday had made the same discovery a year earlier in London and would claim priority.)

Henry constructed electromagnets so strong they could extract iron from ore. He built a powerful battery and an apparatus to send discrete electrical impulses across a wire—proof of a concept that Samuel Morse would apply in the telegraph. Yet to his own frustration, he found it difficult to write up his research and publish it; as with magnetic induction, several of his discoveries would be claimed by others who got them into print first.

Henry's scientific achievements—as well as the impression he gave of being trustworthy, polite, and eager for approval—made him a sensible choice to fill a vacancy at Princeton. Benjamin Silliman vouched for his "intellectual power," "fine tone of character," and "modest and winning manners," while Torrey was sure that Henry would "stand among the first philosophers of the country." Settling in with his wife, Harriet—like Poe, Henry married his first cousin, though Harriet was a mere ten years his junior—he designed a course of natural philosophy. He was relieved "of all the dirty work of the laboratory" by a "coloured servant" or "assistant," Sam Parker, who aided in building, repairing, and demonstrating Henry's electrical equipment—including receiving galvanic shocks. Henry's students were largely future ministers, many of them from the South; he began his course by invoking natural philosophy's utility and religious significance.

In the midst of heavy teaching duties, Henry's visit to Philadelphia was a breathtaking change. Stopping in at Carey and Lea's bookstore, he purchased a book by the science popularizer Dionysius Lardner. He called on his friend Samuel Lukens, who supplied him with scientific equipment; in the shop of

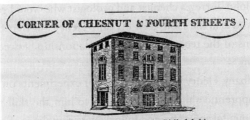

The Old Carey Book-store in Philadelphia.

The Carey and Lea bookshop

the instrument maker Mason, he was impressed by a new set of standard weights. He dropped in on Robert Hare's chemistry lectures and purchased a ticket for a show of automata at the Mechanical Hall on Chestnut, where he was "much diverted with the performance but not much edifyed."

In the evening, at the home of William Strickland, the architect who led Philadelphia's Greek Revival, he attended a "Wistar Party"—a dining club for the city's scientific elite, to which Henry had first been invited by John Vaughan, librarian at the APS (and owner of a fabled wine cellar, to which Henry would, many years later, contribute a 1785 amontillado). There he heard a presentation by the geologist Henry Darwin Rogers on the electrical machine built by the London physicist Charles Wheatstone to measure the speed of light, "composed of an infinite number of minute sparkes which are separated from each other by the action of the revolving mirror."

These dazzling encounters with the nation's scientific and social elite led Henry to the tailor. "I am not altogether pleased with the appearance of my costume," he wrote to Harriet after visiting Philadelphia, "and must endeavor for the sake of appearances to improve it."

On one excursion, he met with Alexander Dallas Bache, whose great-grandfather was Henry's predecessor in electrical research, Benjamin Franklin. Bache had returned home from West Point in 1827, aged twenty-one, to become professor of natural philosophy at the University of Pennsylvania. He invited Henry to his home, showed him "all his experiments," and introduced him to his wife, Nancy, "an interesting little lady who assists him in all his magnetic observations." The Baches were taking recordings of the daily "deviation" of magnetic compasses from true north and their "inclination" on the vertical plane. Their research was tied to the international project to chart the earth's magnetic field across the planet, later called the "magnetic crusade" and championed by Alexander von Humboldt. Henry had already made inroads into the study of terrestrial magnetism and had investigated the magnetic properties of the northern lights.

Bache invited Henry to join him in an exclusive group of "kindred spirits ready to discuss the principles or the applications of science" ranging "over the whole horizon of physical and mechanical research." Bache was convinced that American science could advance only if its most accomplished researchers joined forces to create and lead influential institutions. Other members of the "club" included John Torrey, Henry Darwin Rogers (named by his free-thinking father for Erasmus Darwin, the materialist poet and naturalist), who directed the geological surveys of New Jersey and Pennsylvania, and James Espy, who worked with Bache on steam engines and studied storms.

Bache and Henry began a lifelong friendship that would set the course of American science. In this club of well-trained, highly motivated scientists, Bache was called "the chief."

Alexander's Ambitions

The first of nine children, Bache was blessed with great abilities and personal charm and, like Henry, cursed with a derelict father: Richard Bache Jr. abandoned his wife and children for Texas in 1836 after being removed from the national Post Office for misuse of funds. Avoiding the sins of his father, Bache brought a high sense of duty to his own work. He saw the development of the sciences as inseparable from the strengthening of the nation. In his university lectures, he thrilled students with demonstrations from his research in physics. He joined the hoary American Philosophical Society but soon concentrated his efforts on the Franklin Institute, which was far better equipped—with several lecturers, a museum, exhibitions, and an "Experimental Laboratory and Workshop." Bache reoriented it "to extend, not merely to diffuse, the science of the day."

Founded in 1824, the Franklin Institute was the United States' foremost contribution to the movement for workers' education, which began with Glasgow's Andersonian Institute; George Birkbeck brought the model to

London, while in Paris engineer-scientists of the École Polytechnique taught workers at the Arts et Métiers conservatory. Bache published more than thirty papers in the Franklin Institute's journal between 1829 and 1836; he also prepared an expanded edition of David Brewster's *Treatise on Optics*—a text from which Poe drew for his knowledge of optical illusions and magic lanterns.

Bache's plans for American science were closely aligned with the Whig Party—the enemies of Jackson—and their conviction that the country's affairs were best managed by an educated, dutiful elite. As his West Point training encouraged him to do toward any situation, Bache approached American science as a *system* to be analyzed and improved: "The object of a system is rather to produce uniformity in all parts of a country than to introduce novelty," he wrote while advising his uncle Senator George Dallas on a plan to reform American weights and measures. While maintaining English inches, feet, ounces, and pounds, each measure, he thought, should be established "upon independent scientific principles" and "not be *wholly* borrowed from abroad."

Bache was keen to put science to public and national use. With the multiplying of steam engines, deadly explosions were frequent. Starting in 1830, Bache led a commission with Espy and Hare to study the proper construction and material for boilers and safety flues, and the effects of a rapid pressure increase. Bache secured funds from the secretary of the Treasury, putting him in charge of the United States' "first federally sponsored research project." Citing government policy in France and England, Bache urged the introduction of federally mandated regulations for boilers, later written into congressional law. Bache's vision for American science was already clear: applying internationally approved, expert forms of experiment, observation, and reasoning to matters of public concern, backed by—and reinforcing—a strong and active national state.

Chief among Bache's supporters was Nicholas Biddle, an APS member who wrote up Lewis and Clark's western journey. Born in 1786, Biddle had interests in cartography, natural history, phrenology, and ancient Greek. As director of the Second Bank of the United States and a leading Whig, he faced off against Andrew Jackson in the "Bank War" of the early 1830s. Portraying the bank as a private interest opposed to the will of the people, Jackson ordered government funds removed and placed in his supporters' "pet banks." Biddle huffed that Jackson thought that "because he has scalped Indians and

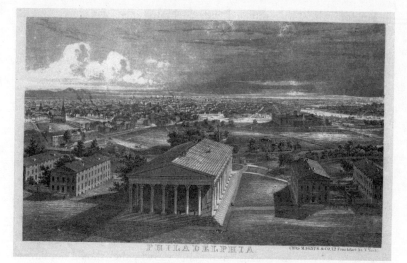

Philadelphia seen from Girard College, facing south;
Eastern State Penitentiary at center right, 1850

imprisoned Judges he is to have his way with the Bank." Biddle overplayed his hand; to show the bank's importance, he placed severe restrictions on credit, which in the end strengthened his opponents' case and led to the end of the bank's charter in 1836.

Biddle remained active in Philadelphia organizations and helped secure Bache's university professorship. He was chairman of the board of Philadelphia's Girard College, a school for white male orphans funded by Stephen Girard, a preposterously rich banker. Laying the cornerstone in 1833, Biddle extolled the need for public instruction in the United States; in other countries it may be "a private misfortune" to be "uneducated and ignorant," but in America "it is a public wrong," because the "general equality of power would be dangerous, if it enabled an ignorant mass to triumph by numerical force over the superior intelligence which it envied." With Biddle's backing, in 1836 Alexander Bache gave up his professorship to become Girard College's first president.

Before the school opened, Girard's trustees paid for Bache to take a European tour to study educational systems—just as Joseph Henry was traveling to purchase instruments and books for his courses. The friends met up in London, where they mingled with celebrated men of science. Assisted by Charles Wheatstone and Michael Faraday, they demonstrated Henry's powerful battery, creating sparks from a thermoelectric current: "Hurrah for the Yankee experiment!" Faraday exclaimed (at least according to one of Henry's students). Faraday introduced Henry to the apparatus he used in his experiments; he showed him the great batteries in the cellar of the Royal Institution that Humphry Davy had connected by wire to the theater above, to stun his audiences with electric light.

Bache smoothly made the rounds at the Royal Society, eased by letters of introduction testifying to his scientific credentials and illustrious ancestor. Henry was pleased to find his scientific reputation preceded him, though he was disturbed by conversations with "English Ladies" who challenged his support for slavery; they cannot, he complained, "understand our prejudices" relative to African Americans "and cannot see why they should not have all the privaledges of a white man."

In Paris, Bache was enthralled by François Arago, an astronomer and secretary of the Academy of Sciences. A close confidant of Humboldt's, Arago leveraged his scientific authority in the National Assembly to advocate for workers and democratic reform and during the Revolution of 1848 would lead the legislative campaign to end slavery in France and its colonies. Henry couldn't speak French; he met few savants and thought that Parisians—with their outdoor dining, specialty boutiques, and gardens with "marble statues in many instances perfectly naked"—were "inhabitants of almost an other planet."

Greater embarrassment awaited him back in Liverpool, at the 1837 meeting of the British Association for the Advancement of Science (BAAS). The BAAS had been launched in 1831 by a group of "gentlemen of science" that included David Brewster, Charles Babbage, and William Whewell, inspired by a similar German organization assembled by Humboldt and the biologist Lorenz Oken. Meeting each year in a different city, the BAAS gathered researchers from all fields of natural science to discuss findings and create amicable bonds.

Henry was happy to be singled out, along with the German chemist Justus von Liebig and the Swiss physicist Auguste de la Rive, as a notable foreign participant. But after his lecture on America's "internal improvements" in railways and canals, in which he recalled a brisk, nine-hour steamboat journey up the Hudson from New York to Albany, one of the session's hosts "jumpt up" and denied such a high speed was possible.

The challenger, Dionysius Lardner—the London-based lecturer on mechanics—sent "the whole room" into a "state of commotion." When a colleague chastised Lardner for disrespecting "a gentleman and a foreigner," Henry pleaded that he deserved no special treatment "on account of my being a Foreigner, that truth and Science should know no country." His noble sentiments were met with applause.

Yet the flap left Henry troubled, even a year after his return home. He was annoyed by "the general low opinion" Europeans showed toward American science. While he was gratified by his warm welcome by many physicists, his European tour showed what a distance his country had to go before participating as a scientific equal on the international stage.

To combat the "great prejudice" against American science, Henry was coming around to Bache's plans for organizing science on the national scale: "I am ever more of your opinion that the real working men in the way of science in this country should make common cause."

Bache began in Philadelphia. When he returned home, Girard College still hadn't officially opened; he accepted a second position as superintendent of Central High School, another experiment in public education. Soon his students were conducting astronomical surveys with the powerful telescope he had installed, while taking daily observations in the magnetic observatory he set up on the grounds of Girard College as part of the Humboldtian "magnetic crusade."

Bache also worked his political connections to secure four thousand dollars from the State of Pennsylvania for thermometers, barometers, and rain gauges, distributed throughout the state to observers who sent their recordings back to the "club" member James Espy, making him America's first-ever state meteorologist. Bache and his friends were getting tied ever more firmly into international networks of observers of the earth and skies.

Political opponents forced Bache to step down from both Girard and Central High. The squabble—prompted by Jacksonian hostility toward state-funded education—convinced him more than ever of the need for "a great system of public instruction, worthy the patronage and support of a free and enlightened people," closely integrated into a well-funded national program of scientific research. Bache returned to the university, biding his time until another opportunity to realize his vision for national science appeared. When it did, he leaped.

Reading Skulls in Public   Henry and Bache saw one enormous obstacle to their goals: they called it quackery. The publishers and lecture halls that turned a profit from diversions such as the "Moon Hoax" or Maelzel's chess player also made room for patent medicines, perpetual motion machines, astrological meteorology, and crackpot theories from archaeology to zoology.

Henry and Bache were convinced that the outlandish, untested, often deliberately provocative claims of quacks—of forgers, frauds, and humbugs—were crowding out the work of serious scientists. As Henry wrote to Wheatstone, "We are overwhelmed in this country with charlatanism. Our newspapers are filled with the puffs of quackery and every man who can burn phosphorous in oxygen and exhibit a few experiments to a class of Young Ladies is called a man of Science." Looking back on their European tour, he told Bache, "The charlatanism of our country struck me much more disagreeably when I first returned than before or even now. I often thought of the remark you were in the habit of making that we must put down quackery or quackery will put down science."

Yet the tools needed to "put down quackery" were sorely lacking. American publications were needed, Henry observed, "in the way of securing priority": to get discoveries into print and make clear who had made them first. Just as damaging to American scientific interests was "the want of an international copy right." Publishers could "filch with impunity" the work of foreign authors, making it impossible for American authors to "receive a proper compensation in money and fame," inhibiting Henry from writing "for my class and the classes of other colleges a text book in which I might set fourth my own claims." Yet Henry thought that the United States was not ready for a public organization such as the BAAS. A "promiscuous assembly of those who call themselves men of science in the country would only end in our disgrace."

Even more troublingly, it was far from easy to define just what distinguished true science from the promiscuous "charlatanism" it opposed. Science was largely brought to the public—usually for a small fee—by popular writers and lecturers in lyceums, libraries, and other public halls. Some of these were indeed the deceptive charlatans Henry and Bache despised. But others were recognized by their peers as "real working men in the way of science." Unlike Bache and Henry, these scientists were willing to use lofty rhetoric, vivid and tactile displays, and moral and political appeals to move their audiences. They were also not above stirring up public controversy and even scandal with their views.

The scientific superstar from Edinburgh George Combe swooped into town at the end of December 1838—just after Poe's arrival—to deliver lectures at the Peales' Philadelphia Museum to paying crowds of hundreds. His topic was phrenology, as systematized in his bestselling book, *The Constitution of Man*. This immensely self-assured Presbyterian, a student of Gall and Spurzheim, was the Anglophone world's leading proponent of the science of bumps on the head. By analyzing one's own character and that of others, and by cultivating positive emotions, he argued, one could develop the organs of the brain to

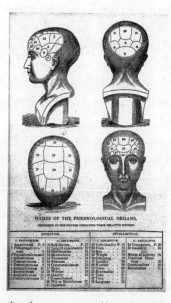

Phrenological organs, as presented by George Combe, 1836

become even more virtuous, reasonable, and productive. His lectures carried a message of self-improvement perfectly pitched for Philadelphia's aspirational audiences.

In Philadelphia, Combe complained about the American custom of applauding with the feet; the much larger crowds in the auditorium above his stomped in appreciation at a performance by the African American composer Frank Johnson and his band. Another inconvenience was the delayed arrival of his collection of sample heads and skulls, which he had shipped from New York. He made do with those of his Philadelphia colleagues, including the craniologist Samuel Morton, who lent him several skulls; Nicholas Biddle lent him a phrenological head purchased during Franz Gall's 1802 tour.

During his lectures Combe kept silent about what he described, in his later published account of his journey, as the "canker" of slavery. He was surprised—much like the "English Ladies" Henry met in London—at the way many white middle-class Americans saw abolition as a dangerous and extreme position. For instance, abolitionists were blamed in the "respectable" press for the 1838 riot in Philadelphia in which the Anti-Slavery Society's newly opened Pennsylvania Hall was burned down just after opening. In its first days, the hall hosted lyceum lectures on meteorology by James Espy, though the lyceum's members were "over-anxious" to distance themselves from any position on slavery. After presentations by William Lloyd Garrison and Angelina Grimké to racially mixed crowds of up to three thousand, a mob set fire to the enormous building. Fire companies arrived; they prevented the flames from spreading to neighboring buildings, but stood by as the hall's roof collapsed.

Despite his objections to slavery, Combe accepted a warm reception from the professor of medicine Samuel Morton. Aided by his former student the Alabama physician Josiah Nott and the international grave robber and self-proclaimed "Egyptologist" George Gliddon, Morton spent years collecting skulls from around the world—from battlefields, tombs, and colonial capture. He cleaned and emptied them, filled them with buckshot and seeds to measure their volume, and put together a ranking of "cranial capacity" that he thought could stand in for brain size and intelligence. Unsurprisingly given Morton's preconceptions, skulls of Europeans came out at the top and Africans' at the bottom. Morton's ghoulish collection of dubiously obtained skulls and "ethnological science" was later used as empirical evidence for a "natural" basis for slavery.

In many respects Combe's and Morton's forms of science could invite charges of quackery or charlatanism, so much were they at odds with the sober, self-directed conduct of science Bache and Henry had in mind. Combe eagerly pursued public acclaim and money with his engaging lectures; phrenology, though accepted by many—for Poe, it had "assumed the majesty of a science"—was subject to multiple interpretations and hounded by opponents. Frederick Douglass would eventually make use of phrenological arguments, but strenuously distinguished the science from the attempts of Morton and his collaborators to build an "ethnological" science of essential racial differences and hierarchies.

Both Combe and Morton were happy to take on potentially scandalous topics. They became notorious less for their conviction that character and intelligence could be read from skulls and arbitrary racial categories (notions largely taken for granted in the era's "average racism") than for their advocacy of a science free from miracles or divine intervention. Both advocated a "uniformitarian" view of nature, believing, with the geologist Charles Lyell, that all natural phenomena—from the rising and falling of water levels to the movements of the planets and the rudiments of character—were ruled by constant laws. Theological doctrines might help with moral questions but were unnecessary when dealing with natural facts.

Morton used his skulls to support the theory of "polygenesis"—that the different races of humanity were distinct species, created in different locations. This view directly contradicted the Bible's account of a single act of creation, "monogenesis." Scripturally minded researchers steered clear of Morton's theories, while the German naturalist Friedrich Tiedemann used the same craniometric methods to argue for the unity of the human species and against slavery. Polygenesis was directly challenged in the 1830s and 1840s in ethnological works by the African American authors Robert Benjamin Lewis, Hosea Easton, and James W. C. Pennington, who grounded their historical appraisals of African civilization—and denunciations of the crime of slavery—in a monogenetic account of creation.

Despite hints of theological irregularity, both Combe and Morton were respected as men of science by researchers worldwide. Morton was Bache's colleague at the university and the longtime president of Philadelphia's Academy of Natural Sciences. Awed by a visit to Morton's "American Golgotha," the Swiss naturalist Louis Agassiz praised its "series of six hundred skulls, mostly Indian, of all the tribes who now inhabit or formerly inhabited America" as itself "worth a journey to America." Agassiz would later become an impassioned advocate of polygenesis and race science, and would be warmly welcomed by Bache and Henry into their clique of elite scientists. And despite Combe's opposition to slavery, he would write an appendix for Morton's overview of racial classification, *Crania Americana*. In the 1830s and 1840s, the lines between legitimate science, political provocation, and crowd-pleasing quackery were exceedingly difficult to define.

Natural Historian for Hire

Edgar A. Poe landed in Philadelphia in 1838. Only three years younger than Bache, he, too, had been raised among his city's elite and received an excellent education. Now, however, he was an impoverished outsider—as Henry had been in his youth—seeking recognition and stability. His work in Philadelphia would closely parallel that of Bache and Henry. All three were striving to place American intellectual life on a stable and unified footing, to allow it to compete with that of Europe. Strikingly, Poe's first publication in Philadelphia—and the one that sold the most in his lifetime—was a scientific textbook.

When Poe arrived with Virginia and Maria Clemm, they were "literally suffering for want of food," living on "bread and molasses for weeks together." Poe's friend James Pedder, well situated at a sugar manufacturer's, purifying the raw goods delivered from Caribbean slave islands, came to their aid; his daughters, Bessie and Anna, visited with gifts for "Sissy" and "Muddy." Pedder was also editor of *The Farmers' Cabinet*, publicizing techniques for improving soils and raising crops—the kind of practical, commercially oriented publication in which much of the era's natural science was reported and discussed. Pedder had studied the beet industry in France and was scheming to introduce beet sugar to the States.

Pedder helped Poe find odd jobs. His old friends in Baltimore Nathan Brooks and Joseph Snodgrass published his occasional pieces in *The American*

Museum of Science, Literature, and the Arts. Hoping for a government post, in July 1838 Poe wrote to the novelist James Kirke Paulding, who had become Van Buren's secretary of the navy; he longed to "obtain the most unimportant Clerkship in your gift—*any thing, by sea or land.*" No luck.

By September 1838 the Poes had moved to a small house on Locust Street, with a garden suitable for the pet fawn that a friend offered as a gift for Virginia: "She desires me to thank you with all her heart—but, unhappily, I can not point out a mode of conveyance." He had to content himself with imagining "the little fellow . . . already nibbling the grass before our windows." Pedder offered a more useful gift: an introduction to a natural historian in need of a writer's help.

Thomas Wyatt, a lecturer and schoolteacher based in Delaware, had published a large textbook on conchology—the classification of shells—with Harper & Brothers. Wyatt's textbook relied on previous works by the French biologists Lamarck and Blainville. He also received assistance from Isaac Lea, the Philadelphia publisher and naturalist.

In the 1830s, geology was one of the most hotly pursued branches of science, not least because of the growing industrial importance of coal. Geology and conchology were closely linked: knowing which shells and rocks went together made it possible to line up geological strata as successive chapters of the long history of the earth. Isaac Lea, whose passion for natural science was stoked by his friendship with the geologist Lardner Vanuxem, wrote that geology and conchology, "its sister science," revealed "objects of the highest importance, a thorough knowledge of our cosmogony," or the origin of the universe. In a more radical vein, the poet and naturalist Erasmus Darwin had taken for his family's emblem the phrase *e conchis omnia*—everything from shells.

Lea's eyes were opened to the wonder of creation when he first examined a crate of shells from China and Ohio: he "did not know what it was to live on God's earth before." Lea published "Description of Six New Species of the Genus Unio"—shells of freshwater mussels—for the American Philosophical Society and became, along with Samuel Morton, a leading light of Philadelphia's Academy of Natural Sciences, founded in 1812 as a less formal, more inclusive scientific society than the patrician APS.

In 1832, like many Americans of his generation pursuing scientific topics in depth, Lea took a tour of Europe. He attended the second meeting of the BAAS. Just as Bache and Henry would four years later, he rubbed shoulders with Faraday, Brewster, and Babbage. The *Bridgewater Treatises'* geologist, William Buckland, told him that "England's prosperity is based upon her coal mines" and "upon their exhaustion she would return to her original barbarism." When Buckland scoffed at America's paltry coal deposits, Lea unrolled his geological maps to demonstrate that "the quantity of anthracite and bituminous coal was almost unlimited in North America"—a promise to outrun the English in wealth and civilization (and coal smoke).

In Paris, Lea was given access to the library of the legendary naturalist Cuvier; he acquired part of Lamarck's shell collection and met the anatomists of the Jardin des Plantes—Blainville, Geoffroy Saint-Hilaire, Ferrusac—who reclassified the museum's *Unio* shells according to Lea's system. On his return Lea published a large folio with full-color prints, *Observations on the Genus* Unio.

Thomas Wyatt built on Lea's classifications in his conchology textbook, merging them with Blainville's and Lamarck's to form a more comprehensive overview of all known shells. Yet as Wyatt gave lectures on the lyceum circuit, his book proved too large and expensive for his listeners, many of them women and children. He needed a handier, more affordable volume but could not publish a similar work without infuriating his publisher, the mighty Harper & Brothers, by undercutting their sales. He needed a new edition: shorter, cheaper, and signed under another author's name.

Enter Poe, newly arrived and eager for work. Thanks to *Pym* and his reviews at the *Messenger*, Poe had a proven record on scientific topics. Fluent in French, he could work through the relevant volumes by Cuvier, Lamarck, Blainville, and Geoffroy Saint-Hilaire, all available for consultation in the Library Company on Locust Street—an august space overlooked by a giant bust of Athena.

Published in 1839, *The Conchologist's First Book* by Edgar A. Poe was slim, portable, and inexpensive, with several plates of engraved shells. The preface and introduction were lifted from Wyatt's book and from Thomas Brown's *Elements of Conchology*, which openly acknowledged their own debts to French precursors. Though later wags accused Poe of plagiarism for this book, all

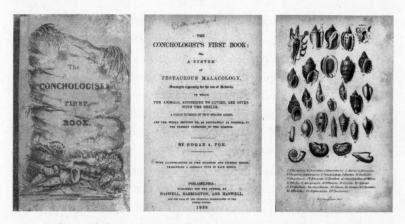

The Conchologist's First Book, by Edgar A. Poe, 1839: cover, title page, plate 12

"new" systems of natural history depended on earlier systems, which were in turn the product of an enormous collective and largely anonymous labor by observers, collectors, and taxonomists around the world. In his preface, Poe thanked Isaac Lea for his "valuable public labors" and for "private assistance" in preparing the book.

Poe introduced important improvements. The biologist Stephen Jay Gould pointed out the book's "progressive, even innovative, arrangement of material": Brown's book had followed the order of description of Lamarck, presenting the shells descending from those taken as most advanced or perfect to the lower, more "primitive" types, but Poe adopted a more widely practiced convention, ascending from "lower" shells upward.

Poe's subtitle, *A System of Testaceous Malacology*, announced a more significant advance. "Malacology," taken from the Greek word for "soft," is the study of small, soft creatures, while "testaceous" means having a shell. Previous works of conchology, Poe wrote, "appear to every person of science very essentially defective, inasmuch as the relations of the animal and shell, with their dependence upon each other, is a radically important consideration in the examination of either." He saw "no good reason why a book upon Conchology (using the common term) may not be malacological." Rather than a study of the ruins left by dead creatures, Poe described both shells and their squishy inhabitants, raising conchology "from artificial description to integrative biology." Poe's book was not just shorter and cheaper but more complete, with innovations that surpassed its predecessors.

Unlike Wyatt—but like Isaac Lea—Poe also included natural theological reflections: "To an upright and well regulated mind, there is no portion of the works of the Creator, coming within its cognizance, which will not afford material for attentive and pleasurable investigation." He quoted the German naturalist Carl Bergmann, who wrote that shells are "medals of the Creation"—lasting records of God's design.

The first edition sold out; a second, published the same year, added "more recently discovered American species," noting that the work had been adopted by several schools. In three editions, the book sold more copies than any other Poe published in his lifetime. It brought him at least fifty dollars, putting food on the table. It also gave him a crucial contact with Isaac Lea, whose Philadelphia publishing house would print Poe's first collection of stories. And the success of *The Conchologist's First Book* gave him something to crow about as he put his skills up for sale.

In addition to his three poetry collections, at age thirty Poe was the author of two very different books: *The Narrative of Arthur Gordon Pym*, a playful, sensational, mysterious tale, full of empirical detail and psychological insight but denounced as an "attempt at humbugging the public," and *The Conchologist's First Book*, a widely read scientific textbook, adding to the best established research of the day, extending a rational classification over a significant domain of nature.

In Philadelphia, a vast public hunger for entertaining novelties was matched by projects to standardize knowledge and ban the speculations of quacks. Standing at the intersection of these two currents, Poe would be right at home.

8

Methods Grotesque and Arabesque

Gentleman's Assistant   Poe found steady employment at *Burton's Gentleman's Magazine* in May 1839, a fashionable periodical owned by William Burton, a stentorian English stage actor. Despite giving *Pym* a damning review, Burton offered Poe a job at ten dollars per week—requiring only "two hours a day, except occasionally"—corresponding with authors, writing reviews, and laying out text. It offered less recognition and autonomy than the *Messenger*, but Virginia and Muddy could not live on molasses.

Poe got down to work penning reviews. He clashed immediately with Burton, who forbade him to print a blistering critique of a Baltimore poet. He patronizingly counseled Poe against "uncalled for severity" and pontificated, "The troubles of the world have given a morbid tone to your feelings which it is your duty to discourage."

As Burton's assistant, making his rounds in a few square blocks inland from the docks on the Delaware, Poe had an entrée to the city's intellectual scene. One evening he dined with the city's editors and journalists at the house of the playwright Richard Penn Smith; Burton made a grand entrance at midnight, with actors and writers in tow. Poe met writers who would become lifelong friends and occasional enemies: Thomas Dunn English, a blustery medical school graduate, and Henry Hirst, a lawyer and absinthe enthusiast who sold exotic birds. He discussed new schemes for railroad ties with the inventor James Herron and techniques of engraving with the painter John Sartain. In *Burton's* editorial office on Dock Street he put in much more than the promised two hours per day, filling columns with more, than a dozen reviews per issue. He dissected novels, plays, memoirs, and travel accounts: all that a gentleman needed to know, including *The Canons of Good Breeding; or, The Handbook of the Man of Fashion*. Yet the well-rounded gentleman also needed more of substance.

Poe used his position at *Burton's* to solidify his reputation for scientific acumen. He reviewed the latest works of science: S. Augustus Mitchell's *System of Modern Geography*; a richly illustrated *Synopsis of Natural History* by Louis-Céran Lemonnier, translated by Thomas Wyatt, Poe's old partner in conchology; Henry Duncan's *Sacred Philosophy of the Seasons*, a natural theology textbook in verse.

Behind *Burton's* cover (featuring a bespectacled Benjamin Franklin) Poe launched a series titled "A Chapter on Science and Art." Drawing on his reading in periodicals from the United States and England, in this series he surveyed the outputs of an age of frenetic discovery and invention. He discussed electroplating, engraving by aqua fortis, a plan for a national "Central School of Natural Science," and a new paper mill that converted rags into books, "the most astonishing machine ever invented": throw a shirt in at one end "and see it come out Robinson Crusoe at the other."

Inventions and discoveries seemed without end. Poe brought them to readers with knowledge, insight, and humor, presenting himself as an expert in various fields—weighing in with equal ease on atmospheric pressure, steam engines, roller skates, and astronomical sunglasses. With his "Chapter on Science and Art," Poe made himself one of America's first science reporters.

In Search of Method   Such reports seemed to offer confirmation that all domains of nature could be explained by science, from the movements of the most distant stars to the motive springs of the mind. The promise of the Enlightenment—to use observation, experiment, and reason to place the world within a uniform grid of quantification and classification—seemed to be coming true. Even though some precincts of reality might remain obscure, they would all, it seemed, *eventually* be brought under the rule of scientific law.

The roots of this faith were long and celebrated. Important developments in the seventeenth century—symbolized by the names Galileo, Descartes, and Newton—fed into the eighteenth century's explorations and encyclopedias. The revolutions in America and France, industrial development, and new systems of explanation in all fields of knowledge gave it further strength. It was in Poe's lifetime, in the first half of the nineteenth century, that claims for the universal progress and benefits of science first came to be taken seriously not only by academics, gentleman scholars, and tinkerers but also by politicians and ordinary people. Scientific understanding was sedimenting itself as an admired, increasingly authoritative baseline of reality. ·

Poe was one of many thinkers who tried to make sense of this colossal change and pin down the nature and effects of scientific thought. Yet the very success of science made it difficult to define. Chemistry, biology, physics, and geology had made rapid advances by concentrating on sharply focused lines of inquiry. Each had its own instruments, language, and questions—familiar to insiders but increasingly inaccessible to nonspecialists. A worry grew that the overall picture of knowledge had shattered into incompatible fragments.

From the time of Aristotle much of what we now know as "science" went under the name of "natural philosophy." The term implied a unified explanation of all physical phenomena according to a single system of causes. But over the eighteenth century it became clear that classifying animals and plants, for example, was different from calculating the trajectories of comets, measuring the output of engines, or defining geological eras. By the early nineteenth century the unity of natural philosophy was breaking down.

Samuel Taylor Coleridge noted the change at the 1833 meeting of the British Association for the Advancement of Science. Listening to a succession of highly specialized reports, Coleridge interjected that the researchers' work was so narrow and disconnected that they hardly deserved to be called philosophers. The host, William Whewell, replied that if "philosopher" is "too wide a term," then, "by analogy with *artist*, we may form *scientist*."

This exchange underlined the need to strengthen the definition of science and the professional identity of researchers, despite the diversity of specializations. In France, Auguste Comte propounded his system of "positive philosophy," which explained the rise of modern science as the result of a law of historical progress, taking humanity from religion, to metaphysics, to empirical science. Others (including Coleridge and Whewell, in quite different ways) drew inspiration from the work of Immanuel Kant and the idealist philosophers who followed him. Kant had argued that knowledge comes from combining sense data with concepts and schemas (such as uniform space and time) that are built-in, or "a priori" mental equipment. What Kant called "ideas of nature"—the existence of God, or the eternal life of the soul—were neither provable nor disprovable by science and thus lay beyond its limits.

In the 1830s, philosophers and spokesmen for science sought to identify a single method that all true sciences shared. Bacon's trusty philosophy of "induction"—generating laws from observations—was updated in John Herschel's 1830 *Preliminary Discourse on the Study of Natural Philosophy*, which interrogated the relationship between observation and theoretical speculation. In 1840, Whewell's *Philosophy of the Inductive Sciences* held that each science grew from certain "Fundamental Ideas," resembling Kant's a priori categories, guiding the "colligation" of facts into laws. These laws were confirmed by their ability to predict new facts (Whewell would cite as an example the surprising discovery of Neptune in 1846, whose location astronomers predicted based on aberrations in the path of Uranus) and by their "consilience"—their applicability to facts of several different kinds, as with gravity's ability to explain tides, falling bodies, and planetary motion.

In the United States, Bacon's induction was reinforced by a "common sense" philosophy that insisted that ordinary reasoning and observation could be trusted to provide knowledge—including about the existence of God. Natural theologians, including the *Bridgewater* authors, were admired in America for showing scientific knowledge to be compatible with—even subservient to—worship of the Almighty. They argued that matter on its own was dead, inert, and passive; it required an omnipotent deity to hold it together and give it life.

The practice of science was also seen to embody Protestant virtues: a disciplined, modest, self-denying search for universal truth. In the United States, this package was all the more attractive when joined to the enlightened ideals of Jefferson and Franklin: utility, efficiency, parsimony. A rising drive to develop and guide science also aligned with unified plans for industrial progress, such as Clay's "American System," which plotted roads, canals, railroads, and communications to integrate the growing national territory.

Technical innovation was seen as particularly "republican," a useful, non-elitist expression of science, raising the intellect and material conditions of all. According to William Ellery Channing in a lecture he gave in Philadelphia in 1841, "Science has become an inexhaustible mechanician; and by her forges and mills, and steam-cars, and printer's presses, is bestowing on millions, not only comforts, but luxuries which were once the distinction of a few." Though sidelined by Jackson, John Quincy Adams's inaugural vision of a unified American scientific and technical infrastructure, including a system of astronomical observatories or "light-houses of the skies," still had strong partisans.

Alexander Dallas Bache, Joseph Henry, and their newfound ally Benjamin Peirce—a bullheaded mathematician and astronomer at Harvard who had helped Nathaniel Bowditch translate Laplace's *Mécanique céleste*—saw great strides being made overseas. The pace was set by scientific organizers such as Humboldt, Herschel, Babbage, and Arago. Working closely with observatories in Berlin, South Africa, Greenwich, and Paris, and maintaining correspondence with hundreds of observers worldwide, these men of science were coordinating the study of the earth's meteorological, physical, geological, and biological processes, gradually linking up the domains of nature. Their work was grounded in precise astronomical coordinates and shared instruments—as in Humboldt's lines of similar temperature and biogeographic distributions of plants, Babbage's push for international standards, and the study of daily and seasonal changes of terrestrial magnetism in the "magnetic crusade."

Bache, Henry, and Peirce wanted American scientists to participate as equal players in these world-spanning networks. To do so, they would need robust, reliable institutions to foster research and training and to keep charlatans out. They painfully felt the lack of such institutions in the United States. On an unprecedented scale, science was on the march, gathering up facts and placing them within a stable, all-embracing picture of the world. America would have to run to join this movement.

Making Darkness Visible   The arrival of photography in 1839, while both Poe and Bache were in Philadelphia, was celebrated as a major scientific advance. Bache wrote one of the earliest American reports on the daguerreotype, photography's first widely accessible form. His report appeared in Philadelphia's *United States Gazette* in September 1839, drawing from French announcements. The invention, by the Parisian stage designer Louis Daguerre with the aid of Nicéphore Niépce, captured images "drawn by light" in a permanent medium.

The daguerreotype was an adapted camera obscura—an enclosed box with a lens on one end. At the other end, the operator inserted a copperplate that had been coated with silver and exposed to iodine. Directed at a scene, "the action of light does the rest." When the plate was removed and exposed to mercury vapor, a stunningly detailed, lifelike impression of the scene appeared "in the most miraculous beauty." François Arago secured a lifetime pension for the inventors and anticipated its uses in natural history, archaeology, and astronomy. Bache explained the process in enough technical detail to allow readers to try for themselves to create "a drawing in which the light and shade is truly represented, and which may be exposed, without change, to the action of light."

With Bache's encouragement, several colleagues in Philadelphia went to work on the process. Joseph Saxton, who had worked in London as an assistant to Michael Faraday and now worked in the Franklin Institute and the Philadelphia Mint, teamed up with Robert Cornelius, a skilled metalworker who could produce silver-coated copperplates. Saxton took the first U.S. daguerreotype from his window at the mint—a rendering of Central High School, one of Bache's institutional bases.

Saxton and Cornelius, working with the university professor Paul Beck Goddard, made chemical improvements that shrank photographic exposure time from minutes to seconds, making rapid portraits possible. Using iodide of bromine as an accelerator, along with stronger light magnified by focusing mirrors, Cornelius took the world's first selfie.

The daguerreotype was a machine that rivaled the work of painters. The *Knickerbocker* editor Lewis Gaylord Clark saw the first daguerreo-

Robert Cornelius, daguerreotype self-portrait, Philadelphia, 1839

types shown in New York, brought by a student of Daguerre's. He noted ghostlike images of the Seine, including the "Quay de la Morgue," where washing women were seen "hanging out their clothes, which almost wave in the breeze." The images resembled the British painter John Martin's panoramas of biblical scenes—and John Milton's vision of hell in *Paradise Lost*: "The 'darkness visible,' the floods of light, the immensity of the space, and the far perspective, in their dim, obscure state." John Herschel said of the early daguerreotypes he saw in Paris that "every gradation of light & shade is given with a softness & fidelity which sets all painting at an immeasurable distance." Yet the painter Thomas Cole, then at work on his allegorical *Voyage of Life* series, was dismissive: "If you believe everything the newspapers say (which by the by would require an enormous bump of marvellousness) you would be led to suppose that the poor craft of painting was knocked in the head by this new machinery for making Nature take her own likeness."

For some, photography appeared as the ideal meeting point of empiricism and technology—the fulfillment of the age's double faith in observable facts and in machines. It was an automatic method, seemingly free from human interpretation or subjectivity, which revealed and replicated the visible surface of the world. An article in *Blackwood's* enthused, "The telescope is rather an unfair tell-tale; but now every thing and every body may have to encounter his double everywhere."

One of America's first indoor daguerreotypes was taken at Philadelphia's Academy of Natural Sciences, probably by Goddard; it shows three men surrounded by animal skeletons and display cases. The man in the top hat is likely Samuel Morton, the skull collector, and the young man in the cap, his apprentice, Joseph Leidy, who would become a major paleontologist. Historians have also argued that the self-possessed figure seated at right, with crossed legs and striped trousers, is Edgar Allan Poe.

Poe did wear sideburns; he hobnobbed with members of the academy and knew Morton well enough to ridicule him in a later tale. Whether the image shows Poe—or, as is more likely, some other smartly dressed scientific enthusiast—he was indubitably at the center of America's "daguerreotypomania."

In January 1840 he wrote one of the earliest American appreciations. He described the daguerreotype as the "most important, and perhaps the

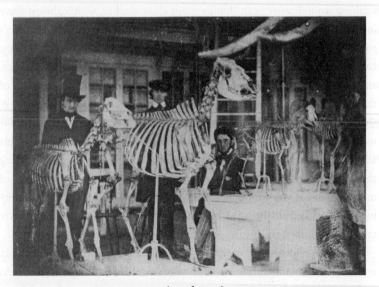

Daguerreotype, Academy of Natural Sciences, ca. 1840

most extraordinary triumph of modern science." But his description went beyond presenting photography as a faithful instrument for capturing the empirical surface of nature; he highlighted its sublime, supernormal effects.

For Poe, the image's shimmering, eerily lifelike qualities pointed to a deeper architecture of reality:

> In truth, the Daguerreotyped plate is infinitely (we use the term advisedly) is *infinitely* more accurate in its representation than any painting by human hands. If we examine a work of ordinary art, by means of a powerful microscope, all traces of resemblance to nature will disappear—but the closest scrutiny of the photogenic drawing discloses only a more absolute truth, a more perfect identity of aspect with the thing represented. The variations of shade, and the gradations of both linear and aerial perspective are those of truth itself in the supremeness of its perfection.

The immediately apparent image suggests a level of perception below what the human eye ordinarily sees. Applying a microscope to the image discloses a hidden layer of truth. An even more powerful lens might reveal another layer, perhaps ad infinitum. The curtain lifts to reveal a curtain; the riddle hides a further enigma.

Poe also hinted that the daguerreotype's underlying action—using the force of light to produce images that were "*infinitely* more accurate" than human art—was nothing short of divine: "The source of vision itself has been, in this instance, the designer." He ended with a prophecy, that its consequences will exceed "the wildest expectations of the most imaginative"; the results of the optical device "cannot, even remotely, be seen." As if to prove Poe's prediction, that year John W. Draper—who had relocated from Virginia to New York University, where he experimented with photography along with Samuel Morse—took a daguerreotype of the moon, revealing shadows and craters in breathtaking relief.

Poe's writing transformed the daguerreotype—a technology embraced by scientists for capturing and freezing the visible surface of the world—into a magical corridor opening an unimaginable (and for the moment invisible) succession of further possibilities. This machine for fixing vision showed

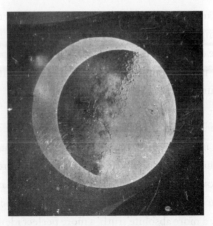

John W. Draper, moon, probably March 26, 1840, sixth-plate daguerreotype

that visible reality was only a first screen, a resting point on the way toward "a more absolute truth." In Poe's hands, the mechanical and material began to unlock the mystical.

A Living Enigma Machine   As portrait studios opened in Philadelphia and New York, and itinerant "daguerreotypists" offered cheaper versions, photography became a popular art. It thus moved like many technical and scientific innovations between public, commercial purposes and more controlled, scientific aims. Likewise, though Poe wrote in the popular press, he understood the projects of science reformers from the inside. His experience as army artificer, his West Point training, and his constant scientific reading put him in a position to publicize, advocate, and adopt their methods.

He even lent his voice to Bache and Henry's cause to "put down quackery." Moonlighting for *Alexander's Weekly Messenger*, a family newspaper, he debunked a local "Weather Prophet, Star Reader, & Fortune Teller." Under the shrill title "A Charlatan!," he excoriated the prophet's book as "a pamphlet of trash" and an "impudent attempt at humbugging the public" (the very phrase Burton used against *Pym*—it stuck in his craw).

As a further contribution to public science, in *Alexander's* Poe launched a popular series of puzzles, or "exercises of the analytical faculties." Rather than improvisation, he asserted that a "degree of rigid *method* enters into enigma-guessing." He suggested that "a set of rules might absolutely be given by which almost any (good) enigma in the world could be solved," alluding to Champollion's recent decipherment of the Rosetta stone.

Injecting a light pastime with one of the era's most important notions— method—Poe was elevating code-cracking to "the majesty of a science." The next month, a reader called his bluff, sending in a message:

850;?9
O 9? 9 2ad; as 385 n8338d— ?† sod—3 —86a5: —8x 8537
95: 370d: o— h—8shn 3a sqd?8d— ?† —og37 —8x8539 95:
Sod—3 o— 9 ?o—1708xah— 950?9n [. . .] ?† 27an8 05:otg38—
9 2038 ?95

Though Poe claimed to have received it the morning the paper went to press, the challenge gave him *"no trouble whatever"*:

ENIGMA
I am a word of ten letters. My first, second, seventh, and third is useful to farmers; my sixth, seventh, and first is a mischievous animal [. . .] My whole indicates a wise man.

The mystery within the enigma was "Temperance" (with sub-solutions "Team" and "Rat").

Other readers joined the game. He solved their hieroglyphs and printed solutions week after week ("More of Our Puzzles"; "Puzzles Again!"). No longer the debunker, Poe now assumed the role of a clever entertainer, much as Maelzel had done with his chess player, stumping a willing audience not with a marvelous machine but with a "rigid method" and "universal rules" for decryption. Poe was accused of "gaggery, or more delicately speaking, of humbug" by readers who assumed "that we wrote our own puzzles and then solved them." But he insisted that cryptology involved neither fakery nor magic: decipherment "is subject to the universal rules of analysis." Some readers saw the performance as miraculous, others as a fraud; "the row increased as the wonder grew."

Poe saw his age's many humbugs and its "men of science" pushing for foolproof forms of scientific authority as secret partners. Together they were establishing the modern matrix of entertainment and science, doubt and certainty. Those who successfully denounced the tricks of charlatans were precisely those with the skills to make themselves believed. Those best equipped to perpetrate a fraud were those who understood how proof worked. Controversies in the press and lecture halls about scientific claims were dialogues about who and what was to be believed, according to what methods and standards.

It was a new and wondrous age: "The simple truths which science unfolds, day after day, are in fact, far stranger, apparently, than the wildest dreams," Poe wrote. "Method" was being offered up as a universal tool for arriving at certainty. But Poe saw that method could easily be taken too far, applied inappropriately, used to mystify rather than enlighten. Poe tweaked the addiction to method in a tale for *Burton's*, "Peter Pendulum, The Business Man," opening with a calling card:

I am a business man. I am a methodical man. Method is the thing, after all. But there are no people I more heartily despise, than your eccentric fools who prate about method without understanding it; attending strictly to its letter, and violating its spirit. These fellows are always doing the most out-of-the-way things in what they call an orderly manner.

Instead, this follower of Franklin applied method to ordinary things. In his "general habits of accuracy and punctuality" he was "not to be beat by a clock"; he demonstrated his "positive appetite for system and regularity" in harebrained get-rich-quick schemes—starting fights, raising cats to sell to the public cat catcher. Poe would later write "Diddling Considered as One of the Exact Sciences," a satirical manual of methods for "diddles," petty scams, and deceptions. Scientific method alone was no guarantee of goodness, truth, or even profit.

Poe also used satire to point out the connection between methods, machines, and the brutality of American expansion. The Second Seminole War had been under way since late 1835. The U.S. Army was deployed to force tribes in Florida off their lands to make way for white settlers, with a long series of bloody and inconclusive confrontations led by General Winfield Scott and Zachary Taylor among others. In August 1839, Poe published "The Man That Was Used Up," a portrait of Brigadier General John A. B. C. Smith, who made his name fighting the "Bugaboo and Kickapoo" tribes. He was now celebrated for his lush speaking voice and enthusiasm for the "rapid march of mechanical invention" in America. "We are a wonderful people, and live in a wonderful age," this antebellum tech bro declared. "Parachutes and rail-roads—man-traps and spring-guns! Our steam-boats are upon every sea." He saw "the most wonderful," "ingenious," and "useful—the most truly useful—mechanical contrivances" springing up like mushrooms and grasshoppers.

While waiting to meet the general in his rooms, a journalist inadvertently kicks over a bundle, from which a feeble voice calls out—for legs, arms, wig, and teeth. The general's servant, Pompey, assembles these parts into the form of a man and inserts "a somewhat singular-looking machine" into the tin man's mouth. Smith's voice again bellows forth with "rich melody and strength."

General John A. B. C. Smith had been scalped, dismembered, and blinded while fighting the "Bugaboo"—a ginned-up, exaggerated terror. Ready-made parts made a new man of him, regular as a triangle. Poe's tale showed the "self-made-man," the free citizen and heroic frontiersman of Jacksonian democracy, as a fragile contraption held together by publicity, stagecraft, and slave labor—the barbarous underside of the march of mechanized civilization.

Poe understood the excitement caused by new discoveries and inventions, by science's promise to bring order to society and the natural world, and by its growing power to tie down ordinary reality and chase away the anomalous, supernatural, and mythical. Yet in Philadelphia he was experimenting with ways to turn this power against itself—to use reason to interrogate reason and explore its hidden shadows. Beyond his work for Burton and at *Alexander's*, he was preparing a collection of short stories, adding new, jaw-dropping tales to his *Folio Club* pieces.

The title of Poe's planned collection, *Tales of the Grotesque and Arabesque*, referred to two modes of deviating from common-sense reality. "Grotesque" tales exaggerated literary and social conventions to comic effect; the term originally referred to the distorted figures molting from the walls of Renaissance grottoes. "Arabesque" tales opened up a dream realm within or alongside regular existence, where supernatural forces could be sensed and harnessed. The romantic critic Friedrich Schlegel had described "arabesque" literature as works that unpredictably shifted through diverse styles and forms. The term also referenced orientalist imagery: the marvels of *The Arabian Nights*, the ornate concentric framings of Islamic manuscripts and Persian rugs, heady potions of perfumes, repetitive rhythms, and hallucinogens wafting in from a sumptuous if vaguely imagined East.

A reviewer would praise Poe's collection as a "succession of richly-coloured pictures in the magic lantern of invention." The image was apt: Poe's literary machine projected images of reality that were amplified, exaggerated, distorted, and revelatory. These were works of a brilliant imagination testing its strength. But to shift so decisively between discrete levels of existence also required him to understand the methods—literary, technical, rational, and imaginative—that were used by "real working men in the way of science" to weave together a stable baseline of reality.

A Passionate Empiricism   The image of science most celebrated in early nineteenth-century America relied on observation, analysis, and classification to dissect—and reassemble—the interlocking parts of nature's machine. Yet this figure of a safe and sober reason was mirrored by a dangerous, ingenious, wildly speculative twin. A philosophical countercurrent was at work in science in these years, focused on organic wholes, aesthetic response, and imagination—and Poe was unusually receptive to its lessons.

Naturphilosophie (German for "natural philosophy") was pioneered by the philosopher Friedrich Schelling, a college roommate of the philosopher G. W. F. Hegel and the romantic poet Novalis. This school fiercely opposed the stable clockwork universe of Newton and the Enlightenment. It assumed that the human mind and the natural world arose from the same underlying principle or force, "The World-Soul" or "the Absolute." This creative power developed by dividing and differentiating itself, starting with a split between matter and consciousness, to form all the entities of the cosmos. The task of a philosophical natural science, Schelling argued, was to use observation, experiment, and reason to rediscover, self-consciously, the original unity between mind and world.

Schelling put forth a research program to investigate the relations among visible forces and grasp the principles uniting them all. Observation and mathematical analysis alone could not unlock nature's secrets. Guiding ideas were needed to orient inquiry, such as the archetypal patterns embodied in living things and the underlying unity of the imponderable fluids (light, heat, electricity, magnetism). Symbolism, poetry, and nature's "hieroglyphics" could offer further insight to the world's processes of becoming.

Naturphilosophie has been called "romantic science." Romantic poetry in the years in which Poe began his career insisted on a world in a state of becoming; as Wordsworth put it, the soul was attuned to "something evermore about to be." Schelling's work brought the speculations and longings of the poets Goethe, Schiller, and Novalis into the realm of empirical and rational science. *Naturphilosophie* infused the empirical study of nature with an aesthetic, surging vitality. It plunged the researcher—senses, emotions, body, and soul—into a whirling storm of alternating powers, reversals, and conversions.

Though Schelling's writings could be dismissed as mysticism, *Naturphilosophie* fed rigorous research. The Danish physicist and poet Hans Christian Ørsted discovered in 1820 that electricity could be converted into magnetism; the Frenchman André-Marie Ampère showed that the conversion also went in the other direction and derived laws of the interaction of electricity and magnetism, research followed up by Michael Faraday—and, in the United States, by Joseph Henry. The comparative anatomists Lorenz Oken, Étienne Geoffroy Saint-Hilaire, and Richard Owen postulated a single ideal animal form behind the variety of visible species. Even Alexander von Humboldt—seen by many as the great champion of empiricism and fact collection—took inspiration from Schelling's conviction of a cosmic unity underlying nature's diverse forms, to be realized again through art and science.

Most American men of science ignored *Naturphilosophie*'s obscure specu- lations, seeing them as remote from practical and utilitarian concerns. But Emerson and the transcendentalists knew Schelling's work through trans- lations and summaries by Carlyle, Coleridge, Germaine de Staël, Victor Cousin, and the chemist and poet Humphry Davy. Poe, too, was familiar with the claims and ambitions of romantic science and philosophy and re- ferred at various points to Kant, the "wild Pantheism of Fichte," and "above all, the doctrines of *Identity* as urged by Schelling."

In *Burton's*, Poe approvingly quoted observations made by Humphry Davy in honor of the geologist William Buckland. If ancient monuments like the Acropolis make us marvel at "the genius of artists" of the past, we should have all the more admiration for "those grand monuments of nature which mark the revolutions of the globe." This was an allusion to the geological theory of Cuvier, which explained the extinction of species by "revolution- ary" catastrophes—floods, earthquakes, temperature changes.

Poe was calling attention to God's sublimity not only as a *creator* but also as a *destroyer*. The earth's history was one of "continents broken into islands; one land produced, another destroyed," while we see "among the graves of past generations—the marble or rocky tombs, as it were, of a former animated world—new generations arising, and order and harmony established." Along with Davy, Poe saw the earth's crags and valleys as evidence of a turbulent creator, subject to violent mood swings. The earth itself was "a system of life and beauty produced, as it were, but of chaos and death."

Like the proponents of *Naturphilosophie*, Poe saw a nature sublimely ani- mated by polarized forces: attractions and repulsions, positive and negative energies, light and dark, in constant movement between order and chaos. Poetry and intuition, as much as observation, calculation, and reason, could be methods for deciphering its design.

Arabesque Operations—Ligeia and Usher

This combination of patient experi- ment and encounters with terrifying, polarized forces made its way into Poe's *Tales of the Grotesque and Arabesque*, published in 1840 in two volumes by Lea & Blanchard. Reviewers found the collection a "playful effusion of a remarkable and powerful intellect" with "vivid description, an opulence of imagination, a fecundity of invention." The tales captured "the light and darkness, the virtue and the vice by which mankind are by turns affected." One London critic had complained that the innovations in *Arthur Gordon Pym*, spread across a large canvas, were "too uniformly extravagant." By contrast, in his most memorable short stories, Poe channeled a steady rise of interest toward a single effect—applying what he saw as the hallmark of powerful poetry to fiction. These were tales as charged as a lightning rod.

In "Ligeia," Poe imagined a beautiful, willful, and supremely learned woman whose studies circle around the concept of identity—the immortal essence that makes one individual distinct from all others. Ligeia was tall, with dark, curled hair and black eyes that suggest an Eastern origin, as with one of Disraeli's Jewish heroines, or a "creole" with African ancestry, but her husband, the narrator, cannot recall where or when he had met her.

Ligeia grows ill and dies. In a state of nightmare-like confusion, her husband remarries. His new wife, Rowena, is pale and blond, the visual opposite of Ligeia. He acquires a remote mansion and decorates it in fantastic style, draping its pentagonal turret chamber in gold and black tapestry with a snakelike censer swaying from the ceiling. This arabesque space wavers between waking and dream, matter and spirit, life and death. As the narrator abandons himself to opium and obsessive recollections of Ligeia, Rowena grows deathly ill; her tomb is prepared. Over the course of a night's long vigil, with strange winds, lights, and sounds filling the chamber, her corpse, laid out on an ebony bed, returns spasmodically to life in a "hideous drama of revivification." In a state of "inexpressible madness" he watches her rise and approach him, but somehow taller, and with dark hair and eyes and a haunting expression of undying will; his departed love, "the lady Ligeia," stands before him again.

In Ligeia, Rowena, and the bizarre equipment and rituals that fuse and separate the two women, Poe personified the polarities and uncanny unifications of romantic science. Like Ørsted, Davy, Faraday, and Ampère—as well as their fictional counterparts Victor Frankenstein and Balthazar Claës, the alchemist in Honoré de Balzac's *Quest of the Absolute*—the narrator carefully arranges materials to initiate a process of conversion, realizing a spiritual transformation in the world of matter. While harking back to magic and ritual, the operation echoes and amplifies contemporary scientific conversions—like those from electricity to magnetism, from steam to work, from distinct elements to an unheralded compound—in which an apparent opposition gives way to underlying identity.

"The Fall of the House of Usher" is often called Poe's greatest tale—a signature work with a staggering unity of purpose in which all elements build envelopingly to a stunning collapse. Poe draws us into the dark, distant region where the Usher mansion stands. We approach it on horseback—an ancient pile with a crack running through its facade, reflected in a small lake. Overtaken by feelings of gloom and foreboding, the narrator wonders, "What was it that so unnerved me in the contemplation of the House of Usher?" As he ponders the question, he is crowded by "shadowy fancies" and looks into the lake, spurring the "rapid increase" of his terror. He is overtaken by a strange, persuasive impression,

> that around about the whole mansion and domain there hung an atmosphere peculiar to themselves and their immediate vicinity—an atmosphere which had no affinity with the air of heaven, but which had reeked up from the decayed trees, and the gray walls, and the silent tarn, in the form of an inelastic vapor or gas—dull, sluggish, faintly discernible, and leaden-hued.

Like an infectious miasma, this hellish, burdensome ether weighs down on the house's inhabitant Roderick Usher, his childhood friend.

Usher appears horrendously transformed by a nervous disorder, with the leaden affect of the "irreclaimable eater of opium." His twin sister, Madeline, is also ill and grows weaker until she dies and is buried behind the metal doors of the family crypt.

During a ferocious electrical storm, Roderick desperately entertains his friend by reading a chivalric legend. With a violent crashing sound, the crypt is torn open, and Roderick, rocking mechanically, cries, "Madman! *I tell you that she now stands without the door!*"

Madeline appears at the entrance, wrapped in her bloody winding sheet—either buried while still alive or resurrected by the convergence of the electrical storm and the sentient ether around the house.

The Ushers embrace and collapse into each other. The narrator flees on his horse. Behind him a "wild light" flares; he sees the rising red orb of the moon blazing through a zigzag crack, widening through the center of the house. The house splits in two, sinking into its own reflection in the lake.

The mansion, doubled by its reflection in the lake, is also mirrored in the song Roderick sings, "The Haunted Palace," an allegory of his own disintegration and that of the Usher family line. "The Fall of the House of Usher" is a hypnotic, fractal reflection on doubling and symmetry—from its images (the house, the siblings) to its sounds (*hushed* and *rushing*, *shudders* and *eddies*) to its form. The story is structured as a chiasmus. Elements from the start are echoed at the close, accelerating toward and away from the tale's central event: Madeline's death, burial, and revival. Usher's sister, Lady Madeline, is a "lady made line": a life doubled in print, reanimated through a fevered reading, disappearing again when the pages close.

From the narrator's musings on the psychological effect of each brick in the mansion, to the final collapse of the tale's tension along with the house, "Usher" is an intensely self-conscious allegory of artistic construction—a meditation on composition and decomposition. Its "mystery all insoluble" was that it was a compound designed to dissolve.

Poe was testing new literary formulas. "Usher" recombined elements from gothic and fantastic tales (much as he had in the earlier "Metzengerstein"), apocalyptic language from the book of Revelation, and imagery from alchemy: the red moon rising at the conclusion has been seen as a reference to the "red king" that marks the accomplishment of the alchemists' Great Work, interpreted by some as the mystical liberation of the soul from the prison of the body. Into this heady mix he added the ethers, atmospheres, and energies of experimental science. The "ancestral pile" of the gothic manor was recharged with the electricity of a "voltaic pile," or battery; the house's life cycle unfolded within its once nutritive, now poisonous milieu. The tale's intense focus on its own technical effects and its reflections on painting, architecture, music, and poetry have made it highly attractive for later repurposing in multiple media; "Usher" has inspired musical and cinematic imitations, from Claude Debussy and Jean Epstein to Alfred Hitchcock and Philip Glass.

Poe also turned his self-consciousness about technique into a satire of his own methods. In "How to Write a Blackwood Article" he speaks as an aspiring authoress, Signora Psyche Zenobia, seeking advice from the legendary editor of *Blackwood's Magazine* for writing one of the zippy, erudite tales of sensation and brushes with death that his magazine made famous and that inspired Poe's own intense, first-person tales. Mr. Blackwood's approach was paint by numbers. First, he tells her, "get yourself into such a scrape as no one ever got into before": fall out of a balloon or into a volcano, or get stuck in a chimney. Next, choose a tone: didactic, enthusiastic, natural, laconic, curt, elevated, metaphysical, or transcendental. Now fill it up with an "air of erudition," sprinkling in similes from Latin, ancient Greek, or German tomes.

Psyche Zenobia follows the formula, writing up her experience of climbing a clock tower and (literally) losing her head. The output is another tale, "The Scythe of Time"—a gothic slapstick lampooning the formulas of Poe's own trademark stories, pushing the arabesque into the grotesque, sending method and mechanism off the rails into lunacy.

Lost in the Crowd   After more than a year at *Burton's*, Poe began another "single connected story," a serial novel set in the eighteenth century, packed with descriptions of the sublime landscapes of the American West. He presented *The Journal of Julius Rodman* as the diary of "the first white Man that ever crossed the Western Wilderness." As in *Pym*, Poe tapped into excitement about exploration. He took inspiration from Lewis and Clark's voyage (written up by Nicholas Biddle) and Washington Irving's *Astoria; or, Anecdotes of an Enterprise Beyond the Rocky Mountains*, a biography commissioned by John Jacob Astor, the fur trader, real-estate speculator, and opium smuggler turned multimillionaire. Poe's first chapters followed Rodman and his small band of travelers up the Missouri toward the Rockies into stunning vistas and deadly scrapes.

Rodman's empire-building plans ended abruptly after six installments in June 1840. Burton had decided to sink his money into a theater, giving himself starring roles. He ran a cagey advertisement for "the best speculation in the publishing way" offered in years, with "peculiar advantages to gentlemen of a literary disposition."

Without warning his readers—or his employees—Burton was putting his magazine up for sale.

Realizing that his situation might change at any moment, Poe acted decisively. He drew up plans for a literary magazine of his own, a high-end publication much like *Burton's* but entirely under Poe's direction. He printed up a prospectus, asking for subscribers in local newspapers.

Burton exploded. A competing magazine would torpedo *Burton's* value. He fired Poe and demanded he repay advances on his salary. Poe replied with cool indignation, as methodically as Peter Pendulum: he detailed his work for the journal, his unpaid writings, and Burton's injustices, including putting the magazine on sale without "saying a word." He left and slammed the door.

That summer, unemployed, he wrote "The Man of the Crowd," a nightmarish vision of the modern city. Its narrator surveys throngs passing the window of his café, classifying each person by their clothing, expressions, and physical habits—from the "division of the upper clerks of staunch firms" and "modest young girls," to "ragged artizans and exhausted labourers," "women of the town," coal heavers, sweeps, "organ-grinders, monkey-exhibiters and ballad mongers"—all ranked and described like the shells and mollusks of his conchology. Only one person evades his taxonomical system: an old man with a desperate, haunted expression of "absolute idiosyncrasy."

The narrator impulsively springs from his makeshift observatory and trails the old man through the city's darkening labyrinths, past gin palaces and dens of vice. At dawn he comes face to face with his quarry, who looks through him, unseeing. "'This old man,' I said at length, 'is the type and the genius of deep crime. He refuses to be alone. *He is the man of the crowd.*'"

This singular specimen is both a general "type" and a unique "genius" beyond the social taxonomy. Absolutely alone while entirely surrounded, he has a heart that "does not let itself be read."

At thirty-one, Poe had begun to establish himself as a hardworking editor and incisive critic, a poet and author of extraordinary fiction, and a well-informed science writer. Now he was adrift without steady employment in the bustle and racket of a city frantic with deals, inventions, publicity, and plans for westward expansion. He was convinced that the nation needed his services—his singular voice and uniquely critical eye—to advance genuine literature and put down pretenders. If no magazine would hire him, he would have to create his own.

9

Dizzy Heights

The General Interests of the Republic of Letters

Cut loose from *Burton's*, Poe set out to define his own ideal publication. His plan was in harmony with the Enlightenment ideals so present in Philadelphia—rationality, universality, impersonality—and with the drive to forge a new national culture. He would do for American literature what Bache and Henry sought for American science: apply rigorous standards, strengthen ties among those active in disparate regions, and concentrate efforts to produce work of equal merit to European productions.

Poe called his planned magazine *The Penn*—after Pennsylvania's tolerant founder and his own sharp instrument. His one-page prospectus was a declaration of literary independence.

The Penn would assert and defend "the rights" of "an absolutely independent criticism." It would end blind obedience to both British models and local prejudices. It would be immune to the manipulations of the puffing system, never giving ground "to the vanity of the author, or to the assumptions of antique prejudice, or to the involute and anonymous cant of the Quarterlies." It would strike "the arrogance of those organized *cliques* which, hanging like nightmares upon American literature, manufacture, at the nod of our principal booksellers, a pseudo-public-opinion by wholesale." His Philadelphia-based magazine promised a "mechanical execution" to surpass New York's *Knickerbocker* and Boston's *North American Review*.

Poe saw *The Penn* as a fearless, independent tribune with a sacred purpose:

to support a unified, genuinely American literature, serving the "general interests of the republic of letters, without reference to particular regions; regarding the world at large as the true audience." By speaking for the entire nation, it could enter the stage of the international and universal.

Poe printed a stack of copies of his prospectus for friends and supporters in New York, Massachusetts, Ohio, Missouri, Maryland, Georgia, and Virginia. Willis Gaylord Clark at *The Philadelphia Gazette* was pleased to see Poe "reigning in his own sphere, where his classic power and genuine good taste, untrammeled by base or palsying associations, shall have full scope and play." The young Philadelphia publisher George Graham praised Poe's "enviable distinction, as an able, vigorous, impartial" if "somewhat over caustic critic," and wished him success in "*paying the printer.*"

Poe also made the rounds in person, as he had done in his campaign for a place at West Point. Judge Joseph Hopkinson endorsed *The Penn*'s high civic ambition to "concentrate in Philadelphia as much literary talent as possible, and be distinguished by works of science and genius." Philadelphia's *Daily Chronicle* ran the prospectus thirty-six days in a row and noted in September the growing "succession of names" on the subscription list.

The Penn became a cause célèbre among the writers, artists, actors, and artisans who met at the Falstaff Hotel on Sixth Street near Chestnut, including the engraver John Sartain and the painter Thomas Sully. Poe's circle included two recent Philadelphia transplants. Jesse Erskine Dow, a naval officer turned journalist, wrote a memoir of maritime life that Poe published in *Burton's*. The novelist F. W. Thomas, a friend of Poe's beloved brother, Henry, made up for his braggadocio and eye for the main chance with loyalty, sensitivity, and tact.

They joined colleagues at the Congress Hall Hotel and the Cornucopia restaurant. This was a raucous crowd, but Poe downplayed rumors Burton was spreading about his intemperance. "I never was in the *habit* of intoxication," he wrote to a friend, though in Richmond he had at times yielded "to the temptation held out on all sides by the spirit of Southern conviviality." For four years, starting in 1837, he "abandoned every kind of alcoholic drink," with the exception of "a single deviation, which occurred shortly after my leaving Burton"—when he turned to "*cider*, with the hope of relieving a nervous attack."

Both F. W. Thomas and Jesse Dow had come to town to work for the Whig Party. Eighteen forty was an election year, pitting Jackson's successor, Martin Van Buren, against the Whig candidate, William Henry Harrison—an ancient hero of the Battle of Tippecanoe in the War of 1812. Though an educated landowner, Harrison could play the down-home man of the people much as Jackson had. In contrast, Van Buren appeared aloof and indifferent in the face of the continuing depression. Harrison's running mate, John Tyler, an old Federalist from Virginia, strengthened the ticket's appeal in the South.

With the "spoils system" now in place, a political supporter could emerge from a presidential election with a plum government post. Despite his general avoidance of politics, Poe joined F. W. Thomas and Dow at a Harrison rally in May 1840, where "some Van Buren men" pelted them with rocks and bricks. Appearing in the midst of this riotous political contest—in which Harrison and Tyler would be victorious—Poe's campaign for *The Penn* had a decidedly Whiggish ring, aiming to overcome regional differences with a rational, procedural authority.

By the end of the year he had nearly a thousand subscribers. He particularly valued the support of Nicholas Biddle—Alexander Dallas Bache's key patron—whose work in science and statecraft exemplified the Whig vision of a strong, consolidated republic led by a wise elite. Prospectus in hand, along with a copy of his *Tales*, Poe paid a visit to Biddle at his estate, Andalu-

T.Birch del.   J.W.Steel

ANDALUSIA the Seat of Nicholas Biddle Esq.

Andalusia, the estate and landscape garden of Nicholas Biddle

sia, a classically adorned mansion designed in part by William Latrobe. Set in an immense landscape garden along the Delaware, the estate was where Biddle had retired after the Bank War. Poe invited Biddle to write an article for the first issue of *The Penn*, with which "I would at once be put in a good position—I mean in respect to that all important point, *caste*, by having it known that you were not indifferent to my success." Poe's charm and learning won over the wizened banker, science enthusiast, and aristocratic defender of the Republic; Biddle took out a four-year subscription.

Though based in a republic in which democracy was widely proclaimed, Poe's national literary institution would be unabashedly elitist. It would advance "a criticism self-sustained; guiding itself only by the purest rules of Art; analyzing and urging these rules as it applies them; holding itself aloof from all personal bias." On seeing Poe's plan for a literary journal, one of his supporters enthused, "Ours has become a mighty nation; but if its institutions are to be perpetuated . . . the minds of the many must be somewhat enlightened." *The Penn* was as much a project of putting down quackery as were those of Henry and Bache. "I am glad to learn," he said, "that you intend to attempt the overthrow of Humbug!"

Lifeline   *The Penn*'s momentum was slowed in December when Poe was confined with "a severe illness." In January 1841 he resumed his efforts, telling Thomas Wyatt of "a most advantageous arrangement" he'd secured with an agent and publisher. All was in place for the debut on March 1.

But on February 4, Philadelphia banks stopped reimbursing paper notes against silver and gold, freezing the city's money supply; it was another panic. Southern banks followed suit, bringing commercial activity to a near halt: "Money is difficult to obtain, even at a high premium." According to another magazine owner, *The Penn* had "such prospects as are seldom enjoyed—an excellent list of subscribers" and "the universal good-will of the public press." Yet with the "present disorder of all monetary affairs," it would have been "madness" to launch. Without the capital to hire a press and printer, Poe had to put his dream of independently running his own magazine on hold.

There was, however, an editorial job available, more or less where he had left it. *Burton's* had found a buyer. The former lawyer George Rex Graham, though not yet thirty, partly owned *The Saturday Evening Post* and a monthly, *The Casket*. He was as amiable, clever, and canny as his features suggested. He purchased the *Gentleman's Magazine* from Burton for thirty-five hundred dollars, along with its subscription list of thirty-five hundred subscribers. In January 1841 he united *Burton's* with *The Casket* under a new title: *Graham's Lady's and Gentleman's Magazine*. It would be "embellished in a style which has not been equaled in this country before," with mezzotint engravings by John Sartain, "the best engraver of the kind," sheet music, and "exquisitely colored" fashion plates.

Graham hired Poe as reviews editor at a decent eight hundred dollars per year. Where Burton had micromanaged and undervalued him, Graham was well aware of Poe's abilities: "As a stern, just and impartial critic Mr. Poe holds a pen second to none in the country."

William Henry Harrison's presidential victory ushered in a new round of government appointments. Stricken with a cold while giving an interminable inaugural speech, Harrison died a month after taking office. His running mate, the Virginian John Tyler, was suddenly thrust into the presi-

Graham's Lady's and Gentleman's Magazine, *May 1841, issue containing*
"A Descent into the Maelström"

dency. Poe's friends F. W. Thomas and Jesse Dow moved to Washington to claim the spoils. Meanwhile, Poe transferred his own civic ambitions from *The Penn* to *Graham's*: with his "additional editorial strength," *Graham's* could enlist the "pride of the American people and writers, in the support of work creditable to National Literature."

The position was a lifeline. One of his first tales in its pages, "A Descent into the Maelström," reads as a reply to the despairing vision of "The Man of the Crowd." It, too, began with an observer plunging into a dangerous, swirling phenomenon. This time, however, observation and reason brought safety and certainty.

Perched on a precipice high above a Norwegian bay, a white-haired fisherman recounts a fateful voyage in which he misread the tide's cues and found his ship caught in the maelstrom: a terrifyingly destructive whirlpool produced when the outgoing tide meets an incoming current.

His ship's speed increased as they whirled around the outer rim of the vortex—a phenomenon that arises, Poe knew from fluid mechanics, when two currents flowing in opposite directions meet, and where, in mathematical terms, every point's distance from the center is inversely proportionate to the speed of its revolution. In other words, as the revolving ship drew nearer to the center, it approached infinite speed—a physical impossibility but a mathematical necessity. Gazing into the whirlpool's center, the sailor was awed by "so wonderful a manifestation of God's power," he imagined "how magnificent a thing it was to die in such a manner." Poe detailed the fisherman's responses to this manifestation of infinity, following his oscillations between the extremes of emotion the politician and philosopher Edmund Burke described as the effect of the sublime: "awe, horror, and admiration."

Amid violent motion and thunderous sound, as the risen moon streamed its light into the abyss, the sailor suddenly reached a strange state of detachment, "possessed with the keenest curiosity." In a ray of light cutting through the nightmarish darkness of the storm, he saw "a magnificent rainbow, like that narrow and tottering bridge which Mussulmen say is the only pathway between Time and Eternity."

Gathering his faculties, he made a series of "important observations": larger bodies fell fastest into the whirling funnel; cylinders fell slower than other objects of the same size.

He lashed himself to a barrel and—against common sense, and against every instinct for survival, but rationally following the implications of his observation—threw himself into the sea.

The boat plunged headlong "into the chaos of foam below," while he floated at its edge until the whirlpool subsided. His circular journey took him from terror and admiration to observation and hope. When his companions dragged him ashore, he was as changed as "a traveller from the spirit-land." His "raven-black" hair had turned white.

Never before had scientific method received such a dramatic, devastating endorsement. The sailor, adopting an emotionally charged but controlled stance of stoic detachment in the midst of chaos, was demonstrating a version of the "objectivity" that the science of Poe's age celebrated as one of its highest attainments. It allowed him to observe important facts and, like Bacon, establish the law behind them, then act. Here detachment brought not horrific isolation, as it did in "The Man of the Crowd," but salvation.

Like the sailor's barrel, *Graham's* was a vessel to see Poe through the storm. Beyond material stability for himself, Virginia, and Muddy, the position offered a highly visible platform for his fiction and critical program. He could console himself that *The Penn* had merely been postponed. Graham even promised to support him in the venture—once *Graham's* was safely launched, of course.

The Science (Shall We So Term It?) of Criticism

At *Graham's*, Poe published a tidal flood of tales, poems, and brilliant, at times ruthless reviews of contemporary publications. Reviewing a welcome satire of American writers, *The Quacks of Helicon*, he declared that "as a literary people, we are one vast perambulating humbug"; he offered his own disinterested judgments as a counter to the cliques and "*coteries*" behind "the *system* of puffery." He applied his judgments to popular works of science as well. He compared Lord Brougham, the most influential science popularizer of the age, to Coleridge, as "one who might have done much, had he been satisfied with attempting but little." Brougham's essays were "immethodical" and "quackish"; they could have been "*better* written by any one of a multitude of living *savans*"—including, presumably, Poe himself.

More effective, he thought, was *Pantology* by the local university professor and West Point graduate Roswell Park. This *Systematic Survey of Human Knowledge* was a "science of everything," a portable overview of the quickly multiplying branches of science. Recalling his own struggle with the "endless, unstable, and consequently vexatious classifications" of conchology, Poe endorsed *Pantology* as an explorer's guide through an era of lightning-fast scientific change. "To human knowledge in general, it is what a map of the world is to geography."

Poe gave a boost to his profile as national literary critic in November 1841 by updating a series he had begun at the *Messenger*, "Autography." It joined woodcuts of the signatures of authors to analysis of their character and work—usually humorous, often appreciative, occasionally sharp. He was feeding on the emerging cult of celebrity and the accompanying craze for phrenology and other methods of character analysis (Poe told a friend that when he had his own head examined by phrenologists, they "spoke of me in a species of extravaganza which I should be ashamed to repeat"). The first installment caused a sensation, with the editor of New York's *New World* borrowing Poe's cut woodblocks for a reprint.

Bostonians, however, took umbrage at Poe's irreverent survey of the nation's literature. Poe had insulted Ralph Waldo Emerson, the former Unitarian minister and leader of the transcendentalist school, who sought to awaken his audiences to their sovereign individuality and their participation in the "World-Soul"; Poe placed Emerson among "a class of gentlemen with whom we have no patience whatever—the mystics for mysticism's sake." Another transcendentalist, Orestes Brownson, he said, "has not altogether succeeded in convincing himself of those important truths which he is so anxious to impress upon his readers." Poe's worst crime against Boston, however, was his accusation of plagiarism against Henry Wadsworth Longfellow—beloved bard and professor at Harvard: "His good qualities are all of the highest order, while his sins are chiefly those of affectation and imitation—an imitation sometimes verging upon downright theft."

In reply, the *Boston Daily Times* denounced Poe's "dogmatism, egotism, and other *isms* equally as offensive"; he was a "literary dictator," a self-appointed "censor general of American authors." The critic Edwin Whipple denounced Poe's literary heathenism: "We would as soon go to a New Zealander for correct views of Christianity as to Mr. Poe for correct criticism."

Poe had developed a strange allergy toward his birth city, Boston—triggered by Longfellow, *The North American Review*, Emerson, and transcendentalism. It sprang from harsh reviews (the *Boston Notion* had derided his *Tales* as a "congregation of nonsense" falling "below the average of newspaper trash"), from what he saw as Boston's cliquish and aloof self-righteousness, and, perhaps more deeply, from his wounded pride at not being welcomed there when he first fled Allan's house.

Yet Poe's attacks on Longfellow never obscured his admiration for the older poet. Likewise, Poe's occasional mockery of transcendentalism hid strong affinities. Like Emerson, Henry David Thoreau, Margaret Fuller, and other contemporaries, Poe wrestled with the strained relationship between individuals and masses, the ambivalent effects of democracy and industry, and the pantheist yearnings of post-Kantian idealism. Poe's eventual cosmology, *Eureka*, shared many ideas with Emerson's *Nature* and "Circles"—the divinity of nature and humanity, the creativity of the senses, and the leap into self-transcendence, what Emerson called the "insatiable desire . . . to forget ourselves." But Poe was impatient with Emerson's allusive, sermonizing style and skeptical of his faith in the expansive individual's power to embrace and renew the universe. Poverty and misfortune further held Poe's optimism in check. Opening a darker current of American romanticism, he shared Nathaniel Hawthorne's conviction of humans' obstinate corruption.

In April and May 1842, Poe devoted several pages to an appreciation of Hawthorne's collection *Twice-Told Tales*, in which he praised not its moral lessons but its aesthetic force. He placed the work in "the highest region of Art—an Art subservient to genius of a very lofty order." Hawthorne confirmed Poe's view that the short story was the ideal field for the demonstration of a writer's power; within its narrow precincts, tone, incident, and style could all be arranged to attain "a certain unique or single *effect*." Hawthorne wielded a pure, powerful style in which "high imagination gleams from every page." With a tone of repose, melancholy, and a "strong undercurrent of *suggestion*," the tales earned Poe's highest endorsement: "As Americans, we feel proud of the book."

In early 1842, he opened *Graham's* reviews section with an "*exordium*" that carefully set forth his agenda for American criticism. The periodical press, he observed, was "beginning to acknowledge the importance of the science (shall we so term it?)" of criticism. A well-grounded, universally accepted criticism could, like any other science, take the place of "the flippant *opinion* which so long has been made its substitute."

American reviewers had long followed British judgments in a "perfect farce of subserviency." More recently they had begun to champion a "national literature" treating American settings, characters, and landscapes—as if "the world at large were not the only proper stage for the literary *histrio*." Reviewers now found themselves "liking, or pretending to like, a stupid book the better because (sure enough) its stupidity was of our own growth." But he saw signs of a more rigorous and principled approach, inquiring into "the offices and provinces of criticism," grounding it in natural principles instead of "a mere system of fluctuating and conventional dogmas."

He feared that this new wave of criticism might only unleash a "frantic spirit of *generalization*"—his view of the vague "conglomerate science" advanced by Emerson, whose "present *role*," Poe said, "seems to be the outCarlyling [of] Carlyle." With Emerson's assistance, Thomas Carlyle had first published his novel *Sartor Resartus* ("the tailor re-tailored") in Boston in 1836; the book was a dense, fragmentary, tongue-in-cheek reconstruction of the thought of a fictional German scholar, Diogenes Teufelsdröckh, and his philosophy of clothes. Poe's 1840 essay, "The Philosophy of Furniture," a winking disquisition on interior design, had more than a patch of *Sartor Resartus* to it. But he saw Carlyle's convoluted and moralizing style as a dangerous influence on criticism.

Poe sought to "limit literary criticism to comment upon *Art*." Critics, he declared, should judge a work for its own sake, in its character as "*art-product*." They should set aside any moral lesson it might preach, the author's life or character, its implications for "the world at large," and focus instead on formal properties—rhyme, rhythm, color, form, harmony, the effect intended and the means used to attain it.

On one hand, Poe was channeling idealist notions of "art for art's sake" advanced by Victor Cousin, Germaine de Staël, and Benjamin Constant, derived from arguments of Kant, Schiller, and Schlegel. On the other, he was establishing the grounds for an "objective," even scientific ("shall we so term it?") analysis of art. For instance, as a guide for the analysis of poetry, Poe wrote a punishingly technical essay—later expanded as "The Rationale of Verse"—to explain the rules and underlying "law" of poetic rhythm, or prosody, through a detailed dissection of "spondees," "dactyls," and "hexameters." The appreciation of poetic meter, he claimed, is universal; it "appertains to no region, nor race, nor era." We have the same kinds of ears as the ancient Greeks, "and a pendulum at Athens would have vibrated much after the same fashion as does a pendulum in the city of Penn."

Poe's presentation of the critic's task as a neutral analysis of an "art-product" was aimed at what he saw as the preachy, cloyingly sentimental literature that cluttered American journals, and the critics who would praise a work for its uplifting message and ignore its feeble execution. Yet leaving aside the difficulty, if not the impossibility, of legislating taste, Poe's position ruled out what some might think of as criticism's central task: to take a stand on questions of truth, right, and the public good. Poe's principles banned any mixture between "pure poetry" and morality or politics. Much like his contemporary scientific reformers, Poe tried to set politics aside—reform, regional strife, abolition—to argue for a criticism that would be impartial and universal, claiming to float above the factional fray.

A *Necessity for Observation*   Poe showcased his analytic powers in his most influential tale: "The Murders in the Rue Morgue"—the first story he published in *Graham's* and the first modern detective story. It introduced the character of C. Auguste Dupin, an inscrutable, idiosyncratic genius, the model for Sherlock Holmes and other erratic, hyperrational detectives.

After an epigraph from the eccentric seventeenth-century scholar Sir Thomas Browne concerning our ability to reconstruct the actions of mythological figures, Poe's tale began with a bit of phrenology. He postulated the existence of an organ for *analysis*, for "resolving thought into its elements." This mental capacity is not, as some have believed, antithetical to the imagination or what the phrenologists call "ideality." As in a mirror, analysis and imagination involve the same intellectual activity, though they unfold in opposite directions. Imagination assembles discrete elements into a complex composition, while a complex composition can be resolved into its constitutive parts by analysis. According to Poe, the same mental process is required to create something as to explain it, just reversed.

The narrator's friend Dupin, an aristocrat fallen on hard times, serves as an outstanding example of analysis at work. The narrator first meets him in a Parisian bookshop where the two are hunting for the same rare volume. Their shared obsession for obscure learning and puzzles leads them to move together into a "time-eaten and grotesque mansion." By day, with shutters drawn, they read and talk. At night, when "the true Darkness" descends, they take long rambles, "seeking amid the wild lights and shadows of the populous city that infinity of mental excitement which quiet observation can afford."

During one expedition, Dupin breaks a long silence by uttering a phrase that, to his friend's amazement, completes the thought he was having at that moment. Dupin explains,

> Observation has become with me, of late, a species of necessity. You kept your eyes upon the ground—glancing, with a petulant expression, at the holes and ruts in the pavement, (so that I saw you were still thinking of the stones,) until we reached the little alley called Lamartine, which has been paved, by way of experiment, with the overlapping and riveted blocks. Here your countenance brightened up, and, perceiving your lips move, I could not doubt that you murmured the word "stereotomy," a term very affectedly applied to this species of pavement. I knew that you could not say to yourself "stereotomy" without being brought to think of atomies, and thus of the theories of Epicurus; and since, when we discussed this subject not very long ago, I mentioned to you how singularly, yet with how little notice, the vague guesses of that noble Greek had met with confirmation in the late nebular cosmogony, I felt that you could not avoid casting your eyes upward to the great nebula in Orion, and I certainly expected that you would do so. You did look up; and I was now assured that I had correctly followed your steps.

Dupin's dazzling leap was based in "apparently intuitive" reasonings, verified by observing his friend's expressions and behavior. It also happened to touch down at key points of contemporary technology, experimental science, and ancient philosophy. With his allusion to "the great nebula in Orion"—a celestial body taken as possible evidence for the nebular hypothesis—Poe also evoked the cosmological theory that traced the chain of events through which the stars and planets came into being. As it was popularized in 1837 by the Scottish astronomer John Pringle Nichol, the nebular hypothesis was linked to materialist theories which held that life and thought—including the exemplary rationality Dupin displayed in the tale—had emerged from earlier, more primitive forms.

One day a shocking news item captures Dupin's attention. The mutilated corpses of two women are discovered: an elderly woman decapitated with a shaving razor, her daughter stuffed, head facing downward, into a chimney in a locked room. Several witnesses heard two voices, one gruff, speaking French, one shrill, speaking an unidentifiable language.

Dupin and the narrator visit the scene; Dupin then places a classified notice in *Le Monde*. To the narrator's astonishment, a sailor appears at their door the next day, confirming all of Dupin's conclusions about the crime, including the identity of the inhuman butcher responsible for the murders.

Dupin's quiet observations, withheld conclusions, and stunning revelations dramatize thought of the highest order. Poe repeatedly contrasted Dupin's fleet analysis with the confused activity of the prefect of police: "There is no method in their proceedings, beyond the method of the moment"; the police's "vast parade of measures" is "ill adapted to the objects proposed."

To illuminate Dupin's method, Poe compared the forms of thought at work in particular games: the "elaborate frivolity" of chess offers fewer challenges than the "unostentatious game" of checkers, while whist—a precursor to bridge—is most demanding. It requires attention not just to the cards but to every aspect of an opponent's conduct: expressions, tone of voice, "the manner of taking up a trick." The rapid steps of logic it requires only appear to be intuitive. Dupin follows similarly swift logical leaps—connecting his close inspection of the window frame, the finger bruises on a victim's neck, the overheard voices, and a detour through *The Animal Kingdom* of the naturalist Georges Cuvier—to identify the culprit.

However outlandish was its solution, "The Murders in the Rue Morgue" offered readers the engrossing spectacle of Dupin working through his "manifold" and "multiform" steps of logic—ruling out possibilities, testing conjectures—and let them share in his enjoyment at astonishing those around him. With its detailing of pathways, walls, and window frames, the story reflected on the architecture and the staging of thought—what is hidden, what revealed, when, and how. The crime is horrible; its solution unimaginable; the means of discovery astonishing, seemingly supernatural, while Dupin's chain of "ratiocination" from clues to solution is irresistible. Reinforcing this sense of inevitability, Poe laid the tale out in a subtly chiasmic structure, with elements of the first half repeated in reverse in the second half, pivoting on the image of the nail in a window frame, *le clou*, the "clue" on which the solution hangs. In *The New-Yorker*, the social reformer Horace Greeley admitted the tale's "deep but repulsive interest."

At *Graham's*, Poe also renewed his cryptographic series, upscaling it as an ancient art with political importance and citing renowned precursors including John Wilkins and Giambattista della Porta. He declared that "human ingenuity cannot concoct a cipher which human ingenuity cannot resolve." In words recalling Dupin's discussion of analysis, cryptography demonstrated that "a rigorous *method*" could be profitably applied in all forms of thought, including "what is considered the operation of pure fancy"; there might, he hinted, even be an algorithm for poetry. Jesse Dow, now the editor of the *Index* in Washington, compared Poe to Champollion: "He can read the hieroglyphics of the Pharoahs [and] tell you what you are thinking about while he walks beside you."

Pocket Universes

Dupin's "peculiar analytical ability" suggests to the narrator the image of "a double Dupin—the creative and the resolvent." Dupin, who is both a poet and a mathematician, unravels and weaves plots with equal ease. While "The Murders in the Rue Morgue" dramatized the sublime mechanics of analysis, other tales in *Graham's* offered up beautiful, self-contained worlds—pocket universes. Many of Poe's imaginative works of this period placed mechanical processes at the heart of creation and suggested a living force within the merely material. They deliberately blurred spirit and matter, dream and reality, weaving worlds out of words.

"Eleonora" began by observing, "They who dream by day are cognizant of many things which escape those who dream only by night." Daytime dreamers travel, "rudderless or compassless," just as easily "into the vast ocean of the 'light ineffable'" as into the "*Mare Tenebrarum*," the "sea of darkness" of the cosmographer Ptolemy.

A daydreamer lives with his aunt and beautiful cousin, Eleonora (an arrangement rather like the Poes'), in an "encircled domain" covered with grass in multiple shades of green, sprinkled with "the yellow buttercup, the white daisy, the purple violet." When the cousins fall in love, "strange, brilliant flowers, star-shaped, burst out upon the trees"; "ruby-red asphodel" replace daisies, while "tints of the green carpet deepened." Flamingos appear, their "scarlet plumage" reflecting the gold and crimson cloud that hangs over the valley, enclosing them within "a magic prison-house of grandeur and of glory." When Eleonora falls ill, the flowers grow dark.

The luminous imagery of "Eleonora"—so popular it was reprinted in at least five other magazines within the year—was a literary version of a new visual entertainment. In early 1840, Peale's Philadelphia Museum hosted *Dissolving Tableaux, or Dissoluble Scenes.* Recently imported from England, this magic lantern effect employed two projectors instead of one, both aimed at the same surface. As the image from the first faded out, it was gradually

Magic lantern landscape, mid-nineteenth century, hand painted

replaced by an image from the second, creating a hypnotic illusion of trans-formation or growth. An advertisement described a brightly colored sum-mertime view of valleys and rivers: these "brilliant landscapes, as if by Enchantment, suddenly and imperceptibly melt away, leaving the eye fixed on a cheerless withering WINTER SCENE," until "the scene suddenly changes" and "nature again puts on its livery of green—trees are seen shoot-ing forth their rich verdure." Next appears a "ROSE BUSH, which, from a small sprout, imperceptibly increases" before the "astonished eyes of the spectators" revealing "the expanding rose in all its bloom, beauty, and color."

"Eleonora" used words to imitate the magic lantern's time-elapsed spec-tacle of the cycles of life. Its polychromatic, kaleidoscopically shifting vision was a literary equivalent of the "dissolving view"—itself a technical approx-imation of the vital powers that Poe, like Shelley and other romantic poets and scientists, discerned within matter itself.

Poe appended a new version of his "Sonnet—To Science" to "The Is-land of the Fay," a story that also conjured a luminously encircled landscape. Slightly modified from its first form ten years earlier, the sonnet's last lines now asked of science,

Hast thou not spoilt a story in each star?
Hast thou not torn the Naiad from her flood?
The elfin from the grass?—the dainty fay,
The witch, the sprite, the goblin—where are they?

Yet the central question still burned: How might a poet love science, "how deem thee wise"?

"The Island of the Fay" was in fact a multimedia artwork. Along with the poem, the story was accompanied in *Graham's* by a mezzotint engraving by John Sartain showing a wooded island in a river, a concentrated blackness set against surrounding light and shade. Placing himself on the grass before this twilight scene, Poe meditated on the thought that forests, rivers, and mountains might be "colossal members of one vast animate and sentient whole"—the earth a being "whose intelligence is that of a God." He saw the universe's planetary systems as "cycle within cycle without end—yet all revolving around one far-distant centre which is the Godhead"; analogously

"The Island of the Fay," mezzotint by John Sartain accompanying
Poe's tale and "Sonnet—To Science" in Graham's, June 1841

he wondered whether the universe contained "life within life, the less within the greater, and all within the Spirit Divine." As in Coleridge's pantheistic musings in his poem "The Eolian Harp," Poe was tilting the language of natural science and natural theology a few degrees, until the metaphysical scaffold separating matter, life, and God began to collapse. If matter were alive—or even *divine*—mankind would participate in this holy substance but would be of no more importance than the "'clod of the valley' which he tills and contemns."

Dreaming with open eyes, he entered into Sartain's image. The island's west side was lit with "a rich, golden and crimson waterfall" of sunlight, the east side "whelmed in the blackest shade." Shadows fell from the trees onto the water, like earth thrown in a grave.

"If ever island were enchanted," he said to himself, "this is it."

He imagined that earth's few remaining fairies had been chased to the island to die, "rendering unto God their existence little by little, as these trees render up shadow after shadow." His fancies became realities: he glimpsed "one of those very Fays" standing in a "singularly fragile canoe," pushing off from the west of the island—joyful in the light, "deformed" by sorrow in the shade. Each time the boat reappeared, after each revolution, the fay was more careworn, with "less of elastic joy." At sunset she faded into the darkness. Poe's reflections fell like leaves from Sartain's print, while his poem's fugitive naiads, sprites, and elves found their way into his daydream.

The tale was no simple lament about scientific destruction. It was an imaginative reply to science's material and mechanical facts. Its vision of a fairy whose life peels away, shadow by shadow, in her revolutions around an island personified astronomical theories about gradual decay in the orbits of planets and comets—a threat to the assumption of God's stable and perfect design. By endowing the "mere bulk" of matter with thought, life, and "elastic joy," Poe was also thinking through the implications of vitalist strands in natural science. Yet this idea of an animating force within matter was shadowed by a counterforce drawing inevitably down, a fissure running through things, a polarity within the "vast animate and sentient whole." Natural discord, he hinted, was a cosmic principle as powerful as harmony.

A delicate interplay between creative and destructive cosmic powers also ran through Poe's "spirit colloquies," dialogues between disembodied spirits. His first, "The Conversation of Eiros and Charmion," described "the wild luxuriance of foliage" that suddenly flourishes when a deadly comet approaches earth's atmosphere before setting the planet ablaze. "The Colloquy of Monos and Una"—a dialogue between two names for "one"—likewise described the earth's destruction, this time at the hands of humans. Poe was one of the first to envision the burning of carbon fuels as the cause of ecological catastrophe; he depicted a near extinction in which "green leaves shrank before the hot breath of furnaces." Yet in a distant future, when "the Art-scarred surface of the Earth" was again free from man-made "rectangular obscenities," the planet would be clothed in "the verdure and the mountain-slopes and the smiling waters of Paradise." The tale finishes with a spirit's description of his own bodily death: the stilling of breath and blood, the sharpening and commingling of the senses, a pulsing sensation of "abstract time" that subsides into perception of "mere *locality*"— until "dust had returned to dust" in the "nothingness" of "immortality."

Poe's pocket universes resembled other contemporary attempts to grasp the cosmos in miniature: Humboldt's synoptic views of ecological systems, handy books such as Roswell Park's *Pantology*, popular astronomy lectures, and the utopian and apocalyptic sermons and tracts of evangelicals, Shakers, Mormons, and Fourierists.

But in a reprint of "A Descent into the Maelström," Poe added a cautionary epigraph, attributed to the philosopher Joseph Glanvill: "The ways of God in Nature, as in Providence, are not as our ways; nor are the models that we frame any way commensurate to the vastness, profundity, and unsearchableness of His works." Poe was convinced that the universe escapes our attempts to frame it. Nevertheless, he kept building models. At times he tried to show his model's limitations *within* the model—working up a mental and literary construction, only to reveal its limits by bringing it crashing down, juxtaposing it with a contradictory reality, or diffusing it into ether. Beyond their heady intrigue, such works held out the intoxicating possibility that the labor of plotting and crafting a miniature universe was of a piece with the creative process of making (and unmaking) the universe itself.

Table Talk with Boz   On March 7, 1842, in rooms at Philadelphia's U.S. Hotel, an intense tête-à-tête was under way. One *tête* belonged to the reviews editor and star story writer of *Graham's*, one of America's most celebrated and widely read magazines. The other was that of the thirty-year-old Charles Dickens, one of the most famous men in the world.

Dickens was stopping off for three days during a whirlwind tour of the United States. Poe's friend the novelist George Lippard had giddily announced the great author's impending arrival: "Sober, quiet, steady Philadelphia has waked up at last! Boston has gone mad—New York crazy—and we suspect Philadelphia is about to become one vast hospital of Boz-Bedlamites." The reason was self-evident: "Charles Dickens is, we are told, the man of the age!"

"Boz" was taken on tours of the Pennsylvania Hospital, the Fairmount Water Works, and the "splendid unfinished marble structure for the Girard College." Lower on Fairmount hill, he stopped at Eastern State Penitentiary, whose Gothic exterior, meant to frighten would-be criminals, contrasted with its modern form of imprisonment: solitary confinement. Though it was introduced as a humanitarian measure, Dickens was repulsed by "the immense amount of torture and agony" this treatment imposed, a "slow and

Portrait of Charles Dickens in Boston, 1842, by Alexander Francis

daily tampering with the mysteries of the brain." A prisoner in complete isolation "is a man buried alive." Dickens saved his social observations for his
novel *Martin Chuzzlewit* and *American Notes*, which he published—to American
indignation—after returning to England.

With Poe, the talk was of literature. Poe had sent Dickens a copy of his
Tales along with his admiring review of Dickens's recent book *Barnaby Rudge*,
which featured a talking raven—modeled on Dickens's own pet bird, Grip.
Dickens replied with inside information about William Godwin's *Caleb Williams*, the book to which Poe had compared his own: "Do you know that
Godwin wrote it *backwards*—the last Volume first," and after arriving at his
conclusion, "he waited for months, casting about for a means of accounting
for what he had done?"

The craftsmen talked shop during "two long interviews." Poe shared
his views about British and American trends, reading him Emerson's poem
"The Humble-Bee." Dickens promised Poe he would try to find an English
publisher for his *Tales*. As Dickens tried to leave on the final day of his visit,

he was besieged by hundreds of fans. The landlord convinced him that refusing a meeting "would doubtless create a riot." He spent several hours
shaking hands.

Poe greatly admired Dickens's plots and characterizations, his vivid descriptions of the streets, shops, clubs, and workhouses of the modern city.
He had nothing like Dickens's celebrity, but his own reputation and critical
authority were on the rise.

As editor and attention-grabbing author in the most sumptuous magazine in the United States—soon to have the largest circulation of any in the
world—Poe was an author to follow and a critic to fear. He had become a
fixture at elegant soirees at Graham's house on Sansom Street. Even his early
poetry was now gaining a reputation, thanks to reprints in *Graham's*.

Poe was scaling giddy heights, making a national and international name
as a modern man of letters. With his scientific writings—his reviews of current research, his cryptography, his realistic tales such as "Maelström," and
his trademark invention, the infallible reasoner Dupin—he was becoming
something more. In line with his age's new possibilities but harking back to
earlier models, Poe was fashioning himself as a universal thinker and natural
philosopher: a Renaissance man who could weave together imagination, observation, logic, and the newest technologies—taking the measure of nature,
mocking mortal folly, and artfully dreaming up worlds.

The Tide Turns

Between Hope and Despair Poe could now afford to keep Virginia and her mother in comparative comfort in a small house, as well as some indulgences, including a gold pocket watch from France. After so much doubt, fear, and hunger, Poe had reached calm seas. Graham admired "how solicitous of the happiness of his wife and mother-in-law" Poe was; other than "the natural ambition of having a magazine of his own—I never heard him deplore the want of wealth."

Yet his relationship with Graham had begun to chafe. Graham had let Poe believe he would help with his cherished project, *The Penn*, but showed no inclination to move forward.

Why kill the golden goose? When *Graham's* started, it had five thousand subscribers. In January 1842, Poe estimated a run of twenty-five thousand. "Such a thing was never heard of before," he boasted. By the spring of 1843, he anticipated a run of "*fifty thousand copies*"—the largest circulation of any magazine in the world, "as many as any steel line engraving will yield" before having to be recast.

Poe had proven beyond doubt his ability to helm a successful magazine. Yet he remained desperate to realize his own vision: "To coin one's brain into silver, at the nod of a master, is to my thinking the hardest task in the world." It soon became harder.

In early 1842, while singing at home, Virginia burst a blood vessel. She began coughing up blood—a sure sign of consumption. There was no cure. The episode was so severe that for two weeks Poe doubted she would recover;

on February 3 he told F. W. Thomas, "It was only on yesterday that the physicians gave me any hope." According to a neighbor, the house was ill-suited for her condition: "She could not bear the slightest exposure, and needed the utmost care . . . yet the room where she lay for weeks, hardly able to breathe except as she was fanned, was a little place with the ceiling so low over the narrow bed that her head almost touched it."

Any negative prognosis was silenced: "No one dared to speak—Mr. Poe was so sensitive and irritable; 'quick as steel and flint.'" He would not hear any talk of Virginia's dying: "The mention of it drove him wild." He told Thomas, "My dear little wife has been dangerously ill."

Poe's nervous state put his job in danger. The day after Virginia's first hemorrhage, he visited Graham to ask for two month's advance salary. Graham "not only flatly but discourteously refused. Now that man *knows* that I have rendered him the most important services." Poe's work over the past months had multiplied the publisher's fortunes. "If, instead of a paltry salary, Graham had given me a tenth of his Magazine, I should feel myself a rich man to-day."

Poe further bristled at Graham's interference with his critical independence; in a few reviews, he admitted, he was "weak enough to permit Graham to modify my opinions (or at least their expression)." As if to compensate, he scrawled a particularly harsh review in February of *Wakondah* by Cornelius Mathews, a New York author and critic.

Virginia's periodic crises were followed by gradual returns to partial health; with her first episode and recovery began "a horrible never-ending oscillation between hope & despair." As Graham recollected, Poe's "love for his wife was a sort of rapturous worship of the spirit of beauty which he felt was fading before his eyes. I have seen him hovering around her when she was ill, with all the fond fear and tender anxiety of a mother for her first-born—her slightest cough causing in him a shudder." At the same time, Poe's impatience with his situation at *Graham's*—the clash between his growing reputation and his enforced subservience—was reaching a boil.

In April he quit. With dire economic conditions still in force—high unemployment, limited credit—this was a reckless, self-destructive move. Poe blamed Graham's insistence on filling pages with expensive, nonliterary material: "My reason for resigning was disgust with the namby-pamby character of the Magazine," its "contemptible pictures, fashion-plates, music and love tales. The salary, moreover, did not pay me for the labor which I was forced to bestow. With Graham who is really a very gentlemanly, although an exceedingly weak man, I had no misunderstanding." There were no harsh words, tense scenes, or accusations as there had been when Poe left *Burton's*—and Allan's.

Poe was wounded by Graham's failure to support *The Penn*. He saw, too late, that their interests had been at odds. "I was continually laboring against myself. Every exertion made by myself for the benefit of 'Graham's,' by rendering that Mag. a greater source of profit, rendered its owner, at the same time, less willing to keep his word with me."

Despite these rationalizations, Poe's departure was an act of desperation. He found the situation with Virginia unbearable; Graham was unyielding; in a state of nightmarish frustration and confusion, Poe wanted to burn it all down. Graham later thought of Poe fondly, if condescendingly. Poe "was quick, it is true, to perceive mere quacks in literature," Graham wrote, but his occasional harshness served a higher cause: "Literature with him was religion; and he, its high-priest."

The Penn still had supporters. Of the nearly one thousand subscribers he had earlier secured, Poe thought that "3 or 4 hundred" would still be prepared to pay for its launch. He had earned widespread goodwill. The news of his departure from *Graham's* prompted *The New World* to describe him "as one of the best writers of the English language now living." Although "Mr. Poe has left Mr. Graham's Magazine . . . in whatever sphere he moves, he will surely be distinguished."

Graham replaced Poe with the Reverend Rufus Griswold, a former editor of the *Boston Notion*. Poe had met the bearded, unctuous minister the previous year while Griswold was assembling a collection of American poetry. Poe wanted to be included, and Griswold needed publicity. Mutual dislike seethed beneath their transactions. Griswold published three short poems and a flat biography of Poe, while Poe's review damned the collection with faint praise: "A thorough analysis of the book might induce many, whose minds are not comprehensive, to think it a bad, instead of what it really is, a good work." Privately Poe saw the encyclopedic tome, crowded with smarmy praise for influential poets, as hackwork. He told

Joseph Snodgrass that the book was "a most outrageous humbug, and I sincerely wish you would 'use it up.'"

Published in 1842, *The Poets and Poetry of America* nevertheless granted Griswold an air of critical authority. Graham offered him the position of full editor and a thousand-dollar salary—two hundred dollars more than Poe's.

Jesse Dow, Poe's friend in Washington, noted *Graham's* rapid decline soon after: "We would give more for Edgar A. Poe's toe nail, than we would for Rueful Grizzle's soul." Griswold for his part spread "malignant, unjust, and disgraceful attacks" on Poe's character to anyone who would listen. Within a few months of Poe's departure, Graham made Poe "a good offer" to return to the magazine, because he was "not especially pleased with Griswold—nor is any one else, with the exception of the Rev. gentleman himself." Poe politely declined.

Fits and Starts   Poe had largely been sober since leaving Richmond, but with Virginia's illness and his departure from *Graham's*, something snapped. In June he traveled to New York to scout a publisher for a new story collection he was calling *Phantasy Pieces*. He ran into a young poet, William Wallace, an enthusiast for mint juleps, and arrived at the offices of the *Democratic Review* and William Snowden's *Ladies' Companion* much out of sorts. A day later he surfaced in Jersey City, looking for an old friend: "He was on a spree, however, and forgot the address before he got across the river. He made several trips backward and forward on the ferryboat," arrived for one cup of tea, then left. Maria Clemm followed, "much worried about 'Eddie dear.'" Eventually he was found "in the woods on the outskirts of Jersey City, wandering about like a crazy man." Clemm took him home.

Poe wrote to the *Democratic Review* with a tale and apologies: "You must have conceived a *queer* idea of me—but the simple truth is that Wallace would insist upon *the juleps*, and I knew not what I was either doing or saying." They refused his story, "The Landscape Garden," whose fantasy of artistic omnipotence—an immensely rich man's plan to resculpt the earth—would nourish Poe in hard times ahead.

He scrambled to publish where he could, piecing together a new Dupin story. "The Mystery of Marie Rogêt" reworked a horrific murder currently in the press: the death of a "beautiful cigar girl," Mary Cecilia Rogers, whose body was found floating in the Hudson. Poe swapped Manhattan for Paris, the Hudson for the Seine. Though its plot was disappointingly haphazard—due to Poe's making adjustments in the tale's second and third installments as new facts were revealed—"Marie Rogêt" was the first detective story to be based on an actual crime.

The series was printed, as was "The Landscape Garden," in Snowden's *Ladies' Companion*, a magazine whose production values were a steep drop from *Graham's*. Poe cringed at the layout, writing to the editor, "Oh Jupiter! the typographical blunders. Have you been sick, or what is the matter?"

By the summer of 1842 debts were piling up for rent and doctor's bills. The Poes moved to a house near the woods on Coates Street, now Fairmount Avenue, promising fresher air for Virginia and lower rent. The landlord's son recalled Poe's fondness for "roving about the country." To a fellow poet Poe later explained, "There are epochs when any kind of mental exercise is torture, and when nothing yields me pleasure but solitary communion with the 'mountains & the woods' . . . I have thus rambled and dreamed away whole months, and awake, at last, to a sort of mania for composition. Then I scribble all day, and read all night, so long as the disease endures." The location inspired "Morning on the Wissahiccon," a description of the region's lush forests, whose narrator, crossing paths with a magnificent elk, imagines himself transported to the Edenic period "when the red man trod alone" before the arrival of the whites and their enslaved Black servants—one of whom appears and claims the elk as a pet.

In May 1842, F. W. Thomas, now living in D.C., suggested a new goal: an appointment in the Philadelphia Custom House, where one of Tyler's supporters was expected to replace the current director. For his loyalty to the Whigs, Thomas had been appointed to the Treasury Department. He painted an enticing picture: "You stroll to your office a little after nine in the morning leisurely, and you stroll from it a little after two in the afternoon homeward to dinner . . . and if you choose to lucubrate [write by lamplight] in a literary way, why you can lucubrate." He was making a tidy salary of a thousand dollars per year.

Poe congratulated him: "I wish you joy. You can now lucubrate more at your ease."

Poe had known Tyler in Richmond, and with some effort he could cast himself as a party man: "I am a Virginian—at least I call myself one, for I have resided all my life, until within the last few years, in Richmond. My political principles have always been as nearly as may be, with the existing administration, and I battled with right good will for Harrison." Further, his literary—and cryptographic—talents could be seen as proof of his usefulness; Tyler's son Robert was a poet and one of Thomas's friends in D.C., and Poe published an encrypted letter from one "W. B. Tyler," intended to court the president's favor.

The customhouse possibility gave him "new life"; it would enable him "to carry out all my ambitious projects." With a government appointment and an investor, he could launch *The Penn* right away.

In August, a new collector was appointed. By then there were "1124 applicants" for only 30 posts. Poe persisted, though the collector "treated me most shamefully": he "scarcely spoke—muttered the words 'I will *send* for you Mr Poe' and that was all."

Thomas commiserated, visiting him in his "rural home" whose rooms "looked neat and orderly" but "wore an air of pecuniary want." He was greeted by Virginia, who had "the most expressive and intelligent eyes I ever beheld," although her "pale complexion, the deep lines in her face and a consumptive cough made me regard her as the victim for an early grave."

Thomas saw his friend's difficulties weighing on him: "His dark hair hung carelessly over his high forehead, and his dress was a little slovenly. He met me cordially, but was reserved, and complained of feeling unwell." Though touched by his "tenderness and loving manners towards his wife," Thomas observed "with deep regret that he had fallen again into habits of intemperance." Poe missed the meeting the two arranged for the next day at Independence Hall, and the customhouse application came to nothing. In November, Charles Dickens wrote to him of his attempt to interest English publishers in his tales; "they have, one and all, declined the venture." (In fact, Dickens had made only one cursory inquiry.)

In December 1842, Poe took advantage of a new act of Congress and voluntarily declared bankruptcy. The legal petition listed forty-five debtors for sums from $4 to $169, for everything from rent, medical visits, and books to piano rental. He owed $10 to J. N. Reynolds, promoter of the Exploring

Expedition, and $20 to Nicholas Biddle, presumably put toward *The Penn* in more optimistic days. The document mordantly listed his only property as "his wearing apparel and a few hundred Sheets, of no use to any one else, and of no value to anyone."

Tales of Guilt and Torture   After Virginia's first crisis, Poe's stories took on a bolder, more sinister aspect. The new series showed terrific focus and formal precision but added a desperate psychological intensity. Exploring the hidden, anxious, and destructive regions of the mind, these tales have left a deep and lasting mark on modern literature and on their readers.

His first story of 1842 was "The Oval Portrait," a haunting fable: a painter's artwork only achieves supreme "life-likeliness" with the death of his wife, who posed for it. Was there some inevitable balance sheet on which each step toward artistic perfection had to be paid in suffering and sorrow? In "The Pit and the Pendulum," Poe confronted a prisoner of the Inquisition with one meticulously designed horror after another: a room of unknown shape with a terrifying pit at its center; metal walls heated to incandescence gradually closing in; a monstrous clockwork, a swinging pendulum sharpened to a blade, slowly lowering.

In "The Black Cat," Poe's legendary study of psychopathology, the torments come from within. In a nerve-racking allegory of addiction, slavery, and the hidden monstrosity of the soul, the speaker details his transformation from a compassionate animal lover to a torturer of the creatures in his care, in the clutches of "the Fiend Intemperance." He turns, inexplicably, against his beloved cat—cutting out its eye, hanging it. Another cat, much like the first, enters his life and leads him to commit murder. His crimes are horribly exposed, as in a daguerreotype, through a chemical process of development that reveals the outline of his victim, and through the cat's hellishly accusing shriek. "These events," says the narrator, "have terrified—have tortured—have destroyed me."

Fresh memories of the cholera epidemic lurked behind "The Masque of the Red Death," set in the late Middle Ages, but speaking urgently to modern fears and inequalities. As the red plague ravages the countryside, killing its victims with "sharp pains, and sudden dizziness" followed by "profuse bleedings at the pores," Prince Prospero gathers his rich friends behind the ramparts of his palace. They abandon themselves to games and revelry, ignoring the suffering of the unfortunates beyond the walls. At a costume ball, each room tinted by a differently colored glass, a new guest arrives. Dressed as a shrouded corpse, he stalks through room after room, until at last, face to face with the prince, he reveals his true identity. The revelers' wealth, power, and self-isolation could not save them: "The Red Death held illimitable dominion over all."

The fall of 1842 saw Poe working on his masterpiece of morbid suspense, "The Tell-Tale Heart." The tale revived the observation drama of "A Descent into the Maelström" with grisly, malevolent twists. Though employing language and imagery from the sciences, Poe gave readers an experience not of certainty and reassurance but of disorientation and horror.

Its famous first lines invoke truth, nerves, disease, hell, and the power of the storyteller. Opening in the middle of a conversation, perhaps an asylum visit or trial, a narrator boasts of the sharpness of his senses:

TRUE!—nervous—very, very dreadfully nervous I had been and am! but why *will* you say that I am mad? The disease had sharpened my senses—not destroyed—not dulled them. Above all was the sense of hearing acute. I heard all things in the heaven and in the earth. I heard many things in hell. How, then, am I mad? Hearken! and observe how healthily—how calmly I can tell you the whole story.

The unnamed speaker has become monomaniacally fixated on the opaque eye of the old man who shares his house: "I think it was his eye! yes, it was this! One of his eyes resembled that of a vulture—a pale blue eye, with a film over it. Whenever it fell on me, my blood ran cold; and so by degrees—very gradually—I made up my mind to take the life of the old man, and thus rid myself of the eye forever." Driven by neither greed nor revenge, he proceeds with cool, objective detachment toward an irrational goal. With patience and precision he stalks the man, peering into his room at midnight with a lantern open "just so much that a single thin ray fell upon the vulture eye." On the seventh night of this deathwatch, the old man awakens, sensing the intruder. After an agonizing silence, he stealthily opens his lantern, "until, at length, a single dim ray, like the thread of the spider, shot from out the crevice and full upon the vulture eye."

At once he hears "a low, dull, quick sound, such as a watch makes when enveloped in cotton"—the beating of the terrified old man's heart, so loudly he fears neighbors will hear.

In its detail and regularity, his description mimics the report of a meticulously designed scientific experiment. Light, long a symbol of knowledge and reason, is here associated with madness. The view of Poe's narrator is filled by a film-covered eye—*not* the perfectly adapted ocular apparatus taken as a sign of God's wisdom in William Paley's natural theology. The darkened room he violates recalls the camera obscura, the technical core of the daguerreotype: an enclosed space into which images of the outside world were projected. Yet this narrator acts by night; the light that pierces the room is artificial; the image projected is not the external world of creation but "the thread of a spider" ensnaring its prey.

The tale restages one of the primal scenes of modern science: Isaac Newton's experiments in the *Opticks*. There Newton detailed the steps through which he carefully cracked open a shutter to let a single ray of light into a darkened room and dismembered it by passing it through a prism, revealing the colors of the spectrum. Poe's tale inverts Newton's experiment, making vision and methodical reason the agents of perversity and death. He flips the heightened sensitivity celebrated in new precision instruments into pathology, allowing his narrator to perceive—perhaps through the vibratory medium of the ether—"many things in hell."

Poe's narrator murders the old man. Methodical as ever, he dismembers the body and buries its parts beneath the floorboards. The police arrive, hearing the victim's shriek; he coolly entertains them, offering refreshments, placing their chairs directly on the guilty floorboards.

He then begins to hear, once again, a constant beating, "like a watch wrapped in cotton"—his dead victim's heart, growing louder and louder. The merciless pounding and the indifferent chatter of the police drive him to a frenzy, until at last he cries out in horror and agony—confessing his crime.

Just as his desire to demonstrate his rational calm led to the uncovering of the murder, the heightened sensitivity of his senses gave him access to sounds, hallucinated or real, that took him beyond the limits of sane experience.

Detailing an act of violent irrationality in the language of scientific method, Poe dramatized the dark side of the Enlightenment—a vision uncannily familiar to Americans enmeshed in a "rational" economic and political order defined by ruthless competition, slavery, and violent settlement. Though usually presented back to them in soothing tones of equality, freedom, and progress, beneath the appearance of calm reason pulsed veins of terror, obsession, and cruelty.

Poe gave the screw another turn. While the heart hidden beneath the floorboards might symbolize the narrator's criminal madness—the monstrous guilt that his calculating reason and glib patter cannot mask—it also echoes the ultimate symbol of method and rationality: a mechanical watch, the embodiment of Newton's clockwork universe and the icon of natural theology in Paley's argument on design.

The tale peels back a rational surface to reveal inexplicable and irrational forces. When another layer is lifted, these forces are revealed as instruments of a deeper mechanism, with a reason all its own.

It also made for a damn good read. The tale has terrified schoolchildren and delighted adults since its publication.

The Stylus

"The Tell-Tale Heart" appeared in the first issue of *The Pioneer*, edited by James Russell Lowell, a young, highborn Massachusetts poet with whom Poe felt "a certain coincidence of opinion & of taste." Poe trumpeted the journal's arrival as a new dawn for American letters. "In these days of self-bepuffed and glorified magazines," he hailed "the PIONEER as the first in the great work of reform." He wrote to Lowell suggesting that they form a writer's "coalition"—a secret syndicate, union, or association of authors, to put the business of literature in the hands of its producers, instead of the owners and publishers.

The Pioneer's second issue included Poe's "Lenore," a poem of grief over the death of a young bride. It also featured a satire by Nathaniel Hawthorne, "The Hall of Fantasy." Poe was given a place in Hawthorne's imaginary club of literary greats "for the sake of his imagination, but was threatened with ejectment, as belonging to the obnoxious class of critics."

Poe's own dream of a journal was suddenly revived at the start of 1843, thanks to Thomas C. Clarke, a rich temperance advocate who edited a family weekly, the *Philadelphia Saturday Museum*, and whose *American Pocket Library of Useful Knowledge* Poe had favorably reviewed. Clarke commissioned a biography of Poe, who furnished the facts, many exaggerated or sheer bunk, while his friend Henry Hirst completed the article, accompanied by an unfortunate engraving. Poe quipped, "I am ugly enough God knows, but not *quite* so bad as that." The biography brought Poe's varied achievements, including his poetry, before the public eye.

Poe managed to convince Clarke of the viability of a prestigious monthly journal. *The Penn* seemed an overly local title, now tainted by failure. The new magazine would be called *The Stylus*—the same as *The Penn*, but "now, under the best auspices, resumed, with no other modification than that of the title." A launch was planned for July 1843.

The partnership with Clarke and an upturn in his wife's health ("Virginia is nearly recovered—indeed I may say quite so—with the exception of a slight cough") reenergized him. He sent the *Stylus* prospectus to F. W. Thomas in Washington, confiding his delight in securing "the great object—a partner possessing ample capital, and, at the same time, so little self-esteem, as to allow me entire control of the editorial conduct."

Clarke was giving him "a half interest" and operating funds; in return, Poe would provide "literary matter" for the first year. It was a tall order. He would "write as much as possible myself, under my own name and pseudonyms, and hope for the casual aid of my friends, until the first stage of infancy is surpassed." Backbreaking literary labor could be borne, however, if he had control of the content, a share in the profits, and his name on the masthead.

Blowout on the Potomac

Poe's excitement about *The Stylus* was already rising to dangerous levels when an announcement came on March 6: the Philadelphia Custom House had a new director, Judge Calvin Blythe. A new wave of appointments was imminent.

Gathering a few dollars owed him, Poe immediately set off for Washington. With F. W. Thomas choreographing introductions, he would seek an endorsement from Whig officials, and subscribers for the *Stylus*—perhaps even federal support.

He took a room in Fuller's City Hotel on Pennsylvania Avenue where he found Thomas in bed with congestive fever, "covered all over with the marks of cupping and blistering." Thomas gave him a letter of introduction to Robert Tyler.

What happened next is not entirely clear, but it involved Poe at one point knocking on the door of the White House.

At Fuller's, Poe met old friends including Jesse Dow, who found him "somewhat excited, having been over-persuaded to take some Port wine." The next day he gave a letter and an air gun to the natural historian John Kirk Townsend. He might have met the daguerreotypist Mathew Brady, visited a Pennsylvania congressman interested in codes, and convinced President Tyler's son Robert along with the new secretary of the navy, Abel Upshur, to write for *The Stylus*.

In the midst of his spree he wrote to Clarke that subscriptions were

Daguerreotype by Mathew Brady of Pennsylvania Avenue, Washington, D.C., 1843, the year of Poe's stay on the street; Metropolitan Hotel and White House indicated at top

pouring in from "*all* the Departments," possibly including Commodore Jesse Elliott of the navy. His frenetic activity was advancing the cause: "I believe that I am making a *sensation* which will tend to the benefit of the Magazine." He was on a winner. "Send me $10 by mail, as soon as you get this. I am grieved to ask you for money, in this way—but you will find your account in it—twice over."

According to Dow, "On the second day he kept pretty steady, but since then he has been, at intervals, quite unreliable." Thomas Dunn English, the physician and author from Philadelphia, was there, sporting a new mustache; Poe christened him "the Don." On Pennsylvania Avenue he borrowed money from a journalist who found him "seedy in his appearance and woebegone." At some point, Poe's body gave out, and he was confined to bed. Thomas called a physician, but the pain was more than physical; "he suffered much from his indiscretion."

Dow begged Clarke to bring him home: "He exposes himself here to those who may injure him very much with the President, and thus prevents us from doing for him what we wish to do and what we can do if he is himself again in Philadelphia. He does not understand the ways of politicians, nor the manner of dealing with them to advantage. How should he?" If Clarke could not come, they would put him on a train, "but we fear he might be detained in Baltimore and not be out of harm's way."

Poe managed to drag himself back to Philadelphia. After a shave and "a warm bath & supper" he visited Clarke: "I never saw a man in my life more surprised to see another. He thought by Dow's epistle that I must not only be dead but buried & would as soon have thought of seeing his great-great-great grandmother." Virginia's "distress of mind has been even more than I had anticipated," he admitted to Dow and Thomas. "She desires her *kindest* remembrances to both of you—as also does Mrs. C."

Shame seeped through his jests about his drunken behavior. He thanked Dow "a thousand times for your kindness & great forbearance, and dont say a word about the cloak turned inside out, or other peccadilloes of that nature."

To F. W. Thomas he poured out his heart: "My dear friend. Forgive me my petulance & don't believe I think all I said. Believe me I am very grateful to you for your many attentions." He asked him to apologize to the landlord "for making such a fool of myself in his house, and say to him (if you think it necessary) that I should not have got half so drunk on his excellent Port wine but for the rummy coffee with which I was forced to wash it down." He asked also for another word to Robert Tyler: "if he *can* look over matters & get me the Inspectorship, I will join the Washingtonians," a militant temperance society: "It would be a feather in Mr Tyler's cap to save from the perils of mint julep—& 'Port wines'—a young man of whom all the world thinks so well & who thinks so remarkably well of himself." Though he tried to laugh away his fragmentary, nightmarish recollections, Poe was mortified.

Later in the month, Thomas told Poe that "the President, yesterday, asked me many questions about you, and spoke of you kindly." Tyler's son Robert had "heard of your frolic from a man who saw you in it, but I made light of the matter." Tyler even sent a letter of personal recommendation to Judge Blythe on Poe's behalf.

The collector was besieged by thousands of men "ready and anxious to take a public office now, who, in ordinary times, would rather trust to their own independent exertions." Poe's appointment never came through.

The Destruction of the World

For many followers of William Miller in upstate New York, the blazing comet arriving in the sky in 1843 heralded Christ's return and the end of the world. On April 1, Poe published a reprint of his earlier spirit colloquy about an earth-ravaging comet under the title "The Destruction of the World." His new preface was only partly reassuring: "From the celestial visitant now present we have, of course, nothing to fear." But "to-morrow its counterpart, or some wonder even more startling, may make its appearance."

He and his family moved again, to Spring Garden, north of the city, to "a lean-to of three rooms." He still planned to launch *The Stylus* in July and continued chasing subscribers and commissioning work from Lowell and Hawthorne for the future journal. Clarke still welcomed him to his home—along with Hirst and Thomas Dunn English, who did not forget

Poe's insults to his mustache. As Clarke's daughter recalled, the former "*bons camarades*" were now at odds: "All three of them happening in early one evening, they had to be kept apart lest they come to deadly strife"— English in the parlor, Hirst in the library, and Poe "shown as usual into the dining-room."

English vented his resentment in a serial story, "The Doom of the Drinker," showing Poe as a "gentlemanly looking personage" whose "fine analytical powers, together with his bitter and apparently candid style, made him the terror of dunces," but who appropriated the ideas of others and was, in sum, "the very incarnation of treachery and falsehood." Griswold would also gossip about Poe's drunken fits, apocryphally dating them to his time at *Burton's*. Rumors reached Baltimore, where Lambert Wilmer whispered to a friend that "it gives me inexpressible pain to notice the vagaries to which he has lately become subject."

In May, Clarke let Poe know, with regret, that he was abandoning *The Stylus*. The hoped-for salvation had flown.

Poe's cousin William Poe, to whom he had described his "many recent reverses" and his "sickness & despondency," warned him against a "too free use of the Bottle," the "great enemy" to their family. The illustrator Felix Darley, a partner to the dashed *Stylus* plan, recalled Poe's stoic bearing, "a refined and very gentlemanly man" who was "interesting always, from the intellectual character of his mind, which appeared to me to be tinged with sadness." The novelist Mayne Reid recalled spending "some of the pleasantest hours of my life—certainly some of the most intellectual" in Spring Garden with Poe and Virginia, though the "rose-tint upon her cheek was too bright, too pure to be of Earth."

Reid also observed Maria Clemm at work. "She was the ever-vigilant guardian of the house, watching it against the silent but continuous sap of necessity." Mrs. Clemm was "the sole messenger, doing the errands, making pilgrimages between the poet and his publishers, frequently bringing back such chilling responses as 'The article not accepted,' or, 'The check not to be given until such and such a day'—often too late." With Poe unemployed and without prospects, Muddy's fastidious economies were all that kept the household afloat.

The Flash of the Bug   March 1843 brought a welcome announcement. *The Dollar Newspaper* was offering a
hundred-dollar prize for a short story contest—"*Very Liberal Offers and No
Humbug.*" Poe had already sold his latest tale to Graham—an unprecedented concoction of cryptological drama, entomological misdirection,
minstrel show, and pirate legend. Poe bought it back and submitted it to
the contest.

On June 14, the *Dollar* announced, "*First prize* of ONE HUNDRED
DOLLARS to 'THE GOLD BUG,'" declaring it "a capital story." Dollars,
gold, and capital were on people's minds: the runner-up was "The Banker's
Daughter," and third prize went to "Marrying for Money."

The protagonist of "The Gold-Bug," Legrand, a ruined aristocrat, lives in
self-imposed exile off the South Carolina coast on Sullivan's Island (known
to Poe from his time in the army). He finds a strange insect on the beach, a
scarab with a golden, metallic-appearing carapace marked with the image of
a death's-head. He wraps it in a nearby scrap of parchment. When he casts
the paper into the fire, writing appears: invisible ink revealed by the flames.
Rescuing and reading the document, Legrand finds a coded message. He
methodically deciphers it and is led, along with his caricature of a servant,
the formerly enslaved Jupiter, to a fabulous buried pirate's treasure—gold
and gems worth millions.

"The Gold-Bug" marked a significant turn in Poe's fiction toward American settings. It was a wish-fulfilling story about using one's wits to exchange
worthless paper with seemingly meaningless writing on it—like the deval-
ued paper currency of the state banks, or the scribbles "of no value to anyone" in Poe's bankruptcy declaration—into gold. Robert Louis Stevenson
would closely follow its trail in *Treasure Island.*

Clarke alerted *Saturday Museum* readers to this "unique work of a singularly constituted, but indubitably great intellect." There was a rush on copies
at *The Dollar Newspaper*'s offices, with the original and a reprint selling out. It
was quickly turned into a play. In the Philadelphia *Daily Forum*, a young journalist denounced the piece and the prize as "A Decided Humbug," baselessly

accusing Poe of plagiarism. Poe prepared to sue for libel, but the issue was settled with a handshake and a retraction.

This fresh burst of celebrity brought Poe a new series of printed tales and a further venture: a lecture tour. The subject: "American Poetry."

Publicity was provided by George Lippard—a long-haired, muckraking novelist (his *Quaker City* depicted the vile appetites of Philadelphia's elites and their machinations against ordinary working people). He recognized Poe's heroic literary efforts and, like many, loathed Griswold. Lippard pronounced him a born poet with a mind "stamped with the impress of genius," likely "the most original writer that ever existed in America. Delighting in the wild and visionary, his mind penetrates the inmost recesses of the human soul, creating vast and magnificent dreams, eloquent fancies and terrible mysteries."

Poe delivered his speech in Philadelphia to a crowded hall, with hundreds "unable to gain admission." A "highly intelligent audience" was riveted by his "great analytical power" and "command of language." He took the show on the road: to Wilmington, Delaware's Temperance Hall, the Mechanics' Hall in Reading, the Odd Fellows Hall in Baltimore, the Franklin Lyceum, and the Philadelphia Museum—the city's most distinguished stage for popular science and entertainment, where George Combe had given his lectures on phrenology and where magic lantern "dissolving views" debuted. Poe's lectures were a stepping-stone to "a sound Magazine, devoted to all the higher objects of American Literature, edited, owned and controlled by Mr. Poe."

Poe sought a lecture at the Boston Lyceum, but its secretary predicted a low "probability of your success." Poe's attacks on Bostonians—on Emerson, Griswold, and Longfellow—weren't doing him any favors in Philadelphia either. Graham commissioned a review of Longfellow's *Spanish Student*; Poe's article was so harsh, Graham told Longfellow, that the editor had to pay him not to publish it. He mentioned an IOU from Poe: "I do not suppose it will ever be redeemed, and I doubt if the writer of it will be"—a literally damning judgment.

Thomas C. Clarke, despite dropping out of the *Stylus* plan, still treated Poe kindly in the *Saturday Museum*. But in January 1844, Clarke retired from publishing. Lippard tried to set the record straight about Poe's contributions to the city's cultural scene: "It was Mr. Poe that made Graham's Magazine what it was a year ago; it was his intellect that gave this now weak and flimsy periodical a tone of refinement and mental vigor."

But the round of recognition sparked by "The Gold-Bug" came too late. One by one, the doors of Philadelphia were closing.

II

The March of Science and Quacks

Science on Parade   Poe's dream of leading a national literary journal, to be guided by impersonal, universal stan-
dards, was now buried, if not dead. Meanwhile, other people's schemes
for lifting the level of the nation's intellectual life were moving forward—
projects to promote, diffuse, and organize research at the national level.
Some strode forth to loud fanfare, while others, behind the scenes, would
have the far greater effect.

In Washington on April 1, 1844, scientists and government officials were
lining up on Pennsylvania Avenue. The National Institute for the Promotion of Science was starting its first public convention with a parade.

President Tyler and members of his cabinet marched from the Treasury
Building to the Presbyterian church. As they filled the pews, the Marine
Corps's marching band played "a solemn air." An "appropriate prayer" was
offered by the Reverend Clement Moore Butler of Georgetown, after which
"a newly invented instrument, combining the organ and piano," sent forth
strains of "dulcet harmony."

The president welcomed the crowd. He was followed by John W. Draper,
who reported on the action of the sun's rays as revealed by the daguerreotype. The astronomer Elias Loomis next presented his analysis of the comet
of 1843.

Interspersed between musical performances, a disorienting range of
lectures were delivered: Lieutenant Matthew Maury on the Gulf Stream,
Virginia's George Tucker on "the Future Progress of the United States,"

lectures on entomology, a petrified forest, the nebular hypothesis, the historical schools of France and Germany, the theory of "One Electric Fluid" (presented by the president's son John Tyler Jr.), meteorological observations by the "Storm King" James Espy, and, from the Reverend Eliphalet Nott, the president of Union College who had designed a coal-burning stove and several other less notable inventions, a discourse titled "On the Origin, Duration, and End of the World."

The event stretched across ten days. In attendance were "members of Congress, strangers, and citizens" with a "large number of Ladies" in the upper gallery. The government officials who organized the convention had promoted it as an "intellectual banquet" of the "broadest popular character."

For this very reason—its broad and popular character—many of America's most accomplished men of science chose to stay home. More than forty leading scientists—including Joseph Henry, Benjamin Peirce, and the Yale professor and editor of *The American Journal of Science and Arts*, Benjamin Silliman—declined the National Institute's invitation.

On reading the program, Henry sarcastically wrote to the chemist John Torrey, "What do you think of my preparing a paper for the Washington Institute on *bubbles* to be presented at the great meeting? Would the title be considered objectional in connection with that of the magniloquent communication to be made by Dr Nott on the origin and destruction of the world?"

The institute was a bold step toward national support for science. But for Henry, Bache, and other like-minded scientists, it was a bubble, a fad, and a step in the wrong direction.

A Cool Half Million   The institute arose as the solution to an enviable problem: What to do with eleven crates of gold? Upon his death in 1829, James Smithson, the illegitimate son of the Duke of Northumberland and an amateur geologist, left half a million dollars to the U.S. government to "found in Washington, under the name of the Smithsonian Institution, an establishment for the increase and diffusion of knowledge." In 1838, Richard Rush sailed to Britain to recuperate Smithson's booty.

No one could agree on what to do with the gold. One possibility was an unkept promise from John Quincy Adams's inauguration speech: a national observatory, a lighthouse of the sky. Another was a national school of science—a plan supported in Congress and backed by the *Southern Literary Messenger*. In 1840, Poe echoed this call, arguing in *Burton's* that Smithson's "whole life is a plain commentary" upon his intention to found not a general university or museum but "a College for the advancement of *Science*."

The success of the British Association for the Advancement of Science inspired American researchers to consider another possibility: a national scientific organization. Tocqueville saw "the spirit of associations," the voluntary organization of groups around shared interests, as a defining characteristic of American democracy in the 1830s. In 1838, an attempt to start an association of America's men of science was blocked by members of Philadelphia's American Philosophical Society, satisfied with their position as the country's oldest (if languishing) scientific society. At the New York Lyceum, John Torrey agreed "with the Philadelphians" that the time was not right: "There is indeed too much Charlatanism in the country—enough to overpower us *modest men.*"

Despite such reservations, a plan for a national scientific body to be funded by the Smithson bequest took shape within Van Buren's cabinet. Its leader was the secretary of war, Joel Poinsett, a world traveler and plantation owner from South Carolina. An ardent promoter of the interests of his fellow slaveholders, he envisioned the extension of U.S.-owned plantations into Mexico, the Caribbean, and South America—a plan later pursued by the oceanographer Matthew Maury.

Poinsett was the leading government supporter of the U.S. South Seas Exploring Expedition, the project initiated by J. N. Reynolds and promoted by Poe at the *Southern Literary Messenger*. Its seven boats had set sail in August 1838 under the command of Charles Wilkes and accompanied, at Poinsett's insistence, by a scientific corps—including the botanist James Dana, the painter Titian Peale, and the zoologist and ethnologist Charles Pickering (Nathaniel Hawthorne's application to document the ship's voyage was declined).

In 1840, as early shipments of specimens from the expedition began arriving in D.C., Poinsett suggested that Smithson's gift be directed to fund a national museum. Such a "national cabinet" could store and display the expedition's specimens along with rocks, plants, and animals from state surveys and local collectors. It would be placed under the control of the National Institute for the Promotion of Science.

A number of government officers, including the secretaries of state and the Treasury, drew up the proposed National Institute's constitution. Poinsett was its director, along with James Kirke Paulding—one of Poe's early supporters, now Van Buren's secretary of the navy. The president and the vice president were also given official roles.

With this board of political appointees, the approaching presidential election of 1840 highlighted the fragility of the institute's conception. Would these politicians—many with strong southern and Jacksonian sympathies—reward or refuse researchers on the basis of political loyalties? Would a new administration support it?

Such worries over the institute's enmeshment with politics proved justified: with Van Buren's defeat to the Whig Harrison in 1840, Poinsett retired to his acres in South Carolina. When the Exploring Expedition's boats returned to New York's harbor in 1842, the institute's secretary, Francis Markoe, a clerk in the State Department, took charge. He began haphazardly arranging the expedition's colorful birds, plants, stones, and ethnographic objects in the Great Hall of Washington's Patent Office (now the National Portrait Gallery). Markoe also began to plan a national convention for the institute, hoping to strengthen its case to receive Smithson's gold.

The institute's organizers abandoned any hope of competing with "the extent and depth" of similar organizations in Europe, such as the BAAS; instead, they aimed merely to "collect" and "diffuse" the "intelligence, science, and practical observations of our countrymen in the different parts of the United States." Membership was wide open—to anyone with "a disposition" to "contribute or derive useful information." Despite this popular vocation, Markoe saw that the institute's success would depend on support from accomplished researchers with established reputations. He invited dozens to attend the convention. He made a particular appeal to members of the Association of American Geologists and Naturalists (AAGN). Recently formed by veterans of state geological surveys, this group held its first meeting at the Franklin Institute in Philadelphia in 1840.

Markoe knew that the AAGN planned to meet again in Washington in May 1844. He hoped to persuade them to abandon their meeting or combine it with the institute's in April. Yet he clearly had in mind "something more popular" than the AAGN's usual business. James Dana, botanist for the Exploring Expedition, felt the AAGN should meet and share their findings "without the aid of a band of music, and with more real science." Some researchers worried that the institute was more concerned with publicizing science than with doing it; according to Bache, Americans "have half a hundred persons engaged in diffusing science for one who is occupied by research."

Joseph Henry thought the institute was unfit to "decide questions of a strictly scientific character," because it was "under the control principally of amateurs and politicians." According to Dana, the modesty of the upcoming AAGN meeting would upstage the institute's extravagance: "If we can have a general attendance, the quiet business-like style of the meetings and the real value of the subjects discussed, and original matter brought out will put them in the background, bass-drum and all."

Henry peevishly wondered what would become of "the mass of diluvium which the Institute has drawn down on itself in an avalanche of pseudoscience." The institute's grab bag of presentations, its ceremonial pomp, and its direction by politicians all smacked of crowd-pleasing charlatanism—anathema to his vision of well-planned, self-directed, sober, and rigorous research.

A Humbug Epidemic   Though Bache agreed, he was too shrewd a politician to snub an organization that might win Smithson's gift. Instead, he used his slot at the convention to deliver a programmatic address titled "The Wants of Science in the United States"—in which he protested, and not subtly, against the institute's direction. He cautioned against "*precipitously* moving forward" with a scientific association without clearly reckoning with America's need for "workers" rather than "talkers." Efforts to impress and entertain the public would be counterproductive: "We need more activity, but not to be stirred up every year by exciting means." He insisted that the country needed not just any scientific union but "*such a* union *as would* repress charlatanism, not the form of association which would deliver us up bound hand and foot as its prey."

Henry and Bache were wary of any popular scientific entertainment, fearing that rigor would be sacrificed for spectacle. They were also insistent that any American version of the British Association must not be a democratic free-for-all. The BAAS meetings were popular and open events, bringing in members from across Britain. Wide participation was good for its coffers, but when "every person has the liberty of making a communication," Henry pointed out, amateurs, dilettantes, and "the profound *savant* are on the same level." The self-selecting leaders of the BAAS, however, were able to keep a grip on its overall direction; "the great body of the members have no voice in the management of the Institution, and in this respect the society is quite as aristocratical as the government of the nation."

Though a firm believer in America's republican ideals—and that the tendency of science was "republican in the proper sense of the term"—Henry thought that any U.S. scientific organization would also need "aristocratical," top-down oversight. The risk of a takeover by charlatans and quacks was far greater in America. He was annoyed at Benjamin Silliman's *American Journal of Science and Arts* for printing far-fetched electrical and physiological theories. When Samuel Morse contacted him with questions about the telegraph, Henry warned that "in the minds of many, the electro-magnetic telegraph is associated with the many chimerical projects constantly brought before the Public." Henry thought an American association would have to restrict its membership to those who "have served, as it were, an apprenticeship to the business of experimenting," echoing his own hard-served apprenticeship as a metalworker; otherwise, "the *third* and *fourth* rate men would soon control the affair and render the whole abortive and ridiculous."

Henry and Bache saw quackery lurking not just in the lyceums and popular press but among their own colleagues. Their friend at the Franklin Institute James Espy had recently earned himself the title of "Storm King" by his popular lectures on meteorology and his book, *The Philosophy of Storms.* Espy argued that storms were caused by heat near the earth, forcing water to rise and form clouds; it then cooled, condensed, and fell as rain—a theory he demonstrated with his "nephelescope," a device that produced a tempest in a teakettle. He proposed a solution to the droughts that tormented western settlements: setting a rash of fires beyond the Appalachians would, he believed, draw down rain. This was one of the first plans of artificial climate modification; fortunately, it was never tried out.

Bache expressed reservations about the "strange course" Espy was taking, and Henry worried about his "want of prudence." Benjamin Peirce bemoaned Espy's "air of self-satisfaction" and judged that "even storm kings are intolerable in a republic." The former president John Quincy Adams diagnosed Espy as "methodically monomaniac"; phrenologically speaking, "the dimensions of his organ of self-esteem have been swollen to the size of a goiter." (Poe would instead question the meteorologist's originality: "The chief portion of Professor Espy's theory has been anticipated by Roger Bacon," he wrote in 1846.) In America's clamorous public sphere, profitable and "ostentatious presentations" on controversial topics could spark thoughtless admiration, dangerous enthusiasm, and envy. The discomfort of Espy's friends showed that even a background in solid science and membership in the exclusive scientific clique around Bache, Henry, and Peirce offered no brakes on the slope to demagoguery.

Henry held a special loathing for the man who would become, by far, America's most successful scientific lecturer in the 1840s, Dionysius Lardner—the very rogue who had challenged Henry at the Liverpool BAAS meeting in 1837. Born in Dublin, Lardner took Anglican orders but devoted himself to evangelizing on mechanics and physical science instead. In London he created a successful scientific publishing concern. He became a fixture on the lecture circuit, ingratiating himself with Charles Babbage through a paean to the calculating engine; backed by Lord Brougham, he was installed as professor at University College London.

A seductive lecturer and a prolific seducer, Lardner had an illegitimate son with a married woman. The husband of another of his paramours, Mary Heaviside, was a captain of dragoons who gave him a thrashing in Paris. Lardner and Heaviside relocated to the United States. Rebuffed in Philadelphia, they moved to New York. In 1841, Lardner began public lectures at Clinton Hall and Niblo's Garden on astronomy, electromagnetism, steam engines, and extraterrestrial life.

Lardner accompanied his lectures with "extensive and splendid illustrations," magic lanterns, the artificial illumination of the oxyhydrogen microscope, and dioramas—a vividly colored, backlit display invented by

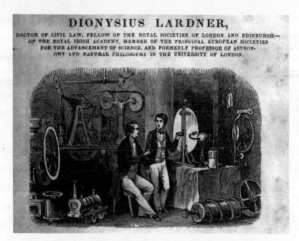

Dionysius Lardner giving a lesson, surrounded by equipment from his lectures

Daguerre that produced illusions of depth, movement, and change. When the cause of his exile got out, New England papers were scandalized. Lardner had "destroyed the peace of a happy family," but "the villain, instead of being scoured from all decent society, is invited to deliver a series of lectures." Henry wrote to Torrey, "I see by the papers that the Gay Lothario Dr Dionysius Lardner is flourishing in New York. I think it is time for you to leave. The Philadelphians have some reason to crow—they gave him no countenance."

Despite the disapproval of leading American scientists, Lardner's easily digested, multisensory spectacles were immensely popular: from Clinton Hall he went on to lecture in lyceums from Boston to New Orleans, St. Louis to Cincinnati; in Philadelphia he entertained audiences of more than one thousand. Lardner, who flaunted the title LLD (doctor of laws), appeared in one of Poe's humorous tales, "Three Sundays in a Week," as "no less a personage than Doctor Dubble L. Dee, the lecturer upon quack physics," whose opinions were gospel to a foolish character. To Joseph Henry's view of science, Lardner was a double threat—both a moral reprobate and a simplifying charlatan.

Yet Lardner's lavish spectacles proved how hungry American audiences were for science. Showmen would rush in to meet that appetite if working scientists did not.

In a spirit close to that of Bache and Henry, the dyspeptic New York physician David Meredith Reese diagnosed charlatanism as a political and moral crisis in his 1838 book *Humbugs of New-York*. New Yorkers, he moaned, "have taken the pills of foreign and domestic quacks by the thousand, with Lobelia, Cayenne pepper, and vapour baths . . . and are now equally busy in bolting down Phrenology and Animal Magnetism." Charlatans make "a pedantic show of learning"; they need only to "decry all existing systems" and "denounce all the learning of the schools" to find "a great multitude of disciples." Though New York was ground zero, the disease was nationwide. Because new scientific schemes were frequently allied with radical programs for political and moral reform, Reese's polemic targeted political and religious movements alongside "pseudo-scientific" systems. "Both Popery and anti-Popery are impostures on the public," he proclaimed, while "Ultra-Temperance" became a public nuisance once it sought to ban all booze.

The full intent of Reese's polemic became clear only in a late chapter taking up one-quarter of the book, titled "Ultra-Abolitionism." The anti-slavery movement, the doctor declared, went beyond the vague hope for a gradual end to slavery espoused by "Jefferson, Franklin, Rush, and John Jay, of the old school." According to Reese, abolitionists who condemned slavery as a sin "*in all circumstances*" and adopted strong actions—such as brandishing a musket against a pro-slavery mob, as had Elijah Lovejoy of Missouri, which got him killed—were violating Scripture (which offered examples of acceptable slavery) and citizens' rights (to their human "property"). Reese declared the Anti-Slavery Society "the most gigantic imposture which ever afflicted either the church or the state." For Reese the age's greatest humbug, and the greatest danger to the common weal, was not phrenology, the water-cure, patent medicines, or even Catholicism—but abolition.

Faced with the volatility of American politics, Bache and Henry were convinced that science depended for its survival on keeping controversial issues such as slavery at a distance wherever possible. The hostility they expressed toward charlatanism was in part a hostility toward any "popular" issues that might inflame passions. Beyond their endorsement of a common background of Protestant natural theology, they largely kept explicit religious claims—and controversies—out of their science. In the 1840s, their moderate, modest stance also meant trying to stand clear of projects of reform for workers' and women's rights, as well as the increasingly divisive issue of slavery and the race science being marshaled to support it. Out of fear of stirring up the public and of creating fissures among the fragile community of researchers, Bache and Henry largely avoided the topic.

This discretion was by no means due to their opposition to slavery. Henry was baffled by the thought that Blacks and whites might one day marry or exercise the same political rights. Their close ally the Harvard mathematician and astronomer Benjamin Peirce was a convinced and public supporter of slavery. Whatever opinions Bache had on the question, they did not interfere with his decades-long friendship with the future Confederate president, Jefferson Davis, or with Louis Agassiz, who would become a diehard proponent of race science.

The "modesty" and "rigor" that Bache and Henry promoted as a scientific and moral ideal led them to avoid topics that might stoke regional antagonisms and interfere with their goal of uniting the nation's "real men of science." For Bache, Henry, and most of their allies, the ideals of scientific disinterest and "objectivity" operated in tandem with a tacit acceptance of the status quo, including white supremacy and slavery—a "polite" silence toward the era's most urgent moral and political issue. Strikingly, Poe's "scientific" approach to literary criticism—which set aside moral and political considerations to evaluate literature as a pure "art-object"—revealed his alignment not only with Bache and Henry's projects of American intellectual reform but also with their complicity in the antebellum era's "average racism."

Bache's "Great Scientific Work"

Bache believed that another way to protect science from political interference was, paradoxically, to entrench it so firmly within the state that it could withstand the vagaries of party conflict and public whim. After his setbacks in Philadelphia—forced to step down from both Girard College and Central High School by Jacksonian politicians—he was on the lookout for a solid, unobtrusive institutional base from which he could quietly build up the nation's scientific infrastructure.

His chance came in 1843, when Ferdinand Hassler died. Hassler was the superannuated superintendent of the U.S. Coast Survey (USCS). Started in 1807 to produce maps of coastlines for fortifications and commerce, in recent years the USCS had lacked staff, equipment, and direction. After Hassler's death, the neglected federal bureau appeared to Bache as an open route. As Henry put it, the Coast Survey was "intimately connected with the scientific character of the country: a great scientific work which should be intrusted to some one with an established scientific reputation." Bache perfectly fit the bill.

Bache tapped his family and influential friends to lobby for the appointment. Benjamin Peirce declared, "I shall allow myself no rest till I have accomplished my object," and drummed up support for Bache among Boston's elite, obtaining references from John Quincy Adams, Whig Senator Rufus Choate, the historian George Bancroft (who, as customs director of Boston, had placed Orestes Brownson and Nathaniel Hawthorne in customhouse positions), and prominent merchants. Joseph Henry also mobilized his connections in New York and New Jersey, though it meant abandoning the natural philosophy textbook he and Bache planned to write together.

The campaign succeeded. President Tyler appointed Bache superintendent of the Coast Survey in 1843; according to Henry, Bache "had the support of all the most prominent scientific men in the country and the Cabinet could not do otherwise." He relocated to Washington, D.C., with his wife, Nancy. There he set about transforming a sleepy government agency into a national research powerhouse.

Under his sole authority, away from public lecture halls, journalists, and college administrators, Bache plotted highly focused expeditions to map the nation's coastlines and harbors. For half the year, he worked in the field. He and his survey teams extended grids of carefully measured triangles over the terrain; they also conducted hydrographic surveys, throwing lead lines over a boat's side to measure the depth and contours of the land beneath coastal waters.

For assistance he called "to the aid of the survey the real talent of the country." Staff were recruited from the army and navy—many from West Point—and included students recommended by his friends. Benjamin Peirce provided a steady supply of tractable Harvard graduates (and his headstrong son Charles, the future philosopher). The "chief" patiently instructed these

A. D. Bache

Alexander Dallas Bache while superintendent of the Coast Survey

young men and a very few young women—the astronomer Maria Mitchell, who in 1847 would discover a comet, was a distant relative of Bache's and spent a summer with the survey—in the ways of science. The survey offered rigorous advanced scientific training and experience of a kind unavailable elsewhere in the United States. It created a generation of researchers loyal to Bache and his friends, trained to their exacting standards.

Bache spent the other half of the year in Washington in offices near the Capitol, computing results, overseeing the production of maps, and composing his annual report. Assisting him was the wizard instrumentalist Joseph Saxton, brought from the Philadelphia Mint to become superintendent of the nation's Office of Weights and Measures, who repaired the survey's apparatus, improving the precision of Humboldtian measurements of all kinds (air pressure, magnetic inclination, distance, and time). Though Bache faced outright hostility from many in Congress who saw federal funding for science as an elitist extravagance, his family connections and political savvy helped him cultivate allies and placate foes. At first he foresaw ten or twelve years of work to cover the entire U.S. coastline, but he managed to expand its mission and annual appropriations. By 1848 the U.S. Coast Survey was one of the best-funded branches of the U.S. government, with a mammoth annual budget of four hundred thousand dollars and no sign of stopping.

Bache transformed the U.S. Coast Survey into a "national scientific project." Under his direction, it was steadily becoming "the general scientific agency of the government," a solid base for research and training, and a model for federally supported but politically independent expertise; it owed its existence to no single group or party.

Yet other "wants of science in the United States" remained unmet. The crowd-pleasing National Institute was, for Bache and Henry, an entirely inappropriate forum for exchanging information and coordinating among localities or allowing the "real working men" of American science to speak with a unified voice. And Smithson's five hundred thousand dollars had yet to find a home.

Just Rewards   Poe's lecture tour of late 1843—through Pennsylvania, Delaware, and Maryland—retraced the path of other itinerant lecturers and performers. He was tilting at his replacement at *Graham's*, Rufus Griswold, by giving his lecture the name "American Poetry"— the title of Griswold's collection, which Poe deemed a "humbug." But he had other rivals on the lyceum circuit.

Most aggravatingly, Dionysius Lardner—Poe's "Doctor Dubble L. Dee, the lecturer upon quack physics"—returned to Philadelphia in December for a series of Christmas engagements. The "Gay Lothario" was quieting the rumors of his immoral conduct by ladling religious sentiment (and music) over his performances.

Lardner had also expanded his special effects. He added the "Planetarium," a room-sized mechanical model of the solar system designed by an artisan from Ohio, and accompanied its display with an organ performance of Handel's *Messiah*. His "telescopic panorama of the firmament," an enormous oxyhydrogen-illuminated magic lantern display with moving slides of the solar system and the comet of 1843, was backed by Haydn's cosmogonic symphony, *The Creation*, while a "moving panorama," a massive scroll of images, slowly unfurled to show St. Peter's and Jerusalem to the sound of Mozart's Masses.

These contrivances adorned what Lardner called his "Bridgewater Lectures". a "view of natural theology" presented with "no sectarian doctrines." He proposed "to demonstrate that the modern discoveries in Astronomical and Physical Science, prove the EXISTENCE and manifest the ATTRIBUTES of the DIVINE AUTHOR OF THE UNIVERSE." Lardner had no license to use the Bridgewater name. He was piggybacking on the reputation of those venerable works of natural theology—much as his former patron in London, Charles Babbage, had done with his controversial *Ninth Bridgewater Treatise*. Lardner's well-oiled productions were calculated to appeal to families, to dazzle and leave them with a satisfied, pious sense of having understood the universe—without having to work too hard or to entertain any dangerous ideas.

Between 1841 and 1845, Lardner's lectures and publications earned him an astronomical two hundred thousand dollars, nearly six million dollars today. The penniless Poe couldn't help but envy such a haul, though he disdained the performances' bad taste and "quack" aspects. Music and mechanical effects could

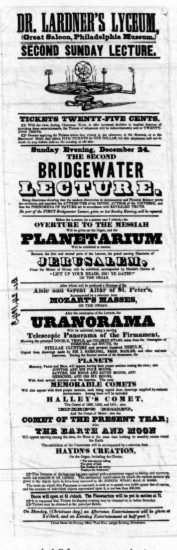

*Playbill for Dionysius Lardner's
"Bridgewater Lecture"
at Philadelphia Museum,
December 1843*

cover a multitude of intellectual sins. The effectiveness of such gimmicks, like the "namby-pamby" embellishments that Poe thought debased *Graham's* magazine, only proved how easily the masses could be captured with noise, excitement, bright lights, and sentimental reassurance. As Poe wrote, "The nose of a mob is its imagination. By this, at any time, it can be quietly led."

Pleasing the crowd, though difficult enough, was too simple a game for Poe. For years he had been writing for two audiences at once. A single tale could entertain a popular audience with suspense, humor, and shocks, while he hoped that its philosophical resonances and literary craft would eventually be recognized by the inquisitive few. His critical writings provided maps and keys for future explorations of his work; in his tales he quietly buried cryptic allusions, dense subtexts, and self-referential in-jokes for later readers to find and decode. As he put it, "Where is the ingenuity of unravelling a web which you yourself (the author) have woven for the express purpose of unravelling?"

Lone Explorer   That autumn, Poe reviewed a special issue of Silliman's *American Journal of Science and Arts* featuring an account of the Exploring Expedition—the national project he had championed at the *Southern Literary Messenger* and took as inspiration for *Pym*. In addition to its botanical, zoological, geological, and ethnographic specimens, the expedition had brought back, unwillingly, a Fijian chief named Veidovi. Captain Wilkes kidnapped him in a brutal show of force in which two American sailors and dozens of Fijians were killed. Falling ill on the last leg of the journey, the captive chief died in New York Harbor. The onboard naturalist Charles Pickering packaged and shipped Veidovi's skull to Philadelphia, where Samuel Morton added it to his "American Golgotha."

Over the four years of the voyage, Wilkes had earned the animosity of most of his crew—not least "the scientifics," whose research he frequently blocked. Wilkes was court-martialed on his return, facing accusations of misrule and excessive force. Cleared of the charges, he went to work in Washington, to help bring order to the makeshift gallery of specimens at the Patent Office.

In July 1843, Wilkes, as was his wont, put himself in charge. He added explanatory signs, rearranged the displays, enhanced the lighting, banned tobacco spitting, and painted above the entrance in gold letters "Collection of the Exploring Expedition." The refurbishment was a hit. Visitors lined

U.S. Patent Office Museum, displaying collections of the U.S. Exploring Expedition

up to gawk at the brightly colored tropical birds and flowers, "gems and gold and iron ores from Brazil; copper and silver ores from Peru and Chili; vast collections of shells and corals"; and ethnological objects, including clothing, weapons, and skulls from Pacific Islanders.

This spectacular display of cargo seized by American naval forces seemed to justify the expedition's enormous expense and complications. It flattered the nation's sense of its own technical progress and superiority by inviting a ranking of the cultures it had collected: "By a walk through the National Gallery, we travel with more than railroad speed over the Pacific, and examine into their various productions and the relative intelligence of the savages. The degradation of the New Hollander stands out in bold relief in contrast with the more advanced, though no less barbarous Feejee." More than one hundred thousand people would visit the exhibit each year over the next decade. It served as a warrant for future outlays by the government for scientific and colonial expansion.

With public approval flowing to the expedition, Poe sought justice for his friend J. N. Reynolds, who had dreamed up the expedition, only to have Wilkes maneuver him out of a place on board. Poe acknowledged "the many able and respectable gentlemen" in the expedition's scientific corps but was outraged by "the scandalous chicanery" that thrust "from all participation in the enterprise the very man who gave it origin, and who cherished it to consummation." He portrayed Reynolds as a visionary, tragically ahead of his time. Though the self-aggrandizing Wilkes tried to claim the glory, "this triumphant Expedition" should be remembered as "The Expedition of Mr. Reynolds."

Reynolds was a kindred spirit—a bold, imaginative adventurer whose courage and toil went unrewarded. In Philadelphia, Poe's literary efforts had not saved him from poverty; neither had his plans for a national journal, even one led by a "coalition" as he had proposed to Lowell. Such an association—or conspiratorial plot—closely paralleled those of the nation's most active and ambitious researchers, whose projects to secure federal support for national institutions had begun to find success. Bache was now in Washington, helming the Coast Survey—a major step forward for his and Henry's top-down, centrally coordinated vision of national science. Congress had granted Samuel Morse thirty thousand dollars for an experimental telegraph line between Washington and Baltimore. Though the National Institute was shunned by leading men of science, its public efforts sharpened the sense of the need for an American association for science, shaped and directed by its elite.

What did Poe have to show for his labors and inventions—his ingenious literary experiments, his tireless reporting and analysis, his relentless campaign to advance American letters? Would any of it be remembered? His plans for a national journal had collapsed; his ventures in Washington and with the Philadelphia Custom House ended in fiasco. He could barely feed his aunt and ailing wife.

A girl who grew up near Poe's home in Spring Garden recalled seeing him "going down Seventh St., into the city" wearing a Spanish cloak, with a "grave and thoughtful aspect." Although "little over thirty he had the appearance of middle age. To his neighbors his name meant very little."

In April 1844, he was forced to leave Philadelphia, the city where such tantalizing possibilities had once opened before him. He left with Virginia for New York. Though his family was destitute, to the point of actual starvation, the country was entering an expansive phase. John Tyler was preparing to run for reelection, and the campaign presented the annexation of Texas and Oregon, up to the 54th line of latitude, as the way forward.

Poe now had to rethink the hopes he had entertained in Philadelphia—of creating a secure, authoritative position from which to weigh in on literary, philosophical, and scientific matters. He could not follow a path like Bache's. Heading to New York, he was approaching the orbit of another, rather different contributor to American science—P. T. Barnum.

Poe was prepared to go to new lengths to bring attention to his writing and ideas—even if it meant going beyond satirical "grotesques" to outright hoaxes. He would continue to pursue in earnest his high aims in fiction, poetry, and criticism, and would push his philosophical and scientific theorizing further than ever. But no city offered greater chances for fame, or a better market for humbug, than New York.

PART IV

New York City

I have reached these lands but newly
From an ultimate dim Thule—
From a wild weird clime that lieth, sublime,
Out of SPACE—out of TIME.

—Poe, "Dream-Land"

THE MODEL.

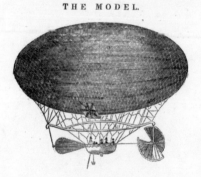

OF THE VICTORIA.

New York Sun, April 13, 1844; woodcut
of "The Victoria" balloon

12

The Market for Novelty

Subdued by Science   On April 13, 1844, two days after the closing of the National Institute's gala in Washington, the front page of the New York *Sun* barked out,

ASTOUNDING NEWS! BY EXPRESS VIA NORFOLK! THE ATLANTIC CROSSED IN THREE DAYS! SIGNAL TRIUMPH OF MR. MONCK MASON'S FLYING MACHINE!!!

Below the headlines the article gushed, "The great problem is at length solved! The air, as well as the earth and the ocean, has been subdued by science." Other exclamations peppered the columns: "God be praised! Who shall say that anything is impossible hereafter?"

It was astonishing news: the first successful air balloon flight across the Atlantic. The chief aeronaut was Monck Mason, an Irish adventurer and science writer who in 1836 had sailed in a hot-air balloon from Wales to Germany.

The flight from England to Charleston, South Carolina, was made possible by Mason's technical improvements: a device to control the balloon's height, a guide rope serving as a regulating ballast, and a rudder of cane and silk. *The Sun* included the onboard journal of another famous man of science, Harrison Ainsworth, detailing technical challenges the "rapturous" journey faced; the remarkable machine was shown in a detailed woodcut. *The Sun* obtained the exclusive story thanks to "the energy of an agent at Charleston"—alluding to its extraordinary means of getting news quickly, including special express trains, carrier pigeons, and the electromagnetic telegraph of Morse.

The article ended with an awestruck prophecy: "What magnificent events may ensue, it would be useless now to think of determining." No need to think: just buy another copy for your friends!

In actual fact, the only real journey was experienced by *The Sun*'s readers: they had been taken for a ride. The story was a hoax, written by Poe, who had arrived in the city a week earlier. But, he later said, there was "nothing put forth in the Balloon-Story which is not in full keeping with the known facts of aeronautic experience"; any part of it, he insisted, might "really have occurred."

In an essay exposing the hoax one month later, he wrote that the public response was "a far more intense sensation than anything of that character since the 'Moon-Story' of Locke." From sunrise until 2:00 p.m. on the day of the publication "the whole square surrounding the 'Sun' building was literally besieged." He quoted a New Yorker waiting for an "extra" edition: "As soon as the few first copies made their way into the streets, they were bought up, at almost any price, from the news-boys, who made a profitable speculation," some selling for half a dollar. The story was republished as a single-sheet broadside "extra"; *The New York Herald*'s reporter was indignant at this "attempt to hoax" for being "blunderingly got up," while the *Saturday Courier* reported that around "50,000 of the extras were sold."

For Poe, the hoax's reception provoked intriguing reflections: "The more intelligent believed, while the rabble, for the most part, rejected the whole with disdain." He saw this as a historical change: "twenty years ago credulity was the characteristic trait of the mob, incredulity the distinctive feature of the philosophic," but the positions had reversed. "The wise are disinclined to *disbelief—and* justly so." Though there were certainly frauds and false claims, the age was so crowded with astonishing discoveries and inventions that the most intelligent course was now to believe first and ask questions later.

Poe's balloon hoax was a perfect calling card for his arrival in New York. There the presentation of astonishing new discoveries, accompanied by controversy, was being tuned to a fine art—as part of a powerful if chaotic machine of publicity, doubt, and belief.

When Doctors Disagree   Luck was on his side when he arrived on April 6. It was raining at the docks; scurrying ashore, he found a boardinghouse and returned with an umbrella to hold over Virginia, who "coughed none at all" as she stepped ashore. Their residence on Greenwich Street was stocked with a table from a fairy tale, as he told Maria Clemm: for dinner they had "the nicest tea you ever drank, strong & hot—wheat bread & rye bread—cheese—tea-cakes (elegant) a great dish (2 dishes) of elegant ham, and 2 of cold veal, piled up like a mountain and large slices—3 dishes of the cakes, and every thing in the greatest profusion. No fear of starving here."

Even so, Virginia "had a hearty cry last night," missing her mother and Catterina, the family's tortoiseshell cat. He reassured Muddy, who would join them soon, "I feel in excellent spirits & have'nt drank a drop—so that I hope soon to get out of trouble."

New York was the largest city of the United States. Its docks heaved with goods for sale and throngs heading west on the Erie Canal. Advertising pitches and sellers' cries came from all directions, along with the racket of coal wagons on cobblestones. Poe was struck by the noise. "Where two individuals are transacting business of vital importance, where fate hangs upon every syllable and upon every moment—how frequently does it occur that all conversation is delayed, for five or even ten minutes at a time," until "the leathern throats of the clam-and-cat-fish-venders have been hallooed, and shrieked, and yelled, into a temporary hoarseness and silence!" Maria Clemm joined them in the boardinghouse, close to the booksellers, magazine offices, and printers on Ann Street, well below the mansions recently sprung up along Fourteenth Street but a short walk to Broadway, City Hall, and the abject slums of the Five Points neighborhood.

Reading rooms and libraries abounded; the city had more journals and publishing houses than any other. Amid the chatter, there was plenty of respectable science in New York. The Lyceum of Natural History, founded in 1817, boasted an excellent collection of preserved animals, plants, and rocks. In 1836 it purchased a large building on Broadway near Prince Street, renting rooms to the Phrenological and Horticultural Societies, its opening celebrated with a geology lecture by Benjamin Silliman. The lyceum's leading lights included John Torrey; his assistant, the botanist Asa Gray; the storm theorist and engineer William Redfield; and John W. Draper, continuing his work on chemistry, physics, and the daguerreotype. The lyceum's building was repossessed in 1843, but Draper helped to relocate it in NYU's medical building.

Yet the voice of sober fact and practical utility could be drowned out in the city's cacophony of entertainments, sensational news, and radical science. At Niblo's Garden on Broadway at Prince Street, one could enjoy dinner, plays, music, panoramas, scientific lectures, and the prestidigitations of Antonio Blitz, a celebrated stage magician. At the Society Library on Broadway—where Thomas Cole's large allegorical paintings *The Voyage of Life* were hung—lecturers in philosophy, literature, and the arts filled the bill.

A few blocks east of Washington Square, the imposing Clinton Hall—where Dionysius Lardner launched his American career—was the headquarters of a vegetarian society, a suffragist group, and one of the largest intellectual concerns in the country: the Phrenological Society. Run by Orson and Lorenzo Fowler, it held a collection of skulls and busts, a lecture hall, printing office, and reading room in which to learn about bumps, organs, and social reform. In Clinton Hall, lectures of confirmed and established science and solid learning vied with new theories and systems, some of them wild speculations, others outright frauds, while still others hovered uncertainly between the factual, the provocative, and the apocalyptically urgent.

The author of *Humbugs of New-York* declared the city "the chosen arena of itinerating mountebanks, whether they figure in philosophy, philanthropy, or religion," in a race to the bottom: "The more ignorant, impudent, and even vicious, such charlatans proclaim themselves to be, the greater power and patronage they may expect." Several journals spoke to and on behalf of the masses. Horace Greeley's *Tribune* published the popular astronomical lectures of Ormsby Mitchel, the West Point mathematician and colleague of Bache's who had founded the Cincinnati Observatory. The *Tribune* also published translations of Charles Fourier, the French utopian socialist. Fourier sought to reformulate society on the basis of a more rational and passionate distribution of labor (though his advocacy of a mind-boggling variety of sexual arrangements was abridged for American sensibilities).

Daily newspapers had first sprung up as commercial bulletins announcing commodity prices and the arrival of ships. *The Sun*, where Poe published his balloon article, was the first large broadsheet read by those who could not afford subscriptions or sixpenny dailies. Since starting in 1833, it grew through the "newsboy system" and the success of Richard Locke's "Moon Hoax."

Another hoaxer, Phineas Taylor Barnum, had recently become the owner of the American Museum on Broadway, just below City Hall Park. The building was adorned with flags, a lighthouse lamp, and a rooftop garden for views and balloon launches. The son of a shopkeeper, Barnum hailed from Bridgeport, Connecticut. With a mop of dark curly hair, large eyes, a booming voice, and a captivating flow of balderdash, he speculated in real estate and launched a lottery network before dedicating himself to popular entertainments.

In 1837 (while Poe was plotting the mystifications of *Pym*), Barnum met

Daguerreotype of P. T. Barnum and Charles Stratton ("General Tom Thumb") by Samuel Root; playbill for Barnum's American Museum, January 1845

one of his heroes at Boston's concert hall: Johann Maelzel, on tour with his chess-playing automaton. Barnum was promoting his own first successful attraction, modestly labeled "The Greatest Natural and National Curiosity in the World." This was Joice Heth, a wizened woman he claimed was a former slave and George Washington's nanny, making her more than a century and a half old. Heth smoked cigars and regaled audiences with the infant antics of "the father of our country." For doubters, Barnum presented a bill of her purchase and an affidavit that "eminent physicians and intelligent men" from around the country "examined this *living skeleton* and the documents accompanying her, and all *invariably* pronounce her to be as represented 161 *years of age!*" Maelzel, "the great father of caterers for public amusement," approved of Barnum's dehumanizing spectacle and assured him of future success: "'I see,' said he, in broken English, 'that you understand the value of the press, and that is the great thing. Nothing helps the showmans like the types and inks.'"

Barnum kept his attractions in print. His rotating list of curiosities included giants, albinos, Native American dancers, jugglers, magicians, automata, stuffed animals, fossils, and monstrous creatures. As he had done with Joice Heth, he counted on viewers' eagerness to gawp at other humans, especially if they were set apart as exaggeratedly different from themselves. One of his most famous performers was Charles Stratton, or Tom Thumb—"the smallest person that ever walked alone"—a child dwarf said to be decades older than he was, trained to talk and move like an adult, drinking and smoking cigars by age five. Some of the wonders were genuine; many were exploitative frauds; all were open to questioning and debate.

Barnum's "Grand Scientific and Musical Theater" hit its stride with an exhibit inspired by the U.S. Exploring Expedition, which docked in New York in 1842 after its four-year voyage around the globe. The onboard death of the captured Fijian chief, Veidovi, was much discussed in the news. When word spread of the lurid plot to preserve his head for Morton's craniological examination, Barnum added a new attraction to his museum's handbills: "the head of the cannibal chief." Visitors discovered, after paying, a plaster cast of the dead man.

To feed the "Feejee fever," Barnum borrowed a Japanese curiosity from a showman in Boston, Moses Kimball; renaming it the "Feejee Mermaid," he announced the upcoming arrival of an English scientist, "Dr. Griffin, agent

BARNUM'S AMERICAN MUSEUM,
ILLUSTRATED.

A PICTORIAL GUIDE TO THAT FAR-FAMED ESTABLISHMENT; CONTAINING MUCH
INTERESTING MATTER, AND HIGHLY USEFUL TO VISITORS.

Open from 8 A. M. until 10 P. M.

Barnum's American Museum, south Broadway, guidebook, 1850

of the Lyceum of Natural History in London, recently from Pernambuco."
The famous doctor—actually Barnum's associate Levi Lyman—appeared at
the New York Concert Hall on Broadway and displayed the mermaid, without mentioning Barnum—who then announced that he had purchased "the
curious critter" and would be showing it at his American Museum.

In the month after the Feejee Mermaid's arrival, the museum's ticket revenues tripled from about one thousand to more than three thousand dollars
per week. Doubts about the chimerical "Dr. Griffin" and the mermaid, "regarding which there has been so much dispute in the scientific world," were
part of the attraction. Some said the creature had been captured alive in the
Fiji Islands; others declared it "an *artificial* production, and its natural existence claimed to be an utter impossibility."

Barnum struck a pose of open-minded neutrality. He "can only say" that
the mermaid has "such *appearance of reality*" as any fish." But who, he asks, "is
to decide when *doctors* disagree"? The paying audience, of course: *you decide.*
Whether it was "the work of *nature* or *art,*" the object was "*the greatest Curiosity*

in the World." Barnum took the mermaid on the road. In South Carolina the controversy led to threats of a duel between a minister-naturalist and a newspaper editor; Charleston residents feared "mob violence."

This ugly fracas—and Barnum's handsome profits—arose from a "mermaid" that closer inspection revealed as the leathery upper body of a monkey sewn onto the tail of a preserved fish. Yet it was not a simple case of an exhibitioner trying to pass off a forgery as an authentic specimen. Barnum staged not only the specimen but also the controversy—the more involved and protracted, the better. At Charles Willson Peale's museum in Philadelphia, natural wonders such as the mastodon's skeleton had been introduced as authoritative facts. True to their father's Enlightenment legacy, Peale's sons presented a specimen similar to the Feejee Mermaid but pointed out its stitches, explaining exactly how it had been made. Barnum's shows, instead, invited viewers to make up their minds for themselves.

Barnum's exhibitions were among the most important routes through which working-class audiences learned about natural history and popular mechanics. Amid the fakes and provocations were wonders and facts: exceptional geological specimens, rare plants and animals, fossils, clever inventions, demonstrations of natural history, chemistry, and astronomy. In this version of antebellum science, knowledge was experienced and felt, at the price of a cheap ticket. It was open to disagreement, debate, speculation, and grinning exploitation.

Barnum was advancing an extreme form of the "charlatanism" that Bache and Henry saw as the enemy of their scientific aims. While he diffused uncertainty and disagreement among a broad audience, they aimed to concentrate scientific certainty and authority in a few hands. For Bache and Henry, any public display of science risked being dragged down to the level of a carnival, jubilee, revival, or town hall meeting. Worst of all from their perspective, Barnum encouraged his low-paying crowds to think that their opinions mattered in questions of scientific truth.

The conditions of persuasion were changing in the United States, through the expansion of commercial culture, high-intensity evangelism, and the regular combat of political campaigns. Much of the testing of beliefs took place in face-to-face public meetings: political speeches, religious revivals, science lectures, stage magic. The press was the indispensable accelerator, reaching unequaled speeds in New York. Bache and Henry sought fortified bulwarks against this rising current, while Barnum gleefully hastened it forward and rowed along. Poe's strategies were at times those of Bache, at times those of Barnum. Pushed by poverty and the threat of starvation, he constantly shifted his positions; in his writings the pursuit of truth was accompanied by the play of glitter and shadow.

Mechanical Paragraphist   In his first steady job in New York, writing "Doings of Gotham" for a Pennsylvania newspaper, *The Columbia Spy*, Poe reported on the city's buzz of gossip, politics, business, and novelty. He saw a city "thronged with strangers, and everything wear[ing] an aspect of intense life." The streets were "insufferably dirty," and cabs, omnibuses, and shrieking cats tormented pedestrians. He roamed "far and wide" on the island Mannahatta, its Native name ("Why," he asked, "do we persist in *de-euphonizing* the true names?"), and was struck by the "air of rocky sterility" in parts of its interior and by shanties built by Irish squatters. A "voyage of discovery and exploration" on a skiff around Blackwell's (now Roosevelt) Island took him past "magnificent cliffs, and stately trees" on the East Side. He predicted that "in twenty years, or thirty at farthest, we shall see here nothing more romantic than shipping, warehouses, and wharves." He was right.

Already he saw preparations being made for the presidential election of 1844. As president, John Tyler had made the advancement of slavery his major cause, upending the Whig Party's careful balance between northern and southern supporters. Seeking a route to reelection, in 1844 Tyler worked out an annexation treaty with settlers in Texas who had declared independence from Mexico back in 1836. He submitted the treaty to Congress; it was at first defeated, but raised the possibility of a war against Mexico and the addition of another slave state. Tensions ran high. Pro-slavery congressmen had tried to keep slavery from being debated in Congress, but John Quincy Adams found ways to circumvent their "gag rule" and read out petitions against slavery in the House of Representatives.

Abolitionists and other foes of slavery were outraged by Tyler's plan for Texas. He ran as an independent, while Henry Clay, who tepidly opposed annexation, became the Whig candidate. Clay's nomination was announced via the first telegraph across state lines, linking the Whig convention in Baltimore to Washington at lightning speed. With Andrew Jackson's blessing, James Polk of Tennessee became the Democratic Party candidate, with George Dallas, Alexander Dallas Bache's uncle, as his running mate. Polk was hell-bent for Texas. He also promised to seize Oregon, then a British territory and part of Canada—a sop to opponents of slavery, to maintain the balance between free and slaveholding states. Tyler withdrew from the race and endorsed Polk.

By May, Poe saw in New York "'Polk Houses,' 'Polk Oyster Cellars,' and 'Polk hats, gloves, and walking-canes,'" which were "already contending with their rivals of Clay." As the election geared up, Poe worried that "the mob-disorder which so lately beset Philadelphia" would explode here, in the race riots led by "Native Americans"—as Anglo settlers called themselves—against freed Africans and Irish.

Poe tracked the intrigues of the New York periodicals, noting the launch by the novelist Nathaniel Willis of *The New Mirror*. At summer's end Maria Clemm appeared at the *Mirror*'s office, "mentioning that [Poe] was ill, that her daughter was a confirmed invalid, and that their circumstances were such as compelled her" to take matters into her hands.

The timing was fortunate; the editors were expanding to two daily editions, launching the *Evening Mirror*. Poe accepted fifteen dollars per week as a "mechanical paragraphist," "announcing news, condensing statements, answering correspondents, noticing amusements." Among his many notices, as in Richmond and Philadelphia, were scientific developments: he wrote to defend a work by John W. Draper that used photography to study the characteristics of plants, and he related the construction of the large telescope for Ormsby Mitchel's Cincinnati Observatory. Like the "monster instrument now nearly completed by the Earl of Rosse"—a telescope fifty-four feet long being built in Ireland, to be used to examine nebulae—its observations would bring "something of that thrilling awe with which man might be supposed to regard an angel."

Though steady work, for Poe it "was rather a step downward, after being the chief editor of several monthlies," to "sit at a desk, in a corner of the editorial room, ready to be called upon for any of the miscellaneous work of the moment." Willis was impressed with "how absolutely and how good-humoredly ready he was for any suggestion, how punctually and industriously reliable," and "how cheerful and present-minded in his work." Poe put his head down and did the job.

Toiling at the center of New York publishing, working furiously on new tales by night, Poe was adapting his sensorium to a supercharged media environment. New York's production cycles were faster, louder, and more cutthroat than Philadelphia's. The pace and demand for sensational novelty was set by the penny papers, which had to sell in great volume, every day, to survive. They lit a fire under other publications.

A vicious culture of literary feuds, celebrity, and personal attacks had also taken root in New York's literary journals—inspired in part by the clashes Poe himself had stirred up in Richmond and Philadelphia. Reputations could be inflated not only by puffs and reprints but also by insults and attacks. Willis at one point refused to respond to a critic who had gone after Poe. "A reply from *me* to Mr. Briggs would make the man," he said. "Notoriety is glory in this transition state of our half-bak'd country."

At the *Mirror*, Poe caught the attention of a clique of ambitious authors calling themselves Young America, echoing contemporary European movements of political and cultural nationalism. Led by Evert Duyckinck, a high-minded journalist with wispy blond hair and a Vandyke beard, they turned their ire upon authors who imitated British "silver fork" novels—with particular animus toward *The Knickerbocker* and its editor Lewis Gaylord Clark, who feted his authors with sumptuous dinners. Where Clark and *The Knickerbocker* imitated European fashions, the Young Americans called out for properly American topics: "an image of rural life, of men in cities." Cornelius Mathews's poem, *Wakondah*, based on Indian legends, and Harry Franco's *Big and Little*, depicting New York City life high and low, answered the call.

At first, the movement was aligned with the northern branch of Jacksonians; Duyckinck was the literary editor of the *Democratic Review*, run by John O'Sullivan, who coined the slogan "Manifest Destiny" (Poe dubbed him an "ass"). Yet Polk's anti-intellectual and militarist campaign of 1844, and the recognition that American literature needed federal support, drove them toward the Whigs.

Just as the Whigs advocated tariffs to protect American manufactures, the Young Americans argued for an international copyright to protect American literature—as Joseph Henry sought for American science. U.S. law did not cover works by foreign authors; because there was no obligation to pay writers from abroad, printers pirated novels of Dickens, Scott, or Edward Bulwer-Lytton (as well as scientific texts) with impunity, running extremely cheap editions in vast numbers. Works by American authors, printed in smaller batches, had to be priced much higher to turn a profit. Readers had a choice: an inexpensive book by a well-known British author, or a costly book by an unknown American. Cheap, foreign, and famous swamped expensive, local, and obscure.

The cut-and-paste habits of magazine production also made it difficult for American authors to establish a recognizable identity. Cornelius Mathews—earnest, amiable, eggheaded, and bespectacled, widely considered a bit dim—anguished over this "false and lawless state of things." Cutting and reprinting "tend to unsettle the lines that divide nation from nation; to obliterate the traits and features which give us a characteristic individuality." An international copyright would give authors more control of their own works and careers. Mathews was a favorite of Duyckinck's, and their cause was joined by Harry Franco—the pseudonym of Charles Briggs, who in January planned to launch a new weekly, *The Broadway Journal*; James Russell Lowell was a comrade in arms.

Ironically, when Poe arrived in New York, the Young Americans saw the critical positions he had previously taken—against literary cliques and puffing—as a reason to bring him into their literary clique and puff his work. Poe was ready: the biography he had helped write for the *Saturday Museum* was being revised by Lowell for a new introduction to his work, along with a new engraved portrait and appreciations of his poetry, tales, and criticism. He had already posed for a daguerreotype (and would do so at least five more times); he was one of the first authors to use photography to craft his public identity.

Yet just as Poe entered New York's publicity machine, he mercilessly satirized it with a tale he published unsigned in the *Southern Literary Messenger*: "The Literary Life of Thingum Bob, Esq." Raised by a barber who shaves the editors and poets in the city of Smug, Thingum Bob shoots to the top of the literary heap with a couplet praising his father's patent hair treatment—the restorative "Oil-of-Bob." To crown his success, he becomes editor and owner of a periodical uniting "all the literature of the country in one magnificent Magazine."

Poe's tale slashed maniacally at the tricks of literary fame—tactical pla-

giarism, reciprocal puffs, high-toned attacks, ginned-up controversy, political opportunism—and ridiculed Poe's own most cherished ambitions. He
oiled the gears of his story's reception by planting unsourced gossip about
the stir it was making: "The query is put to *us* especially, here in the North—
'who wrote it?' Who *did?*—can any one tell?"—a publicity strategy worthy of
Thingum Bob himself.

In the *Mirror*, Poe wrote a facetious exposé of a new entertainment
brought to America by P. T. Barnum: the "Swiss Bell-Ringers" (actually from
Lancashire, England), who performed at Niblo's Garden and the Society Library. By means of "numerous bells, of various sizes," these seven mustached
men produced "the most delicious music." Poe's notice revealed them to be
"ingenious pieces of mechanism" driven by the "power which operates in the
Electro-Magnetic Telegraph." A battery under the stage, he explained, "com-
municates by a hidden wire with each of them," sending shocks "regulated
and directed by the skilful musician and mechanician who secretly manages
the whole affair." This was also a leading theory to explain how Maelzel had
controlled his chess-playing automaton.

Though his position at the *Mirror* was humble, it put him at the center of
New York's literary buzz. He could point out the hidden wires controlling
the disorienting jangle of New York's media—even as he began to use them
for himself.

We Live in a Wonderful Age

Poe was incomparably productive
in 1844, bringing a dozen stories
into print. While his previous tales were often set in London, Paris, Venice,
Göttingen, or misty regions off the map, he now favored American settings.
The new tales registered the blaring jingoism and thirst for expansion in the
Polk-Clay presidential clash. They also answered the era's trite paeans to
industry and nation:

> Through ponderous looms the rapid shuttle flies,
> And weaves the web which shines with varied dyes.
> Here, gliding cars, like shooting meteors run,
> The mighty shuttle binding States in one.

Many of Poe's tales of this period satirized such exultations about American ingenuity and progress. "The Thousand-and-Second Tale of Scheherazade," a new *Arabian Night*, exaggerated modern technical feats into fabulous wonders. The United States was presented as "a nation of the most powerful magicians" who send their voice from one end of the earth to the other (the telegraph), direct the sun to make a picture (the daguerreotype), and tame "a huge horse whose bones were iron and whose blood was boiling water" (the steam train).

In "Some Words with a Mummy," Poe turned the tables. This story featured actual figures known to him from Philadelphia—George Gliddon, the grave-robbing "Egyptologist," and Samuel Morton, the race scientist and skull collector. Poe caricatured Morton as "Doctor Ponnonner" (who defended his crank theories "upon honor"). In his rooms, Ponnonner attaches a voltaic battery to a mummy supplied by Gliddon. They reanimate the Pharaoh, "Allamistakeo," who is understandably annoyed to be disturbed.

Gliddon and Ponnonner condescendingly explain to the African king "the marked inferiority of the old Egyptians in all particulars of sciences, when compared with the moderns, and more especially with the Yankees." The Pharaoh refutes their boasts with proof of Egypt's knowledge of chemistry, astronomy, artesian wells, steam power, and architecture. The only "modern" discoveries the Egyptians lacked, it appears, were democracy and quack medicine, innovations that the Pharaoh places on a par. The American experiment might turn out to be an illusion—or "all a mistake."

The tales Poe produced in 1844 dredged up the contradictions and hypocrisy of America's "great age of progress." The apparent hero of "'Thou Art the Man,'" Charley Goodfellow, appears "open, manly, honest, good-natured, and frank-hearted"—pioneer virtues that allow him to get away with murder. "The System of Doctor Tarr and Professor Fether" relates a visit to an asylum where a new, humane, and lenient "system of soothing" has been introduced and where the inmates are revealed to be running the place, to the mangled melody of "Yankee Doodle."

This series of tales often featured such surprising twists. Audiences demanded (and, paradoxically, expected) novelty and surprise—the fiction writer's price of entry into a marketplace crowded with thrills. "The Angel of the Odd," another of Poe's zingers from 1844, reveled in "the extravagant gullibility of the age," people's readiness to believe reports of inventions and bizarre occurrences. "The marvellous increase of late in these 'odd accidents' is by far the oddest accident of all," the narrator observes. In bust-to-boom New York, exceptions were the rule; surprise was the only sure thing.

Reasoning with Chaos   In these dense, antic tales—full of reversals, puns, grimacing humor, and occasional horror—Poe poked at weighty philosophical questions. Is nature predictable and orderly? Is there a reliable way to know it? Or does chance rule the universe? His detective tales, featuring C. Auguste Dupin, confronted such questions head-on.

Poe published his third Dupin mystery, "The Purloined Letter," in 1844 in a year-end anthology, *The Gift*; like the first two it reflected on the variety of methods for grasping a hidden design or cause. Where "The Murders in the Rue Morgue" highlighted the methodical and logical (if "multiform" and "manifold") steps behind seemingly intuitive leaps of knowledge, "The Mystery of Marie Rogêt" focused on probabilistic reasoning. It closed with Dupin's observation that there is no reason to assume that two similar series of facts must have the same causes: "the most trifling variation in the facts" might cause two otherwise identical series to diverge.

The "doctrine of chances" or "calculus of probabilities" was just beginning to shift its meaning and implications. Originally this form of inquiry was used to overcome the uncertainty involved in judgments based on limited knowledge and to reduce the errors in astronomical observations. Announcing a notion that would become widespread in physical science in subsequent decades, Poe suggested that chance might not only explain human error but also be part of the structure *of nature itself*. In "Marie Rogêt," alongside his discussion of "the calculus of probabilities," Poe noted, "It is no longer philosophical to base, upon what has been, a vision of what is to be. Accident is admitted as a portion of the substructure."

This view, of chance as part of the fundamental structure of nature, harked back to the ancient materialist philosophy of Epicurus and Lucretius, which saw all of nature as the result of the chance interaction of tiny particles, triggered by the occasional "swerve" that set an atom on a new course. Lucretius's *De rerum natura* (*The Nature of Things*) laid out this cosmology in verse for Roman readers; Erasmus Darwin's *Botanic Garden* and *Temple of Nature* updated it for eighteenth-century freethinkers. Darwin's materialist cosmology found a vast and enthusiastic readership in America in the early nineteenth century, though some worried about its implication of either atheism or pantheism. "We see not how the . . . pupil of Darwinian metaphysics can stop short of Spinoza's creed," one theologian warned. The death of the chemist John Dalton in 1844 prompted new appreciations for his "atomic theory," while four new translations of Lucretius's poem appeared after 1799. The most influential of these, by the Reverend John Mason Good, spurred Poe—along with Percy and Mary Shelley, Hawthorne and Melville—to reckon with the implications of understanding life and thought as by-products of chance interactions.

The Epicureans' materialist vision of a universe ruled by chance was anathema to natural theologians, convinced that nature followed the predictable laws maintained by a benevolent deity. Yet in New York's hardscrabble offices and alleys—where a trivial encounter might bring glory or ruin—Poe had reason to suspect that in both life and dice, mere luck divided winners from losers. The science of statistics, grounded in probabilistic reasoning, was being introduced by states to track populations and set policy; it was offered as a tool for traders in New York's much-read *Hunt's Merchants' Magazine*, which Poe called the "absolute authority in mercantile matters." While statistics helped master the slings and arrows of an expanding market economy, in "Marie Rogêt" Dupin had used probabilistic reasoning to uncover the causes of a brutal crime.

When Poe revived Dupin in "The Purloined Letter" in 1844, he took a more psychological approach. To foil the plot of a mischievous minister—blackmailing the queen with a compromising letter—Dupin enters into the mind of his adversary. He relates the story of a boy who always won a guessing game by adopting the facial expression of his opponent, then waited to see "what thoughts or sentiments" arose in his own mind or heart. Likewise, to anticipate (or reconstruct) the moves of the minister, Dupin has to empathize, to think and feel along with him.

Poe sharply contrasts this intuitive approach with the narrowly empirical methods of the police. The prefect, knowing the letter to be in the minister's rooms, examines every possible nook, checking drawer after drawer, applying a "most powerful microscope" to all the joints and cracks of the house. His men impose a numbered grid on its entire surface and examine "each individual square inch." They find nothing.

For Dupin, this "probing, and sounding, and scrutinizing with the microscope and dividing the surface of the building," was simply the application of the narrow "set of notions regarding human ingenuity, to which the prefect, in the long routine of his duty, has been accustomed." The prefect assumed the minister would hide the letter as most people would—by concealing it in some out-of-the-way place. But the minister, Dupin knew, was both a mathematician and a poet (much like himself, and like Poe). Adapting his thought to the capacities of this exceptional mind, Dupin knew to look for the letter, folded upon itself, hiding in plain sight.

Dupin's empathic method—entering into the thoughts and feelings of what one seeks to understand—is difficult to square with the objectivity or detached, impersonal observation then being championed by established men of science. Measuring, dividing, calculating, and mapping a well-delimited space were the methods that Poe had learned at West Point and that Bache now applied in the Coast Survey.

Dupin suggested that such approaches could only discover facts hidden by an intelligence of a mundane sort. He complained that "the mathematician argues, from his *finite truths*, through habit, as if they were of an absolutely general applicability—as the world indeed imagines them to be."

Poe made Dupin his spokesman for an attack on the rising empirical and mathematical sciences. They weren't *wrong*, only *narrow*, and insultingly confused the intelligence of the creator with that of a plodding bureaucrat.

Mesmeric Revelations   According to proponents of a new science feverishly discussed in New York— mesmerism or "animal magnetism"—the true boundaries of knowledge were being widened every day.

The lights in the hall went down and a man took the stage, dressed like a doctor or a preacher. He began with a sober discussion of the new science:

THE
HISTORY AND PHILOSOPHY
OF
ANIMAL MAGNETISM,

WITH PRACTICAL INSTRUCTIONS FOR THE
EXERCISE OF THIS POWER.
BEING A COMPLETE COMPEND OF ALL THE INFORMATION
NOW EXISTING UPON THIS IMPORTANT SUBJECT.

BY A PRACTICAL MAGNETIZER.

BOSTON:
PUBLISHED BY J. N. BRADLEY & CO.,
OFFICE OF THE DAILY MAIL, 18 STATE STREET.

Cover of History and Philosophy
of Animal Magnetism, *Boston, 1843*

its known facts, its points of agreement with recent discoveries in medicine, electricity, and physiology, and the mysteries still in play.

He then brought out his "subject," a woman said to possess a remarkable sensitivity. They sat in chairs facing each other. Filling his mind with a "strong determination of will, united with benevolence of feeling," he gazed at her without blinking, applying steady pressure to her hands until she closed her eyes. He then conducted "passes," careful movements of his hands close to her skin, from the crown of her head to the tips of her fingers, to her stomach, to her feet.

With a series of questions he tested the depth of her trance: "Are you asleep?"—"Yes—no; I would rather sleep more soundly." Soon she was in the *magnetic state*.

The subject answered questions and obeyed commands; she identified objects and read passages of a book through a blindfold or in a distant room. Audience members were then brought onstage and likewise placed into the state; some had intractable pains relieved; some observed distant events; some awoke without any memory of their acts under the magnetizer's power.

Such "experiments" were held in the same lyceum halls as the lectures of Combe, Espy, and Lardner; they produced uncanny effects comparable to magic lantern shows, with elements of magic and ritual. Though more controlled than evangelical revivals, they hinted at mysteries beyond the world of ordinary sense. In Nathaniel Hawthorne's *Blithedale Romance*, set within the intoxicating currents of the 1840s' utopian philosophies, a character attending a mesmeric performance in a village lyceum sees this new science epitomizing "the mysticism, or rather the mystic sensuality, of this singular age." Ralph Waldo Emerson at first took mesmerism as a fad for minds made unhealthy by sensational media: "I suppose that as the marketplace & the alleys need to be stimulated by the raw head & bloody bones of a murder or piracy with wood-cuts so our wise cotemporaries [*sic*] are glad to be made to wonder by something that is wonderful to the senses." Later, looking back on its heyday, he saw mesmerism affirming "unity and connection between remote points," part of a historic "breath of new air."

Franz Anton Mesmer was a German doctor who had toured Europe performing hands-on cures for chronic illness and pain. Connected to Masonic lodges, Mesmer claimed to be developing a Newtonian science of attraction and cosmic balance, working with an invisible substance akin to electricity that was carried by the ether and circulated through bodies. To many, Mesmer's "cure" was a panacea; to others, an absurd, licentious fraud. The Paris Academy of Sciences organized a Commission on Mesmerism in 1784; its members, including Pierre-Simon Laplace and Benjamin Franklin, concluded that Mesmer's effects, though often genuine, were due only to his patients' imaginations.

In the 1820s in Paris, a new wave of *magnétiseurs* appeared, working with "clairvoyants" and "somnambulists" who claimed to see events happening far away, to read books in languages they didn't know, and, at times, to communicate with the dead. The new wave reached London, where the physician John Elliotson used mesmerism to reduce surgical patients' perception of pain. It appeared in New York in 1829, first presented by a French instructor at West Point, Joseph Du Commun, and took off thanks to Charles Poyen, a Frenchman who left his family's plantations in the Caribbean. Poyen first practiced in Lowell, Massachusetts, in 1836, offering mesmerism as a way to maintain shop-floor discipline among female factory workers. Mesmerism was embraced by followers of Swedenborg, the natural historian and mystic who saw correspondences between the natural world and the hidden realm of spirits and angels.

At the same time, its practitioners eagerly employed the methods and language of science. Silliman's *American Journal of Science and Arts* reported on the topic; it was pursued by respected researchers including the chemist Robert Hare and John Kearsley Mitchell—a Philadelphia physician who treated Virginia Poe (and who briefly owned Maelzel's chess-playing automaton).

Mesmerism proposed an empirical and rational approach to phenomena that were both mental and material, constantly evoking notions from physics such as electromagnetism and the luminiferous (light-bearing) ether. Yet it also challenged the fixities of space, time, and matter upon which scientific consensus was being built, by seeming to demonstrate invisible sympathies between bodies at great distances.

Magnetic performances often appeared as battles of will and reason among magnetizers, subjects, and skeptics; they were intensely entertaining. Even performances that failed to meet expectations lent the science a sense of authenticity; other séances produced experiences so strong and strange that even skeptics had to concede that *something* was happening. An 1843 manual from Boston cited the mighty Boz's conversion: "Mr. Dickens, in a letter to a friend, says, that having witnessed the experiments of Dr. Elliotson in London, 'he should be untrue to that gentleman and to himself, if he hesitated to declare, that he is a believer in the science, and that he became so against all his preconceived opinions."

In the United States, new theoretical frameworks sprang up around mesmeric phenomena: John Bovee Dods advanced a system of "electrical psychology"; Phineas Quimby's "mind cure" placed the source of all illness in the mind; and Orson and Lorenzo Fowler—the New York phrenologists who promoted new sciences and cures, including the healing benefits of octagonal houses—were convinced that phrenology and mesmerism were "twin brothers by nature." Stanley Grimes's *Etherology* likewise brought "the facts of mesmerism and 'phrenomagnetism' into harmony with the known and admitted laws of electricity and magnetism"—an effort Poe praised for its "ingenuity and plausibility."

Mesmerists shared Poe's obsessions with the limits of empirical science and the shadowy relations between matter and spirit, observation and imagination. Poe's woozy 1844 "Tale of the Ragged Mountains," set in the countryside around Charlottesville, imagined a mesmeric rapport crossing decades and continents. The main character, Bedloe, enjoys hiking through the woods after his morning dose of laudanum. While taking a walk in "the strange *interregnum* of the seasons which in America is termed the Indian Summer," he is seized by a vision and relives the experience of a British soldier slain during a rebellion in Calcutta. With a "shock as of a galvanic battery," his consciousness leaves his body and returns to his "original self." With its mirrored, chiasmic structure, the tale shows minds communicating across time and space, as barely suppressed memories of colonial repressions against "Indians"—in America and Asia—struggle to rise again to consciousness.

A *Philosophy from Beyond*   Poe plunged deeper into these mysteries in "Mesmeric Revelation," published in *The Columbian Magazine* in August 1844. It took the format of a medical article, echoing case reports in *The New York Dissector* and the Fowlers' *American Phrenological Journal*. But in Poe's hands, mesmerism revealed strange truths about God, the cosmos, and human suffering.

"WHATEVER doubt may still envelop the *rationale* of mesmerism," he began, "its startling *facts* are now almost universally admitted." Only "doubters by profession" deny that a skilled mesmerist can impress a subject into "an abnormal condition" resembling death, during which the patient perceives "matters beyond the scope of the physical organs."

Poe presents himself as P., a magnetizer with a patient under his care, Mr. Vankirk. One day Vankirk summons him, overcome by anxiety about "the soul's immortality." Reading works of the spiritualist philosopher Victor Cousin and the transcendentalist Orestes Brownson, Vankirk follows remarkable chains of thought while mesmerized in which "reasoning and its conclusion—the cause and its effect" are simultaneously apparent. On awakening, he forgets it all.

They agree on an experiment: P. will ask Vankirk questions while in the state of "sleep-waking." The patient dreamily spells out a "train of convincing ratiocination," beginning with the nature of God, who is neither spirit nor matter as usually understood. Matter has gradations, from the gross to the very fine; at its limit, it is an extremely subtle fluid made of atoms so tiny that they are "without particles—indivisible—one." This weightless, etherlike matter "not only permeates all things but impels all things—and thus is all things within itself. This matter is God." Mesmeric revelation, indeed.

Vankirk explains that this "rare matter" is too fine for our sense organs to perceive until it condenses to form "nebulæ, planets, suns, and other bodies"; only angels—which lack the organs of ordinary existence—perceive its most refined state. Our senses, which perceive by contact with gross matter, are adapted to the atmosphere and entities of earth, while

> there are many things on the Earth, which would be nihility to the inhabitants of Venus—many things visible and tangible in Venus, which we could not be brought to appreciate as existing at all.
>
> But to the inorganic beings—to the angels—the whole of the un-particled matter is substance; that is to say, the whole of what we term "space" is to them the truest substantiality.

Venusians, he claims, have organs adapted to their surroundings, just as we earthlings have organs adapted to ours. Angels alone sense the most subtle matter, which we call space.

Death changes everything. "There are two bodies—the rudimental and the complete; corresponding with the two conditions of the worm and the butterfly. What we call 'death,' is but the painful metamorphosis" through which we realize "the full design" of our being, when we are freed from the limitations of ordinary matter.

But why, P. asks, is it necessary to pass through the imperfect, painful, rudimentary life?

Vankirk murmurs: "Pain, which in the inorganic life is impossible, is possible in the organic."

> P. But to what good end is pain thus rendered possible?
> V. All things are either good or bad by comparison. A sufficient analysis will show that pleasure, in all cases, is but the contrast of pain. *Positive* pleasure is a mere idea . . . The pain of the primitive life of Earth, is the sole basis of the bliss of the ultimate life in Heaven.

Poe's dialogue is a mesmeric contribution to "theodicy"—the term coined by the philosopher Leibniz for the reasons why a benevolent and omnipotent God would allow evil and suffering. Poe's mesmerized patient argues for the "comparative" or "relative" nature of pleasure and pain: to experience the bliss of eternity, we must first know the pains of having a body.

Poe was sketching out a new natural philosophy: a systematic explanation of creation, metaphysics, and life. Everything was matter, he claimed, but matter could reach such refined states that it was indistinguishable from electricity, magnetism, or spirit. This was a materialism that accounted for the powers of the soul, spirit, and mind. God himself was indistinguishable from this all-encompassing subtle ether. Informed by Swedenborg's writings on the senses, his providential design required an uninterrupted creation of organic beings with sense organs adapted to their circumstances. Their pain was a prelude to the bliss that would follow their butterfly-like metamorphosis at death.

Like the balloon story he wrote on his arrival in New York, "Mesmeric Revelation" was a work of imagination disguised as a factual report. But this was no hoax; Poe was testing out serious propositions about the nature of reality. If his conclusions challenged the fundamental methods and assumptions of science and theology, well, then theology and science would have to change.

A Scandalous "Scientific Romance"

A more momentous challenge to established science and religion hit the press that year: the anonymous *Vestiges of the Natural History of Creation*. First published in Edinburgh in 1844, the book went through seven editions in the next three years, taking readers in the United States and Britain by storm.

Vestiges was called a "scientific romance." With engaging language and vivid imagery, it told a unified history of the origins of the solar system and life, weaving together current findings from all the sciences. It replaced an essentially static Christian universe—where between creation and apocalypse one saw only minor adjustments—with a universe undergoing constant development. This "evolution" unfolded according to uniform mechanical laws laid down at the beginning.

The romance (or novel) began with the nebular hypothesis: the idea that the solar system had formed out of clouds of diffused matter. This was the prime example of a "law of development" running through the history of the universe; it suggested that "the formation of bodies in space is *still and at present in progress,*" with nebulae seen in the night sky far from earth gradually condensing into their own planets and suns. The book went on to detail geological history, the emergence of microscopic "infusoria" and "animalcules," and the steady, ongoing emergence of more and more complex life—through plants, mammals, and primates, through to the "numerous nations" of the human race.

A single process of life, unfolding in multiple series according to a relatively restricted set of physical and organic laws, spread all the "animated tribes" of living things across the planet. The book declared that the "difference between mind in the lower animals and in man is a difference in degree only; it is not a specific difference." Instead, its author explained, "LIFE is everywhere ONE. The inferior animals are only less advanced types of that form of being perfected in ourselves."

The central position *Vestiges* gave to the nebular hypothesis exposed the theory to new scrutiny. Though one of its originators, Laplace, was suspected of being an atheist, William Whewell had offered the hypothesis as a proof of divine creation: the solar system might have formed from nebular vapor, but, Whewell asked, "what but design and intelligence" could have prepared the process? In 1837, the Glasgow astronomer John Pringle Nichol's *Views of the Architecture of the Heavens* had further popularized the hypothesis, praising Laplace for revealing "that primeval Creative Thought which originated our system and planned and circumscribed its destiny." Nichol saw the theory as a confirmation of uniform natural law and used it to support a radical, implicitly materialist agenda. Joseph Henry was delighted by its "exact account of the phenomena of the constitution of the solar system" and found the hypothesis a "grand but simple view of the process which the Deity has adopted in framing the universe." Even the theological *Princeton Review* hazarded that the nebular hypothesis might be "the true theory of the universe."

Yet the new book, *Vestiges*, took the nebular hypothesis into dangerous territory. It directly linked the theory to materialist theories of transmutation, suggesting that life arose spontaneously through the interaction of ordinary matter and electricity. For evidence it referred to experiments by Andrew Crosse in which tiny organisms appeared to be generated by applying an electric current to silicate of potash and copper nitrate. Updating eighteenth-century deism, the book stressed that creation—an ongoing, "progressive act" obedient to a uniform law of development—was not finished. The author ventured the "speculation" that humanity as it now exists is preparing "the possible development of higher types of humanity"—a "species superior to us in organization, purer in feeling, more powerful in device and art."

Gathering together the dispersed and specialized sciences, *Vestiges* rewrote the story of the universe. No longer was God actively involved in directing each moment of creation; instead, laws laid down long ago unfolded gradually, over time, bringing each new chapter forward according to a complex program revealed in stages. On earth, the struggle of the individual's soul for salvation no longer played the central role; it had been replaced by the striving of distinct "species" and "life" as a whole to realize the aims of progress.

The notion that even nature's most abrupt changes obey "uniform law" had been urged by the phrenologist George Combe, by the geologist Charles Lyell, and, more provocatively, by the mathematician and inventor Charles Babbage in his *Ninth Bridgewater Treatise*—an uninvited addition to the *Bridgewater* series of natural theological tracts. To explain the appearance of law-bound change in nature, Babbage used the example of his own calculating engine (the same amazing device Poe compared, unfavorably, to "Maelzel's Chess-Player"). Babbage could program his machine to make an unexpected leap in a sequence of printed numbers. In the same way, he argued, a surprising novelty in nature that might appear to be a "miracle" is better understood as the result of a preestablished mechanical law we don't yet understand.

Vestiges quoted Babbage's book approvingly. According to one American reviewer, *Vestiges* was "the most elaborate attempt, which has been made in modern times, to establish a mechanical theory of the universe." But because such arguments placed the power to create species within the laws of the universe itself, they could be difficult to distinguish from atheism. The Yale chemist and preeminent editor Benjamin Silliman complained that in England all the book's errors were "swallowed up by the upper classes, to whom every thing boldly asserted and in captivating style is gospel." He found the book's message "false in religion and philosophy."

For *Vestiges*' second U.S. edition, the publisher Wiley and Putnam made the extraordinary decision to include an introduction by the Congregationalist minister George Cheever (whose poems Poe called "undeniably *médiocres*"). In effect, Cheever advised readers to ignore the arguments they had paid good money to read, preemptively dismissing the book as an "elaborate attempt" to "exclude God utterly by law, from his own world." Francis Bowen, editor of Boston's *North American Review*, saw the book's "true character and tendency" as a revival of the materialist atomism of Democritus, Epicurus, and Lucretius. Tayler Lewis, professor of Greek at NYU, summarized it as "blank atheism, cold, cheerless, heartless, atheism."

The book sent established scientists scrambling to respond. Charles Lyell denounced it; John Herschel devoted his 1845 BAAS address to a refutation. The geologist Adam Sedgwick warned, "If the book be true, the labours of sober induction are in vain; religion is a lie; human law is a mass of folly, and a base injustice; morality is moonshine."

In the United States, James Davenport Whelpley, physician and geologist at NYU, rejected the book's "false conclusions" and mocked its claims of "men originating by slow degrees from monkeys with a crowd of like absurdities following, pell-mell, at the heels of a nebular hypothesis, and ending like that in mere vapor."

The anonymity of *Vestiges*' author was carefully guarded, adding to the brew of questions surrounding the book. Readers had to cast about to guess at its origin and to know how to interpret its aims. Was the author a quack, an amateur, a provocateur, or a legitimate scientist, freed by anonymity to advance his or her true views?

Just as Poe had asked about "Thingum Bob," readers on both sides of the Atlantic were eagerly asking, "Who wrote it? Who can tell?" They were also snapping up copies. *The Broadway Journal*, Charles Briggs's new literary weekly, observed, "It is admirably calculated for the millions and being sold at a low price, it must . . . meet with a rapid sale." Barnum could hardly have done better.

One of the earliest American notices of this "scientific romance" appeared in the February 1845 issue of *The American Review: A Whig Journal of Pol-*

itics, Literature, Art, and Science. The same issue included a startling new poem that also inquired into mysteries of life and death. Its author, too, was unknown; it was signed with a pseudonym.

Like *Vestiges,* the poem made an immediate sensation.

Who wrote it? Who can tell?

"The Raven" would make Poe a household name.

13

A Man of Wonders

Nameless Nevermore   Poe spent the later part of 1844 with Virginia and Muddy in a farmhouse on the west side of Manhattan, near Eighty-Fourth Street. He was preoccupied with the complex internal rhymes of a poem he had begun in Philadelphia, after meeting Charles Dickens, whose most recent book had featured a trained raven. He might also have mused on his subject at the Library Company, where a giant bust of Athena overlooked scholars and their books. At the end of the year, he announced to the poet William Wallace, "I have just written the greatest poem that ever was written."

"Have you?" said Wallace. "That is a fine achievement."
"Would you like to hear it?" said Poe.
"Most certainly," said Wallace.
Thereupon Poe began to read the soon-to-be-famous verses in his best way . . . When he had finished it he turned to Wallace for his approval of them—when Wallace said:
"Poe—they are fine; uncommonly fine."
"Fine?" said Poe, contemptuously. "Is that all you can say for this poem? I tell you it's the greatest poem that was ever written."

When George Graham refused to buy it, Poe sold the poem for ten dollars to the recently launched *Whig Journal*. The Whigs got a bargain. In the journal's "ramshackle Nassau Street office" the long-haired editor George

Colton read it to a group of witnesses "before yet it had gone into type" and, "as he closed with oratorical effect the last refrain, declared with an emphasis that shook the whole mass of his flaxen locks—'that is amazing—amazing!'"

Very quickly "The Raven" amazed others. It appeared at the end of January—first on January 29 in the *Evening Mirror* and soon after in the February issue of the *Whig Journal*. There it was placed between an article urging the creation of panels of scientific experts to evaluate patent claims and an essay by Evert Duyckinck on the situation of literature in 1845, which drew attention to Poe's past and forthcoming works.

The poem's intricate, engrossing stanzas worked their spell from the first lines:

> Once upon a midnight dreary, while I pondered, weak and weary,
> Over many a quaint and curious volume of forgotten lore,
> While I nodded, nearly napping, suddenly there came a tapping,
> As of some one gently rapping, rapping at my chamber door.
> "'Tis some visiter," I muttered, "tapping at my chamber door—
> Only this, and nothing more."

> Ah, distinctly I remember it was in the bleak December,
> And each separate dying ember wrought its ghost upon the floor.
> Eagerly I wished the morrow;—vainly I had tried to borrow
> From my books surcease of sorrow—sorrow for the lost Lenore—
> For the rare and radiant maiden whom the angels name Lenore—
> Nameless here for evermore.

The effect of "The Raven" lay in the combination of its complex, seemingly inevitable rhymes and its stark, mysterious dramaturgy. A lonely, bereaved scholar is visited by a black bird trained to repeat a single word, "Nevermore." It croaks out these three morbid syllables in response to the scholar's increasingly fierce questioning: Is the bird a messenger from the land of the spirits? Does it have a message from the poet's dead love, the lost Lenore? Or is it just a mindless beast, repeating its training with no sense of what it says—an animal automaton acting out an empty program?

Through eighteen stanzas, in the same mesmerizing, half-mechanical rhythmic scheme, with wilder claims and more daring rhymes at each new turn, the bird remains; it pierces the heart of the poet, refusing to leave. In the flickering light of the study, with purple curtains perfumed by incense, the bird offers nothing but the maddening refrain: "Nevermore," twisted into a denial, a prophecy, a torment—a permanent shadow of certain sorrow and unanswered cries:

> And the raven, never flitting, still is sitting, still is sitting
> On the pallid bust of Pallas just above my chamber door;
> And his eyes have all the seeming of a demon's that is dreaming,
> And the lamp-light o'er him streaming throws his shadow on the floor;
> And my soul from out that shadow that lies floating on the floor
>   Shall be lifted—nevermore!

"The Raven" was signed "BY QUARLES"—the name of a seventeenth-century English poet known for enigmatic and emblematic poetry.

In the *Weekly Mirror*, "The Raven" was accompanied by Willis's superlative praise: "It is the most effective single example of 'fugitive poetry' ever published in this country; and unsurpassed in English poetry for subtle conception, masterly ingenuity of versification, and consistent sustaining of imaginative lift and 'pokerishness.'" "Pokerish" captures the sense, as one reads, of a bet made, a bluff daringly maintained—with the poem raising the stakes by satisfying the complicated rhythm and rhyme scheme (modeled in part on "Lady Geraldine's Courtship" by Elizabeth Barrett) despite ever-increasing difficulty.

Each stanza adds a nuance, heightening the intensity of the narrator's questions. The black bird, returning from "the night's Plutonian shore," perching ominously on an all-white bust of Athena, Greek goddess of wisdom, resonated with traditional Christian imagery; it contrasted the hope of revelation with things seen only "through a glass, darkly." The literary scholar Betsy Erkkila has suggested that the work also taps into a racial imaginary—pure white femininity and reason threatened by a dark, bestial intruder. The raven may also be the embodiment of guilt and shame, the return of the repressed, whether individual or national.

The poem juxtaposes knowledge—Athena, the scholar's books—with the unknown and unknowable. What does this strange creature have to say about the afterlife, about the limits of nature and instinct, about the person who trained it? Is there a mind within the bird, or is it mechanical matter? Does "Nevermore" mean the speaker will not see Lenore again, or is it a word without meaning? Can humans ever unravel such mysteries?

Above all, the poem's melody captured its readers, in its incantatory snares and its sense of growing necessity and impending revelation. The actor James E. Murdoch gave an impromptu reading at the offices of *The Broadway Journal*; for one witness it was "the most cherished memory of my life that I heard the immortal poem read by one whose voice was like a chime of silver bells."

Quickly it made the rounds of New York papers: on February 3, Horace Greeley's *Tribune* praised it as "a poem which would have enriched Blackwood"; the *Morning Express* said it "may well defy competition in its way from the whole circle of contemporary verse writers; though Alfred Tennyson might, perhaps, enter for the prize." It was both a technical feat and an insightful marvel, a novelty suited to an age of invention: "As a piece of versification it is as curious as it is, psychologically, a wonder."

Soon Poe's name was out. "The Raven" vindicated not only his earlier poetry but also his current criticism. Charles Briggs, editor of *The Broadway Journal*, wrote to Lowell, "You will see in this week's Journal a grand poem by Poe, which I think you will like. You will see that it is framed according to his notions of poetry. A mere beautiful something entirely free from didacticism and sentiment."

It was reprinted dozens of times around the country. Parodies proved its impact. "The Owl" warned against the ills of drink:

> But the owl he looked so lonely, saying that word and that only,
> That a thimble-full of whiskey I did speedily outpour
> In a tea-cup on the table, which, as well as I was able,
> I invited him to drink of, saying there was plenty more—
> But the owl he shook his head, and threw the whiskey on the floor,
>
> > Plainly saying, "nevermore!"

The New World printed "A Vision," in which the journals of New York were "personified, by goblins strange and tall." "The Gazelle (After the Manner of Poe's 'Raven')" was penned by a "new-found boy-poet of fifteen"; "The Turkey" followed "The Whippoorwill." A lawyer in Quincy, Illinois, forwarded "The Pole-Cat" to his colleague Abraham Lincoln, who answered stiffly but gamely, "I have never seen Poe's 'Raven'; and I very well know that a parody is almost entirely dependent for its interest upon the reader's acquaintance with the original. Still there is enough in the pole-cat, self-considered, to afford one several hearty laughs."

There was "great curiosity to see the writer of that wonderful poem." Poe made a dramatic entrance in New York's salons. At the home of Mrs. Caroline Kirkland, he met "the New York literati, not one of whom had ever before seen him"; his dignified bearing made "a most favorable impression." At a gathering in the home of the physician John Francis, Poe entered, "a pale, thin, and most grave-looking man, whose dark dress and solemn air, with the Doctor's own look of ceremonious gravity, produced an ominous silence where, a moment before, all was hilarity." When Francis announced the stranger—"The Raven!"—Poe "bowed slightly and slowly; with a fixed, and, it almost seemed, a portentous gaze, as if complacently accepting the character thus thrust upon him."

The poet Elizabeth Oakes Smith first heard the poem read by Charles Fenno Hoffman before its author was known. It affected her so much she "arose and walked the floor, and said to him, 'It is Edgar Poe himself.'" Hoffman summarized the inner sense of this "production of genius": "It is despair brooding over wisdom."

The poem seeped into public consciousness. According to Smith, Poe told her that "my Raven is really being talked about a great deal. I was at the theatre last night, and the actor interpolated the word 'Nevermore,' and it did add force to the sentiment that was given, and the audience immediately (he looked so pleased when he said this), evidently took the allusion."

By March 1845, Poe and his "Raven" were being lauded and puffed like Thingum Bob and his hair ointment, Oil-of-Bob. It was an earworm so enticing that it could be used to sell medicated soap, as in the jingle "The Craven: BY POH":

Once upon a midnight dreary, while with toil and care quite weary,
I was pondering on man's proneness to deceitfulness and guile,
Soon I fell into a seeming state 'twixt wakefulness and dreaming,
When my mind's eye saw a scheming fellow counterfeiting Soap—
Yes! counterfeiting GOURAUD's matchless Medicated Soap;

". . . Craven, wilt thou now give o'er,
And never counterfeit my Soap or Poudres any more?"
Quoth the craven—"Never more!"

The dubious "Dr. F. Felix Gouraud" sold his "incomparable *Poudres Subtiles* and marvellous Grecian Hair Dye" near Poe's home. Warning customers against "counterfeits," Gouraud refurbished Poe's poem to flog a patent medicine—the paradigmatic humbug of the nineteenth century.

Though Poe sold it for only ten dollars, the poem was a priceless calling card. Writing to his old friend F. W. Thomas, Poe crowed, "'The Raven' has had a great 'run,' Thomas—but I wrote it for the express purpose of running—just as I did the 'Gold-Bug,' you know. The bird beat the bug, though, all hollow."

He Was the Young American

The smash debut of "The Raven" had been well prepared by New York's publicity machine. Just two weeks before it appeared, Lowell published a glowing biography of Poe in *Graham's*. He positioned Poe in terms of the chief concern of the Young Americans: the absence of a truly national literature.

Like all his contemporaries, Lowell said, Poe suffered from the "anomalous," lawless dispersion of American letters; the literary productions of Boston, New York, or Philadelphia were "more distinct than those of the different dialects of Germany." As possibly "the most discriminating, philosophical, and fearless critic upon imaginative works who has written in America," Poe had it in him to forge a genuinely national criticism and to pave the way for a truly American literature. All he needed was "a magazine of his own, in which to display his critical abilities." To date, "he has squared

Engraving of Poe in Graham's, *February 1845*

out blocks enough to build an enduring pyramid, but has left them lying carelessly and unclaimed," scattered throughout the nation's periodicals.

In Lowell's essay, the previously fragmentary Poe appeared complete, a well-established author with an oeuvre, alongside an inviting portrait. Poe's friend Snodgrass thought it a good likeness, though "rather wanting in that *nervousness* of expression so peculiar to Mr. Poe." Lowell spotlighted Poe's poetry, presenting analytic power as the distinguishing trait of not only his criticism but also his fiction and verse.

Lowell's biography focused other critics' attention. Margaret Fuller, a former editor of Emerson's *Dial*, now in New York, noted Lowell's "frank, earnest" essay and praised Poe's poems "The Haunted Palace" and "To Helen" for possessing "such distinguished beauty in thought, feeling, and expression, that we might expect the life unfolded from such a bud to have the sweetness and soft lustre of a rose." Duyckinck alerted *Morning News* readers to this newly discovered force in American letters: "Whenever his name is mentioned it has been with the comment that he is a remarkable man, a man of genius." Though few "knew precisely what he had written," Duyckinck reckoned, "his influence has been felt while the man was unknown."

"The Raven" raised the curtain; Poe had arrived. Duyckinck welcomed

him as the critic who would separate original and genuine works from derivative or "quack" literature, alerting readers to Poe's investigations into "American Cribbage," the rampant plagiarism in magazines. Proving the Young
Americans' point about the greater esteem given "foreign" works, a Boston
magazine reprinted "The Purloined Letter" only after it had been reprinted
in Chambers' Edinburgh Journal, and Duyckinck set the record straight; the story,
by an American, "has attracted far less attention at home than abroad."

Poe took up the Young Americans' causes. In "Some Secrets of the
Magazine Prison-House," he exposed the exploitation faced by "poor devil
authors"—their scandalously low pay, their lack of control over publication,
the extortionate profits that went to publishers simply for owning a press.
The lack of an international copyright law pushed America's "very best
writers into the service of the Magazines and Reviews," whose owners constantly reprinted works without attribution and at best offered a pittance
for original work, simply for appearances' sake; it "would *not do* (perhaps
this is the idea) to let our poor devil authors absolutely starve." A strong
American literature needed strong federal support—like any other growing
industry, like the nascent sciences, and like technical inventions that benefited from the government's patent law.

Lionized! Poe's alliance with the Young Americans forced him to ad-
just some positions: after picking apart Cornelius Mathews's
Carlyle-esque "essayism" and "tomahawking" his long poem, *Wakondah*, in
Graham's, he now excused his "impudent and flippant critique," suggesting it
was merely "a laughing matter." He also favorably reviewed the long poem
Orion, by Mathews's friend in London Richard Horne—a member of Tennyson's circle—and offered soaring praise of Elizabeth Barrett, another of
Mathews's British allies. Horne's heavy moralizing made him guilty of the
crime of "didacticism," while Poe's fickle effusions to Barrett, followed by
astringent qualifications, left her baffled.

Whatever the cost of these compromises, in 1845 the Young Americans advanced Poe's career with dizzying speed. Duyckinck invited him to
publish a new story collection for Wiley and Putnam's Library of American
Books. Better yet, Poe joined Briggs and his silent partner, John Bisco, as

The Broadway Journal's co-editor and "3rd pecuniary owner." Poe was back in the editorial chair, with a controlling stake in a promising new magazine.

He was invited to give a lecture at the Society Library, advertised as "novel, ingenious, and a capital antidote to dullness." Willis promised a literary anatomy lesson: "the critical blade of Mr. Poe" had already brought many authors to a fate resembling the "decapitation of the criminal who did not know his head was off till it fell into his hand."

Poe's lecture attacked "the pernicious influence of coteries"—most of all Boston's, "the chief habitation, in this country, of literary hucksters and phrase mongers," including Griswold and Longfellow. Duyckinck compared Poe's principles of "poetical composition" to those of Hazlitt and Coleridge; his manner was "that of a versed and resolute man, applying to a hideous sore a keen and serviceable knife." The Daily Tribune's reviewer—probably Horace Greeley—praised his "acute and fearless criticism."

Bostonians were less pleased. The Atlas would "prefer the dancing dog, or somersetting monkey" to such a reckless critic; they taunted Poe to "come before a Boston audience with such stuff, they would poh at him at once." The editor of the Evening Transcript, Cornelia Wells Walter, carved an epitaph for the "would-be-critic": "His name was Poe / His life was woe / . . . echo, answering, saith—'Poh!'"

In The Broadway Journal, Poe was reprinting his earlier tales and poems, gathering the dispersed "blocks" of his earlier works into a "pyramid" of literary glory. Would it be an immortal monument or a house of cards?

In its pages he launched a new round of attacks on Longfellow. Polemics between authors were a well-tested publicity stunt, often starting as a good-natured joust among colleagues or a way for an obscure David to make a name against an inflated Goliath. But they easily got out of hand: feelings were hurt, reputations destroyed; lawsuits and duels ensued. While at the Messenger, Poe had been one of the first American journalists to engineer literary feuds of the sort developed by Blackwood's in Edinburgh. The practice was picked up in New York by Park Benjamin in The New World and soon spread. Given the number and proximity of authors, a tiny tiff could spark a blaze. Poe now stoked the flames more dangerously.

At Graham's, Poe had pointed out vague resemblances between poems by Longfellow and Tennyson. In January 1845, Longfellow published The Waif,

a collection of works by little-known poets, many unsigned; Poe suspected
that some of the pieces were Longfellow's own lesser works and questioned
Longfellow's choice not to include any poetry by Lowell (a bizarre objection,
because Lowell was already well known). In reply, the *Evening Mirror* published a letter, signed "Outis," defending the elder poet against Poe's charges
while arguing that some imitation, an unconscious plagiarism, was part of
the natural working of the imagination.

A flame war ignited between Poe and "Outis," who might have been
Poe himself. Just as Barnum invented mouthpieces in the press to attack
his Feejee Mermaid, Poe might have been working both sides of the controversy, creating a stir and raising the question that had not occurred to
anyone (except him) to ask: Who speaks best for American poetry, Longfellow, the famous Harvard professor, or Poe, the new arrival? Poe's editor,
Willis, saw such controversies as a promotional tool and was certain that
the "joust" between Poe and Longfellow's defenders "will serve Longfellow
in the end."

In April the "Longfellow War" continued in *The Aristidean* with a gutter-
worthy review of Longfellow's new collection. The screed was unsigned—
likely written by Thomas Dunn English in consultation with Poe. It opened
with the ad hominem suggestion that Longfellow's reputation was due to
the "influence of his social position" as professor "at HARVARD, and an access
of this influence by marriage with an heiress"; it derided his Boston-based
supporters, the "Longfellow Junto," as a "small coterie of abolitionists, transcendentalists and fanatics in general," and ridiculed Longfellow's "Poems
on Slavery" as "incendiary drivel."

In a notice he published in another magazine, Poe tried to distance himself from this "coarse" review: "There are in it some opinions which, by implication, are attributed to ourselves individually, and with which we cannot
altogether coincide." Whoever was responsible for this piece, Briggs and
Lowell cringed as Poe's attacks on Longfellow began to turn New England
against him.

The controversy did not dim his celebrity in New York: "Everybody
wants to know him; but only a very few people seem to get well acquainted."
He became a fixture at the gatherings of the poet Anne Lynch, in her town
house on Waverly Place: "Poe had always the bearing and manners of a

gentleman—interesting in conversation, but not monopolising; polite and engaging", "always elegant" yet "unaffected, unpretentious, in his manner." Among Lynch's visitors were the poet Fitz-Greene Halleck (man-about-town and adviser to John Jacob Astor), the Kentucky-born abolitionist Cassius Clay, and "the man in the moon," the reform-spirited journalist Richard Locke.

Lynch's regulars included powerful women such as Margaret Fuller, the probing, erudite critic and philosopher who, after making her name as an editor of Emerson's *Dial*, wrote *Summer on the Lakes*, an account of a voyage west where she met seers and Native Americans; in New York she became a contributing editor for Horace Greeley's *Tribune*. Lynch and Fuller were joined by the poets Mary Hewitt, Estelle Lewis, and Elizabeth Ellet, whose poetry Poe had reviewed years earlier, less enthusiastically than she would have liked. In this crowd, a favorite topic was mesmerism, which was discussed along with the mystical and angel-filled philosophy of Emanuel Swedenborg. Other habitues of Lynch's salon were the physician and philanthropist Dr. John Francis, who gave a lecture for the reopening of the New York Lyceum of Natural History, and George Bush, a professor of Hebrew at New York University and an avid Swedenborgian (as well as an ancestor to two presidents).

Poe was entering a fascinating, tempestuous scene. Women in nineteenth-century America were largely excluded from business and politics. If married, they had limited power over their own property. They faced subtle opposition and open scorn if they tried to gain an independent reputation as artists and thinkers. The "woman question" was a key issue for reformers including Fuller, whose *Woman in the Nineteenth Century* was a groundbreaking investigation of the politics of the sexes under America's comparatively egalitarian mores. Educated white women also exerted considerable influence as participants and organizers of salons.

In Poe's circles in New York, women were respected and successful poets, authors, and editors, active in movements of reform—against slavery, in favor of more equitable systems of labor and property, medical education, and women's rights. These influential women, some of whom reclaimed the insulting epithet of "bluestocking" for themselves, organized much of the intellectual life of the age, though frequently off the printed page. Poe published and praised their work, pointing out flaws as well as achievements. They taught him more than he readily admitted in the domains of poetry, philosophy, and social savvy.

Poe was friendly with Mary Gove, author of *Lectures to Women on Anatomy and Physiology*, which held that it was "not moral obliquity, but want of information, that causes the many abuses" women faced. She lectured frankly about sex, contraception, childbirth, and women's inalienable right to their own bodies, even if that meant (as it had for her) choosing divorce over a passionless marriage. For ills from indigestion to cancer, she urged vegetarianism and the "water cure"—consuming great quantities of water and covering oneself with dampened towels, sweating, and applying washings "externally and internally." The regimen was also propounded by Marie Louise Shew, another poet who befriended Poe and Virginia.

Gove's philosophy drew upon Sylvester Graham's dietetics, Swedenborgian mysticism, and the doctrine of passionate attraction of the French reformer Charles Fourier—as presented by Albert Brisbane, a driving force behind the utopian experiment at Brook Farm and a man, according to Gove, "entirely subjected by the spirit of Fourier." Gove's criticism of the inhumanity of "civic society" and its failure to understand either the body or the heart led her through a variety of utopian and reformist circles, especially after her marriage to Thomas Low Nichols, a principled libertine.

Encircled by these passionate thinkers, Poe and the much-admired poet Frances Sargent Osgood grew mutually enthralled. When Willis gave Osgood a copy of "The Raven," its effect on her, she said, "was so singular, so like that of 'wierd [*sic*], unearthly music,' that it was with a feeling almost of dread, I heard he desired an introduction."

They met at the opulent Astor House hotel. Osgood recalled "a peculiar, an inimitable blending of sweetness and hauteur in his expression and manner." Poe greeted her "calmly, gravely, almost coldly; yet with so marked an earnestness that I could not help being deeply impressed." Thomas Dunn English, who had also moved to New York, recalled a soiree hosted by Lynch: "At my feet little Mrs. Osgood, doing the infantile act, is seated on a footstool, her face upturned to Poe, as it had been previously to Miss

Frances Sargent Osgood, ca. 1848

Fuller and myself. In the center stands Poe, giving his opinions in a judicial tone and occasionally reciting passages with telling effect."

Witty and spirited, Osgood was estranged from her husband, the painter Samuel Osgood (who painted the portrait of Poe now in the National Portrait Gallery). The couple had lived in London, where she published two books of poetry. In New York in 1841 she compiled *The Poetry of Flowers and Flowers of Poetry*, gathering floral-themed poems from across the centuries, placing her own verses alongside those of Lucan, Erasmus Darwin, Byron, and Byron's friend Thomas Moore, with commentary on flowers' appearance, meanings, and uses. It included full-color prints, a dictionary of flowers, and at the end "a simple treatise on botany" condensed from Linnaeus. This was a work of natural history in the classical sense, with a resemblance to Erasmus Darwin's Epicurean *Loves of the Plants*—the story of flowers told from all possible angles: descriptive, symbolic, scientific, emotional.

Poe had earlier singled out Osgood's work in his "Poets of America" lecture. In print he admired the music and sweetness of her "songs." Some spoke of children and the home, while others addressed illicit passion and marital discontent in stunningly frank terms:

Frontispiece from Osgood's,
The Poetry of Flowers and Flowers of Poetry, *1841*

Oh! hasten to my side, I pray!
I dare not be alone!
The smile that tempts, when thou'rt away,
Is fonder than thine own.
The voice that oftenest charms mine ear
Hath such beguiling tone,
'Twill steal my very soul, I fear;
Ah! leave me not alone!

Her verse possessed a simple, seemingly artless clarity or grace, "a term applied, in despair," Poe wrote, "to that class of the impressions of beauty which admit neither of analysis nor of comprehension." Many sought her attention, including Rufus Griswold. Virginia encouraged her friendship with Poe (at least according to Osgood), even as Poe and Osgood flaunted their growing attachment in flirtatious poems with hidden messages placed acrostically in their lines.

A member of Anne Lynch's set wrote to a friend in Providence, the poet Sarah Helen Whitman, about the new addition to their salons: "People seem to think there is something uncanny about him, and the strangest stories are told, and, what is more, *believed*, about his mesmeric experiences, at the mention of which he always smiles." Poe's sphinxlike response hinted at depths of sadness and experience—heightened by his allusions to his beloved consumptive wife, to wide and obscure learning, and to the strange new science of animal magnetism. He did nothing to dispel the heady rumors. In fact he urged them along.

Facts in the Case   In late June 1845, Poe's second story collection, *Tales*, was published to great acclaim. A glowing four-page appreciation in the *Whig Journal* gave particular attention to "Mesmeric Revelation," spurring a round of reprints. The phrenologist Orson Fowler signaled the importance of the tale's "*subject matter*," without either "endorsing" or "repudiating" it: he laid it before readers, to "*think it over fully*, and form their own conclusions." Strikingly, Poe's growing reputation as a fiction writer was for Fowler a reason to believe: it was "written by Edgar A. Poe, a man favorably known in the literary world; so that it may be *relied* upon as authentic." Yet the magazine of Brook Farm—the Fourierist commune whose travails Hawthorne distilled into *The Blithedale Romance*—dissented: it dismissed Poe's stories as "clumsily contrived, unnatural, and every way in bad taste"; "there is no health about them; they are like the vagaries of an opium eater."

London's *Popular Record of Modern Science* reprinted "Mesmeric Revelation" as "The Last Conversation of a Somnambule," noting its publication in journals of "respectability and influence in the United States" and its "internal evidence of authenticity." Poe gleefully informed his readers of the "to-do about the tenability of Mr. Vankirk's doctrines" made by mesmeric journals and Swedenborgians who took it as "absolutely true, although at first they were very strongly inclined to doubt." Professor George Bush assured Poe of the work's originality and probability, while *The New World* warned Poe not to "trifle" on "so serious a subject," but invited readers to "draw their own conclusions."

In the months after the debut of "The Raven," "Mesmeric Revelation" charged the atmosphere around Poe with hints of occult power. It revealed him as a philosophical and scientific investigator grappling with serious mysteries and as a writer giving them compelling shape. Solidifying his claims to magnetic expertise, in April Poe reviewed *Human Magnetism* by W. Newnham. He noted the abundance of counterfeit reports of mesmerism but placed the theories of the English magnetizer Chauncy Hare Townshend among "the most truly profound and philosophical works of the day."

As with his ambiguous smile when he was asked about his "mesmeric experiences," he kept the status of his "Revelation" flickering: Was it a work of pure imagination? A true reporting of a false philosophy? A true philosophy advanced through fiction?

Poe followed the run of "Mesmeric Revelation" with a new mesmeric tale in December, "The Facts in the Case of M. Valdemar." Here he wove a tighter and bolder fabric: more story, less speculation, and a robustly characterized patient at the center, Monsieur Valdemar, the "well known august personage with white beard and black hair."

Valdemar is dying of consumption. In the throes of what appears to be a fatal crisis, P—— comes to his side and brings him into the magnetic state. For weeks the patient holds on at the edge of death, his body's collapse postponed by his mind's concentration.

After seven months he begs to be allowed to die. When P—— ends the trance, Valdemar's tongue vibrates with an impossible, horrific phrase: "*I am dead.*" His body collapses into a "liquid mass of loathsome—of detestable putrescence."

In Poe's hands the story itself—with its gradual movement from the familiar to the unearthly via rhythmic, incantatory steps—became a mesmeric talisman, drawing the reader on an intensely concentrated journey to a shocking, repulsive conclusion in which the obscure becomes palpable. The vile matter of a diseased body rushes grotesquely to the fore as the mind and spirit fly.

"Valdemar," too, was believed. Greeley noted that "several good matter-of-fact citizens" took it seriously, though only those with "the bump of Faith large" were entirely taken in. From London, Elizabeth Barrett wrote to Poe that "Valdemar" is "going the round of the newspapers," raising "dreadful doubts." Readers devoured it in *The Times* under the title "Mesmerism in America: Astounding and Horrifying Narrative," *The Popular Record of Modern Science* argued that the "angry excitement and various rumors" it had prompted in New York were "sufficient to show that *something* extraordinary must have taken place."

The leading Boston mesmerist, Robert Collyer, interrogated Poe: "M. Valdemar's case has been universally copied in this city, and has created a very great sensation." He was certain its facts were plausible but begged Poe to "put at rest the growing impression that your account is merely a *splendid creation* of your own brain." Poe reprinted Collyer's letter in *The Broadway Journal* but refused "to offer one word" about "the truth or falsity" of the account.

Beyond their philosophical and narrative interest—or rather, precisely because of it—Poe's mesmeric tales were powerfully effective media stunts. In "Mesmeric Revelation" he tested out a new philosophy, while in "Valdemar" he launched a deliberate hoax based on familiar theories and experiences.

Poe's interest in mesmerism's practical mysticism was a point he shared with many literary figures; it also brought him close to new popular religious movements. Andrew Jackson Davis, "the seer of Poughkeepsie," entered a visionary trance when he met the phreno-magnetist Stanley Grimes. In New York City between 1845 and 1846 he was regularly mesmerized; with his transcriptions, *The Principles of Nature, Her Divine Revelations, and a Voice to Mankind*, he became one of the founders of spiritualism. When Poe paid him a visit, Davis informed him that even if "Mesmeric Revelation" was "poetically imagined," its main ideas "concerning ultimates" were "strictly and philosophically true."

The fusion of mesmeric practices and Swedenborgian visions in Poe's "Revelation" resonated with Davis's philosophy. Their sensational cosmological theories also bore a distinct resemblance to *Vestiges of the Natural History of Creation*. Against a static universe forged once and for all, these radical philosophies envisioned a progressive unfolding of life and thought over time.

The Power of Words

Even before the furor surrounding *Vestiges*, Poe had been preoccupied with cosmology.

He reflected on the subject in his spirit colloquies and in entries to a series he called "Marginalia," first appearing in the *Democratic Review*—pithy epigrams and reflections inspired by his wide and varied reading. In one installment of the series, Poe took literally the notion of "cosmopolitanism," originally advanced by the Stoics. "An infinity of error makes its way into our Philosophy," he declared, "through Man's habit of considering himself a citizen of a world solely—of an individual planet—instead of at least occasionally contemplating his position as cosmopolite proper—as a denizen of the universe."

This literal "cosmopolitanism" meant not only recognizing the absurdity of identifying with one's city, nation, or tribe against others but also trying to understand the universe's design and the thoughts of its creator. In his first "Marginalia" entry he developed a fundamental objection to the version of natural theology in the *Bridgewater Treatises*. By explaining every natural object by a single prompting necessity, the *Bridgewater* authors had overlooked the defining trait of God's creation. Human constructions are simple: "A particular cause has a particular effect—a particular purpose brings about a particular object." However, "the great idiosyncrasy in the Divine system of adaptation," he claimed, is that every effect is also a cause: "The object is either object or purpose, as we choose to regard it." In other words, every element of the creation can be seen to be serving the whole creation and is served by it. No single cause can explain this "reciprocity" or "mutuality of adaptation."

He also thought the natural theologians erred in insisting on the harmony of creation and God's benevolence; such assumptions made it impossible to explain evil, suffering, and imperfection. These were dismal facts all around him and inside him. Already in "The Black Cat" he had argued that moral philosophers failed to take account of "PERVERSENESS"—the inclination to act against one's own interests. "Who has not, a hundred times, found himself committing a vile or a silly action, for no other reason than because he knows he should not? Have we not a perpetual inclination, in the teeth of our best judgment, to violate that which is *Law*, merely because we understand it to be such?" Perverseness, he argued, was "one of the primitive impulses of the human heart—one of the indivisible primary faculties." Poe was, paradoxically, proposing a natural law driving humans to violate the *equally* natural law of self-preservation. He was describing a built-in violation of the creator's alleged benevolence—an Epicurean "swerve" within the straight line of morality and self-interest.

Even on aesthetic grounds, Poe revolted against an order that was too perfect. A later "Marginalia" installment elaborated on Francis Bacon's dictum that "there is no exquisite beauty without some strangeness in the proportions":

> Take away this element of strangeness—of unexpectedness—of novelty—of originality—call it what we will—and all that is ethereal in loveliness is lost at once. We lose—we miss the unknown—the vague—the uncomprehended, because offered before we have time to examine and comprehend. We lose, in short, all that assimilates the beauty of earth with what we dream of the beauty of Heaven.

This preference for the strange, unexpected, novel, or seemingly inharmonious would be one of Poe's decisive contributions to modernist aesthetics.

Poe's questions about the design underwriting nature intersected with his musings on the tremendous power of words. He described "experiments" he undertook at the limits of waking consciousness, alluding to a "class of fancies, of exquisite delicacy, which are not thoughts, and to which, as yet, I have found it absolutely impossible to adapt language." These mental phenomena arise only when the soul is in a state "of most intense tranquility," in those "points of time where the confines of the waking world blend with those of the world of dreams"—when he is "upon the very brink of sleep, with the consciousness that I am so." He described his attempts to bring back lucid reports from this borderland: "Now, so entire is my faith in the power of words, that, at times, I have believed it possible to embody even the evanescence of fancies such as I have attempted to describe."

His faith in the power of words went further. Charles Babbage, in his provocative *Ninth Bridgewater Treatise*, developed the implications of the claim in modern physics that every act and motion makes a "permanent Impression" through its impact on the ether: the air becomes a recording device, "one vast library, on whose pages are for ever written all that man has ever said or even whispered." Babbage, an abolitionist, illustrated this notion with a gruesome anecdote taken from court reports of a slave trader who forced his human cargo overboard; Babbage imagined a record of the victims' cries inscribed on the ether, a permanent witness to his sin.

Poe ran with Babbage's sublime suggestion in a spirit colloquy, "The Power of Words." In it, two spirits discuss the ether, that "great medium of creation"; they observe, like Babbage, that the vibrations from every act, from every sound, leave a physical trace. As proof, one of the angels points out a "wild star"—a planet newly formed by the angel's tears: its "brilliant flowers *are* the dearest of all unfulfilled dreams," and its "raging volcanoes *are* the passions of the most turbulent and unhallowed of hearts." Through words and tears, thought and feeling literally impress themselves upon matter. They alter the course of nature's development, forging and remaking worlds.

Just as in the metaphysics sketched in "Mesmeric Revelation," and the experiment in "Valdemar," Poe was extending factual, material processes to the point at which they blurred into the speculative, ethereal, and spiritual, teasing life and thought beyond their known limits. Challenging simplistic understandings of design, he was chasing the edge of creation, where "unthought-like thoughts" and unstable, formless entities might be converted into felt and spoken things.

14

The Imp of the Perverse

Poe's workload for *The Broadway Journal* was crushing. New pressures forced him into missteps, of which his attack on Longfellow was the worst judged. He wrote in May 1845, "For the last three or four months I have been working 14 or 15 hours a day—hard at it all the time." Despite his growing fame, "I am as poor now as ever I was in my life—except in hope, which is by no means bankable." Thomas Dunn English, his mustached frenemy, said of Poe, "He never rests. There is a small steam-engine in his brain, which not only sets the cerebral mass in motion, but keeps the owner in hot water."

The intense pace of literary production demanded no less. In 1825 there were approximately a hundred periodicals in the United States; by 1850 there were six hundred. Poe thought the sudden rise, "within a few years, of the magazine literature" was a sign of the times: the "whole tendency of the age is Magazine-ward." Tastes and habits had changed: "We now demand the light artillery of the intellect; we need the curt, the condensed, the pointed, the readily diffused—in place of the verbose, the detailed, the voluminous, the inaccessible."

Poe was sharply aware that he was living through a media revolution, driven by new technologies for papermaking, steam-powered presses, and transport by rail. The telegraph and Morse code promised even greater acceleration. Poe was not convinced "that men at present think more profoundly than half a century ago, but beyond question they think with more rapidity, with more skill, with more tact, with more of method." This new speed, condensation, and quantity of information—feeding a ceaseless hunger for facts and distractions—transformed the labor of a "magazinist," as he called himself, into the scramble of the sorcerer's apprentice. He had to sprint to keep abreast of events, in tune with public tastes, a step ahead of creditors.

Eighteen forty-five and 1846 were years of Poe's growing recognition as one of his country's most distinguished story writers, poets, and critics. Rushing like a cyclone as an editor and publisher, he also filled the columns of *The Broadway Journal* and sold work to other magazines, including criticism that asserted the importance of technical mastery and deliberation to create literary effects.

One of the searing ironies of Poe's life was that during this rise to fame, as he developed an ideal of the quasi-omniscient author in total control of the creative process, his life was falling apart—his career, his relationships, and his very mind—a victim of bad luck, alcohol, and self-sabotage. He elaborated his theory of rational artistic mastery in the very months in which his life grew most chaotic and his sanity was most thrown into doubt.

The Pulse of the Machine

Enthusiastic reviews for his *Tales* poured in through the summer of 1845. They underlined the analytic force of his "tales of ratiocination," his powers of imagination and description, and the scientific and philosophical acumen of stories such as "A Descent into the Maelström" and "Mesmeric Revelation."

Preparing a new collection of poetry, he returned with gusto to his critical theories. In defiance of the reforming passions of many of his friends, Poe denounced the "heresy of didacticism." Poetry should not teach, and especially not "dogmas or doctrines about what they call PROGRESS." Instead, "the sole legitimate object" of poetry is *beauty*—or, more exactly, "a thirst for a wilder beauty than the earth supplies." This quest for otherworldly beauty activates that "divine sixth sense, which is yet so faintly understood." It is described by phrenologists as "the organ of ideality," what the spiritualist philosopher "Victor Cousin appeals to, as do all attempts to understand the purest aspect of God."

Though his ideal of beauty soared toward the divine, Poe had for years also been developing a worldly view of art as calculating, rational, and mate rial. He sought to supplant Coleridge's distinction between the *fancy*, which merely combines, and the *imagination*, which creates. Poe saw this as a distinction "without even a difference of *degree*" because, he believed, there was no such thing as human creation out of nothing: "All novel conceptions are merely unusual combinations. The mind of man can *imagine* nothing which has not really existed." Even a fantastic creature such as a griffin, a winged lion, "is a mere compendium of known limbs and features—of known qualities." Everything that "seems to be *new*—which appears to be a *creation* of intellect" is ultimately "re-soluble into the old."

Just before the debut of "The Raven," Poe similarly argued that originality is not "a mere matter of impulse or inspiration" by a godlike poet but an artisanal craft. If the human mind could create something new, out of nothing "it would create not only ideally, but substantially as do the thoughts of God." Instead the highest office of the human imagination is to combine elements—in a "chemistry of the intellect" whereby "the admixture of two elements will result in a something that shall have nothing of the quality of one of them."

He refuted "the old dogma, that the calculating faculties are at war with the ideal." As he had argued in "The Murders in the Rue Morgue," the solving of material, rational puzzles and the creation of novel forms of beauty proceed hand in hand: "The highest order of the imaginative intellect is always pre-eminently mathematical, or analytical; and the converse of this proposition is equally true."

Poe's insistence that imagination is equivalent to analysis— and that both involve a material process of combination—might seem to lower art to the level of brute labor. But the status of labor was rising in this period. Where it had once been seen as a lowly curse and abhorred by the aristocracy, work was increasingly seen as a positive virtue. In Europe, workers' movements championed the dignity of labor, while in America, as Tocqueville observed, all men were expected to have a profession. With the Young Americans, Poe was arguing that literary workers, "poor devil authors," deserved as much consideration and government protection for their work as other artisans then claiming their rights. Poe's references to chemistry and mathematics, however, were distinctive. He was arguing that the work of the artist

requires patient labor, yet like science it enlists
the highest analytic faculties of the mind.

Not all agreed. Some of the critics Poe most
respected accused him of overemphasizing the
role of reason in poetry and hinted that his own
work had too much of the "merely mechanical"
to it. James Russell Lowell's biographical article observed that Poe faced the mysteries of
life and death. But while a mystic "dwells in the
mystery, is enveloped with it," Poe stood outside, as a spectator:

He analyzes, he dissects, he watches

—— *"with an eye serene,*
The very pulse of the machine,"

for such it practically is to him, with wheels
and cogs and piston-rods all working to
produce a certain end.

James Russell Lowell,
daguerreotype,
Philadelphia, 1841

The quoted couplet, from Wordsworth's "She Was a Phantom of Delight,"
is significant: Wordsworth and Coleridge had defined poetry in *Lyrical Ballads*
as "the spontaneous overflow of powerful feelings"—an approach Lowell saw
lacking in Poe's highly deliberate poetry.

Similarly, in his 1845 book of criticism, *Conversations on Some of the Old Poets,*
Lowell referred to "a man who is a scholar and an artist—who knows precisely how every effect has been produced by every great writer." This man's
planned effects miss their mark: "the heart passes by his pitfalls, and traps,
and carefully-planned springes," more easily caught "by some simple fellow
who expected the event as little as did his prisoners."

In a review of *Conversations*, Poe objected to Lowell's suggestion that preplanned poetic effects inevitably fail. Perhaps feeling himself personally targeted, he argued that *conscious deliberation* was intrinsic to art. Powerful effects
don't happen by chance: "If the practice fail, it is because theory is imperfect.

Margaret Fuller, daguerreotype by John Plumbe, 1846

If Mr. Lowell's heart is not caught in the pitfall or trap, then the pitfall is ill-concealed, and the trap is not properly baited and set." A few days later the *Herald Tribune*, in an article likely by Margaret Fuller, rushed to defend Lowell and the "natural" in poetry: "We do not believe that traps of any kind, bait them as you may, can ever succeed in the long run, or impose on the heart."

Exasperated, Poe replied with an article titled "Nature and Art" in the *Evening Mirror*. He refused the very distinction that Lowell and the *Tribune* insisted upon. Art is "nothing more than the arranging, the methodizing, the rendering easily available so as to carry into successful application, the suggestions, the laws, and the general intentions" of nature. Rather than two opposed terms, art models itself on and draws its materials from nature, extending nature's hints and tendencies. At its best, Poe held, *art is the continuation of nature by other means.*

The Grip of the Imp

A repeat performance of Poe's lecture on American poets was scheduled at the Society Library for April 18. Under "mingled rain and hail and sleet," the event was canceled. According to an office worker at *The Broadway Journal*, "It was a little thing, it is true, but he was a man easily upset by little things." The next day Poe appeared at work "on the arm of a friend, intoxicated with wine."

After months of sobriety and astounding productivity in New York, Poe drank. Each spree was followed by an excruciating, guilt-racked convalescence. He would then abstain, remorsefully, before drinking again.

The amount of alcohol consumed yearly—and even daily—by early nineteenth-century Americans might earn the admiration of even Faulkner or Bukowski. Drinking played a particularly important role in the life of magazinists: public houses and clubs were places to trade gossip, news, favors, and opportunities. Poe's pathological susceptibility to alcohol made these sites dangerous. He had long periods of nearly total abstinence, including four extremely productive years in Philadelphia. But the new pressure and sociability brought by "The Raven" gave strength to "the Fiend Intemperance."

Meanwhile, Virginia's health grew worse. Poe's friend Thomas Holley Chivers witnessed Virginia "attacked with a terrible paroxysm of coughing," which Maria Clemm ominously decreed "is not a cold." Anne Lynch warned Poe, "Life is too short & there is too much to be done in it, to give one time to *despair*. Exorcise that devil, I beg of you." In the summer of 1845—with endless work and Virginia's ongoing illness, as well as the intoxicating, disorienting prospect of his dreams of literary success coming true—Poe became dangerously unhinged. His life became a pattern of fragile sobriety shattered by disorienting binges that weakened his body and eroded his judgment.

Drinking began to cost him friends. When Lowell arrived in New York to meet his pen pal in person, he found Poe "a little tipsy," with "that oversolemnity with which men in such cases try to convince you of their sobriety. I well remember (for it pained me) the anxious expression of his wife." Maria Clemm sought to dispel Lowell's "wrong impression of my darling Eddie. The day you saw him in New York *he was not himself.*" Poe sniped back defensively, telling Chivers that Lowell was "not half the noble-looking person that I expected to see." Lowell later satirized Poe's "science of verse" in *A Fable for Critics*:

Here comes Poe with his Raven, like Barnaby Rudge—
Three-fifths of him genius, and two-fifths sheer fudge.

By then, inspired by his wife's convictions, Lowell had become the editor of an abolitionist journal. Poe's review of the *Fable* accused Lowell of regional bias, "a bigotry the most obstinately blind and deaf" and described him as "one of the most rabid of the Abolition fanatics" for praising only Boston poets and ignoring southern authors. As David Reese had in *Humbugs of New-York*, Poe here treated Lowell's abolitionism as narrow-minded, belligerent, and uncharitable . . . to slave owners.

Lowell had been warned of Poe's increasingly erratic behavior by Charles Briggs, the co-owner of *The Broadway Journal*, who was tempted to "haul down Poe's name; he has lately got into his old habits." In one slough of despair, while "dreadfully unwell," Poe told Duyckinck of his resolution "to give up the B. Journal and retire to the country for six months, or perhaps a year," to regain his "health and spirits."

Invited to read a new poem at NYU, Poe froze; whether for inspiration or escape, he "remained in a state of intoxication during the week." On the scheduled day, Chivers found him "in bed pretending to be sick" to "avoid the delivering of the Poem"—reading a volume of the orator Macaulay. That same week, Chivers ran into him on Nassau Street, "tottering from side to side" while a man called after him in pickled admiration, "*The Shakespeare of America!*" Before Chivers could steer him home, they ran into Lewis Gaylord Clark, who had recently reviled one of Poe's articles in *The Knickerbocker*. Poe swore, "while attempting to rush away from my hold, that he would attack him."

Poe confided to Chivers that his "*amour*" with Frances Osgood was growing in intensity. *The Broadway Journal* had published Osgood's poem "So Let It Be," describing the affection between Poe and his young wife, "that fair, fond girl, who at your side / Within your soul's dear light doth live." Virginia surely could not "have the heart to chide" either Poe or Osgood for "the ray that Friendship well might give."

Frances Osgood now wanted Poe to meet her in Providence to discuss a nasty rumor being spread by one of her admirers about Poe's passing a forged check. It was a baseless slander, and Poe eventually received an apology. But on July 2, Chivers found him "drest in his finest clothes, going down towards the *Broadway Journal*" on his way to Providence, with "not a dollar in the world—borrowed ten from me."

Meanwhile, Briggs was trying to seize control of the *Journal* by buying out the third owner, John Bisco. With Poe swanning about Providence, the issue due on July 5 did not appear.

The *Journal* returned on July 12, "like a giant refreshed." Briggs abandoned his share, and Poe signed a new contract with Bisco making himself "sole editor" and receiving "half of the entire profits." With Briggs's departure Poe's already enormous labors doubled; his behavior grew even more unpredictable. Even the stalwart Duyckinck was troubled by Poe's binges. He wrote in his journal, "There is Poe with coolness, immaculate personal cleanliness, sensitiveness, the gentleman, continually putting himself on a level with the lowest blackguard through a combination of moral, mental and physical drunkenness."

In *Graham's* that month, Poe published a mordant diagnosis of his own self-destructive behavior in a tale, "The Imp of the Perverse." As in "The Black Cat," he started by noting phrenologists' failure to locate the organ responsible for the human inclination to do the thing most contrary to one's interest—such as the impulse to leap when admiring the view from the top of a cliff.

This drive to self-destruction, he insisted, was a moral fact, just as strong as and sometimes stronger than the will to do good and stay alive. Though Poe made few overt statements about Christianity, his writing on "perverseness" reveals a Calvinist sense of predestined damnation for most if not all of humanity. He found the optimism of social reformers and improvers naive; the technical and scientific progress of recent decades could never change the human mind's deep-rooted tendency toward cruelty and self-destruction.

In "The Imp of the Perverse," Poe illustrated his point by detailing a murder carried out with aesthetic perfection. The tale resembles De Quincey's 1827 *Blackwood's* essay, "On Murder Considered as One of the Fine Arts"—until Poe's brilliant assassin succumbs to an irrepressible urge to brag about his plan. Perverseness will out.

Whether a fanciful explanation, an apology, or a sincere attempt to overcome deeply conflicted motives, "The Imp of the Perverse" showed Poe using philosophy—and luridly imaginative storytelling—to try to make sense of his own irrational, seemingly unstoppable impulse to set his life on fire.

Poets Behaving Badly   In late summer, he told his cousin Neilson
that Virginia "has been, and is still, in pre-
carious health." Meanwhile, he was single-handedly editing *The Broadway Journal* and writing most of its copy, borrowing money to keep afloat. His flirtatious exchange of poems with Frances Osgood continued as he reviewed the proofs of his forthcoming poetry collection. He also moved his family from East Broadway to Amity Street near Washington Square. He accepted an invitation to give a lecture in October at the Boston Lyceum for fifty dollars. Juggling so many plates, he found himself unable to produce a new poem.

The night of the speech, the Boston Lyceum was packed. Poe's "prose-writings had been eagerly read, at least among college students, and his poems were just beginning to excite still greater attention." The evening began with a two-and-a-half-hour-long historical lecture by the statesman Caleb Cushing. Finally Poe took the lectern, opening with "a long and prosy preface" against didactic poetry—his chief complaint against Boston poets. He then recited a "rather perplexing" poem. Though he called it "The Messenger Star," it turned out to be "Al Aaraaf," his youthful poem of angelic astronomy, originally prefaced by "Sonnet—To Science."

Some audience members, confused by its obscure stanzas and unearthly sounds, left early, having had "*poetry* enough for one night." Yet walking in the dark back to Cambridge, a group of Harvard undergraduates "felt that we had been under the spell of some wizard." A review the next day praised the "elegant and classic production, based on the right principles, containing the essence of *true* poetry, mingled with a gorgeous imagination."

Afterward, Poe was "persuaded" by his hosts to drink champagne. He admitted that the poem had already been published—when he was seventeen. Word got out. Cornelia Wells Walter, editor of the *Evening Transcript* and a friend of Longfellow's, carped, "A poem delivered before a literary association of adults, as written by *a boy!* Only think of it! Poh! Poh!"

He might have left it there, writing the evening off as a politely received if uninspired performance. Instead, seized by the imp, he recast his lecture as a deliberate provocation or hoax. Back in New York he taunted, "We have been quizzing the Bostonians, and one or two of the more stupid of their editors and editresses have taken it in high dudgeon." Walter snapped back that his poetic abilities must have been deteriorating since the age of ten, "his best poems having been written before that period."

Poe lashed out at Walter as an "adorable creature" and a "pretty little witch" seeking "revenge for something that we did to Mr. Longfellow." He then let loose on "the Frogpondians" (for the croaking that comes from Boston Commons): "The Bostonians are very well in their way. Their hotels are bad. Their pumpkin pies are delicious. Their poetry is not so good." After several more rounds of ignoble sniping—with papers from New York to Charleston taking sides—several Bostonians pressured Duyckinck to call Poe off from this "perfectly damnable" course against "one of the most respectable young ladies in Boston." Though Poe continued to lampoon "the Frogpondians" and the fog of transcendentalism he attributed to them, he ceased to attack Walter.

Fresh hells awaited him in New York. Bisco was giving up: he offered to sell Poe his share of *The Broadway Journal* for $150. Poe scraped the sum together after desperate pleas to English, Duyckinck, Greeley, John Pendleton Kennedy, Chivers, and even Rufus Griswold.

The new masthead of *The Broadway Journal*, printed on October 25, listed Edgar A. Poe as "Editor and Proprietor."

Poe's dream had come true. He at last had his own magazine.

The triumph was bitter and brief. A young Walt Whitman met Poe at the office at 304 Broadway and found him "very kindly and human, but subdued, perhaps a little jaded." Poe told Chivers, "I have had a tough time of it—making all kind of maneuvres." He was overwhelmed: "I have to do *everything* myself—edit the paper—get it to press—and attend to the multitudinous *business* besides . . . The moments I now spend in penning these words are gold themselves."

By the start of December, Poe was forced to find a new backer and complained that "every body is at us—little dogs and all," including a Princeton theological student who reviewed "The Imp of the Perverse" as a "humbug philosophical" that chased its idea "from the wilderness of phrenology into that of transcendentalism, then into that of metaphysics." The imp struck again in late December, with "one of his fits of drunkenness" forcing him to print the *Journal* with a column and a half left blank.

His financial backer, Thomas Lane, decided to "close the publication entirely." The final issue appeared on January 3, 1846. Cornelia Wells Walter danced on its grave: "To trust in friends is but so so / Especially when cash is low / The Broadway Journal's proved 'no go'— / Friends would not pay the pen of POE."

Soon he was drawn into a more personal hornet's nest, full of "jealousies and heart-burnings." On a visit to Poe's home, Elizabeth Ellet overheard Frances Osgood and Virginia laughing together—about, she believed, an ardent letter Ellet had sent to Poe. Following the perverse principle of accusing others of one's own misdeeds, Ellet then spread rumors about compromising letters from Osgood to Poe. Her jealous insinuations "became the town talk, at least among literary people."

Ellet persuaded Margaret Fuller and Anne Lynch to join her in confronting Poe. They appeared on his doorstep and demanded he return all of Osgood's letters to him, ostensibly to protect her reputation. Poe returned the bundle, with a parting shot that Ellet "had better take care of her own letters," some of which "compromised herself."

Newly indignant, Ellet sent her brother, a colonel, to demand justice. Poe rushed to the rooms of Thomas Dunn English, asking to borrow a pistol for self-defense. English refused, denying that Poe had ever received any letters from Ellet. With tempers flaring and honor at stake, Poe lunged at English, who "dealt him some smart raps on the face." The colonel was pacified, but Poe and English became sworn enemies. Though Poe and Frances Osgood still exchanged poems in the press, they never met again.

Throughout these entanglements Poe followed up miscalculations and missteps with defiant self-justifications, dangerously raising the stakes until he dramatically overstepped a moral line. Alcohol, anxiety, and a fragile grandiosity played their part. As he put it in a moment of clarity, "It is our intense sense of self which debases, and which keeps us debased." In the public dramas of American letters, accelerated and magnified in the fishbowl of New York, Poe was learning the dangerous lesson of Byron and later celebrities. There's no success like failure, and people will gladly pay (whether applauding, jeering, or weeping) to watch an artist self-destruct.

For Valentine's Day 1846, Virginia pleaded for relief in a sweetly affectionate poem she wrote for her husband:

Ever with thee I wish to roam—
Dearest my life is thine.
Give me a cottage for my home
And a rich old cypress vine,
Removed from the world with its sin and care
And the tattling of many tongues.
Love alone shall guide us when we are there—
Love shall heal my weakened lungs;
And Oh, the tranquil hours we'll spend,
Never wishing that others may see!
Perfect ease we'll enjoy, without thinking to lend
Ourselves to the world and its glee—
Ever peaceful and blissful we'll be.

Underlining her devotion, Virginia spelled out "Eddy's" name in the first letter of each line. Poe took the hint: he moved the family from Amity Street to Turtle Bay on the East River; by May, they relocated twelve miles north to the countryside at Fordham, now the Bronx.

The country air of their "little cottage at the top of a hill," rented for a hundred dollars per year, would be a salve to Virginia's declining health. "There was an acre or two of green sward, fenced in about the house, as smooth as velvet and as clean as the best kept carpet. There were some grand old cherry-trees in the yard that threw a massive shade around them. The house had three rooms as well as a 'piazza' that was a lovely place to sit in in summer." To Virginia and Mrs. Clemm's relief the new residence removed Poe from literary circles and gossip.

Trouble found him still: a trip that spring to Baltimore ended in a spree, followed by sickness. Mary Hewitt wrote to wish him a quick recovery: "Our charming friend Mrs. Osgood, and myself, indulge often in talking of you and your dear wife . . . All Bluedom misses you from its charmed circle, and we often ask when we are to have Mr. Poe back again among us."

A young poet saw him waiting for the train back to Fordham one dreary afternoon. "A heavy shower had come up suddenly, and he was standing under an awning. I had an umbrella, and my impulse was to share it with him on his way home, but something—certainly not unkindness—withheld

me. I went on and left him there in the rain, pale, shivering, miserable . . .
There I still see him, and always shall,—poor, penniless, but proud."

Resolutions and Rebuttals   During the very months of Poe's
sudden fame and the turmoil that
accompanied it, the press was filled with the controversy over *Vestiges of
the Natural History of Creation* and its anonymous author. *The Broadway Journal* alone mentioned the work at least five times in 1845. Poe defended it
against detractors in April, taking a hostile reviewer to task: "We should
be inclined, for example, to think far more highly than he, of the 'Vestiges
of Creation.' If not written by Dr. Nichol, this work is at least worthy
that great man." Besides the astronomer John Pringle Nichol, a great advocate of the nebular hypothesis, several other prominent figures were
suspected of being the book's author, including the phrenologist George
Combe, the publisher Robert Chambers, the geologist Charles Lyell,
the philosopher Harriet Martineau, Charles Darwin, and the mathematician Ada Lovelace. In December, Poe noted the publication of a new
book, *On the Harmony of the Comprehensible World*, by another of the work's
suspected authors, a Tory politician named Richard Vyvyan. *Vestiges* was
the talk of Broadway.

Remaining anonymous, the author of *Vestiges* answered critics in 1846
with a sequel, *Explanations*. One objection to *Vestiges'* theory of the evolution of
the solar system had come from new astronomical evidence. In 1845 the Irish
astronomer William Parsons, the Earl of Rosse, began observations with the
"Leviathan of Parsonstown," a fifty-four-foot-long telescope he had built,
with a six-foot-wide reflecting lens. As a demonstration of its power, at the
BAAS meeting in 1845, Rosse passed around stunning drawings of observations of a nebula in spiral form, later called the whirlpool nebula; Herschel's
less powerful telescope in South Africa had shown the same object as simply
two disconnected clouds.

Rosse claimed that when his colossal instrument was directed at the
Orion nebula—one of the central examples used by proponents of the nebular hypothesis—one saw "groups or clusters" of stars, and not the expected
clouds of gas. He declared that "there can be little, if any, doubt as to the re-

SPIRAL NEBULA.

Spiral nebula, based on Lord Rosse's drawing, with black and white inverted, 1848

solvability of the nebula." Many seized on Rosse's confidence in the "resolution of the nebula" into stars as a refutation of the nebular hypothesis—the
backbone of *Vestiges'* evolutionary cosmology.

In response, *Explanations* argued that Rosse's resolution of *some* nebulae
in no way implied the resolution of *all* nebulae. The only observations that
might put the hypothesis in danger would be of "nebulae comparatively
near, which no increase of telescopic power" had yet resolved into stars.
Explanations went on the offensive against established "men of science."
It was true "that nearly all the scientific men" rejected *Vestiges'* theory. So
what? The "ability of this class to give, at the present time, a true response
upon such a subject, appears extremely challengeable," the author declared.
Excessive specialization had made scientists unable to judge any general
theory: they were "engaged, each in his own little department of science,
and able to give little or no attention to other parts of that vast field." As a
result, "the whole scientific class are disqualified." Instead, the author appealed to "another tribunal," going over scientists' heads to the court of
public opinion.

With *Explanations*, the scientific "class" was even more outraged than before.

In Britain, David Brewster fumed at the author's "unwise and ungenerous expressions." The botanist Asa Gray, formerly superintendent of the New York Lyceum and, since 1842, professor of natural history at Harvard, defended "the professional establishment of science," which condemned *Vestiges'* view of humanity as evolving from electro-vitalist slime. The "unprofessional reader" ought to trust these professionals' testimony "in respect to facts which they are the most competent witnesses of, and inferences of which they have the best means of judging." He appealed to the authority of European experts (listing Cuvier, Whewell, Lyell, and Murchison) against those without credentials, institutions—or a name.

Joseph Henry was annoyed by his students' fascination with *Vestiges*, insisting that a machinelike universe would still have to be "the production of an intelligent agent." While a young Unitarian theologian admired *Vestiges'* scope and found the nebular hypothesis a "magnificent cosmogony," he warned against science's "subtilizing materialism," which steadily reduced the role of God and spirit.

Boston's *North American Review* summed up the central issue: "We have only to decide whether it is more likely that the complex system of things in the midst of which we live—the beautiful harmonies between the organic and inorganic world, the nice arrangements and curious adaptations that obtain in each . . . was built up, and is now sustained by one all-wise and all-powerful Being, or by particles of brute matter, acting of themselves, without interference or control."

Many saw *Vestiges'* use of religious language as a cover for its true aim: to undermine religion and support materialism. A professor of sciences at Dickinson College saw the author's allusions to God as "subterranean approaches to the citadel of his reader's faith"—a typical "atheistic strategy ever since Epicurus." If *Vestiges'* cosmos had no need for an active Governor and Preserver, it might do without God altogether: "Why then should God exist for no purpose but to cause the existence of that which could have existed eternally as easily as himself? There is then no God, or matter is God."

Either possibility—atheistic materialism or pantheism—was abhorrent. Protestant scientists and theologians saw the universe moved by the loving spirit of the Creator, who watchfully governed and preserved it, adjusting it at times the better to realize his purposes. But was it actually, as *Vestiges* suggested, a cold, purely material machine, designed and abandoned once it was built? Was it perhaps merely the result of mindless chance? Or might God and the universe somehow be identical?

The ongoing *Vestiges* controversy also raised unsettling questions about the place and politics of science. Who should be allowed to address such fundamental issues, or decide on the answers? With what methods, with what standards of evidence, and in what forums? And before which jury or judge?

Opening Poetry's Black Box   Now in semi-exile in Fordham with Maria Clemm and Virginia, Poe was removed from the rounds of editorial offices, salons, and public houses. He could look back on the previous months in bewilderment and horror.

His attempts to master the machinery of literary publicity and New York society had veered out of control and backfired. His editorship and ownership of *The Broadway Journal* had ended disastrously, costing him time, health, and dignity. His bouts of drinking, followed by obnoxious behavior, turned allies into enemies. The affection for him in "Bluedom" exploded into a fracas of insult and resentment.

Yet he would not easily accept defeat. He kept writing and publishing. He defended and extended his intellectual territory in a mode he knew well: poetic criticism.

Poe's first poetry collection in more than a decade, *The Raven, and Other Poems*, was published in November 1845, five months after his *Tales*. Reviews were strong and positive. Yet critics repeatedly raised one complaint: that in his poems, inspiration and feeling were subservient to reason and technique. Lowell had ambivalently described Poe's poems' "wheels and cogs and piston-rods all working to produce a certain end." *The Harbinger* thought the *Poems* "have more of effect, than of expression" and "succeeded in the art rather at the expense of nature." Thomas Dunn English, writing before their brawl, observed that Poe seeks, "by means of the mechanism of verse, to throw beauty around a simple narration," where "much of the effect depends upon the mode of construction." One critic felt "The Raven" was "more remarkable for its mechanical construction than for its spirit of poetry."

This impression—of analysis dominating inspiration, machinery replac-ing soul—was reinforced by the selection Duyckinck had made for Poe's re-cent collection of tales; Poe complained that the editor "has what he thinks a taste for ratiocination, and has accordingly made up the book mostly of analytic stories." A more representative selection, he thought, would have shown his works' "wide *diversity and variety*." The related critique of the *Poems*—that Poe's poetry was more of the head than the heart, more of the mechanic than the bard—irked him even more.

In April 1846, in *Graham's*, Poe replied to these complaints with a critical manifesto: "The Philosophy of Composition." In it he appeared not only to agree with his critics but also to go even further. Presenting a detailed method for writing a poem—much as he had, jokingly, for a short story in "How to Write a Blackwood Article"—he cast himself as a poetic engineer, even as a poetic automaton. The essay has become legendary: an object of baffled fascination, mocked by some and revered by others.

Poe begins by inviting the reader to have a "peep behind the scenes" into the workshop of the poet. He promises to explain the process by which he com-posed his most famous work, "The Raven." Against poets who "prefer having it understood that they compose by a species of fine frenzy," he offers a view of "the wheels and pinions" of the poetic machinery. He will show his critical principles at work: every part of a poem should contribute to a single, powerful effect; the poem should be judged not for its message but as an aesthetic object.

The essay starts by dismissing "as irrelevant to the poem *per se*" the circum-stances that led Poe to create a popular poem (grief, longing, or poverty, for instance?). He then presents the considerations that arose at each point and his means for resolving each challenge. The poem must not exceed one hun-dred lines. It should treat the ideal "Beauty." It should have a refrain adaptable to "produce continuously novel effects," concentrated in a single melodious word; thus he arrives at "Nevermore," with "long *o* as the most sonorous vowel, in connection with *r* as the most producible consonant." Its tone should be melancholy, "the most legitimate of all the poetic tones." Keeping in mind "the object *supremeness*, or perfection," he inquired, "What, according to the *universal* understanding of mankind, is the *most* melancholy?" The answer, "ob-viously," is death, and better yet the death of a beautiful woman: "unquestion-ably, the most poetical topic in the world." As the optimized formula unfolds, the author-machinist at each step selects the element perfectly suited to the effect, until "The Raven" springs irrepressibly to life.

With "The Philosophy of Composition," Poe gave "The Raven" a new injection of novelty—much like a behind-the-scenes, "making-of" feature released after a successful film. By insisting on the rule-bound, mechanical aspects of his creation, Poe was refuting the romantic definition of poetry, advanced by Wordsworth and Coleridge and defended by Lowell and Fuller, as "the spontaneous overflow of powerful feelings."

By walking the reader through the "wheels and pinions" of his craft, however, Poe was also taking a stance similar to that of Maelzel, the exhibitor of the chess-playing Turk Poe had debunked back in Richmond. Maelzel showed his viewers the turning gears inside the automaton and went to great lengths to make his machine appear more mechanical than it was, to distract them from the human hiding inside it.

Was Poe's essay, then, a hoax akin to Maelzel's chess-playing automaton? Was he exaggerating the mechanical and deliberative aspects of his creation—giving it an "*air* of method"—to divert attention from the fallible, improvising, emotional, and inspired creator hiding inside it? Many have read it this way. Poe's French champion Baudelaire held that Poe was making himself out to be "much less inspired than he naturally was," but "genius is always allowed a bit of charlatanism." We might see Poe as "pokerishly" throwing his critics' jibes back at them in exaggerated form: "You think my poetry sounds mechanical? Wait until you hear how I engineered it!"

Yet at the same time, Poe was stating views on poetic composition that he had earlier expressed with undoubted sincerity, though now more starkly than before: the necessity of method, analysis, and "writing backward" from a chosen end to produce a unified effect.

Poe seemed dead serious and at the same time fully conscious that what he was proposing—a completely preplanned, rule-bound, self-writing poem, produced as if by a machine—was preposterous, nearly unthinkable.

In an uncanny anticipation of Poe's essay, in July 1845 *The Illustrated London News* introduced a "Machine for Composing Hexameter Latin Verses," displayed at London's Egyptian Hall. Built in Bridgewater by an heir to the Clark's shoe fortune, it resembled a "small bureau book-case." At the pull of a lever, cylinders turned within it and letters fell into line, forming Latin verses that

* The machine is described by the inventor as neither more nor less than a practical illustration of the law of evolution. The process of composition is not by words already formed, but from *separate letters*. This fact is obvious; although some spectators may, probably, have mistaken the *effect* for the *cause*—the *result* for the *principle*, which is that of Kaleidoscopic evolution; and, as an illustration of this principle it is that the machine is interesting—a principle affording a far greater scope of extension than has hitherto been attempted.   The machine contains *letters* in alphabetical arrangement.   Out of these, through the medium of *numbers*, rendered tangible by being expressed by indentures on wheel-work, the instrument selects such as are requisite to form the verse conceived ; the components of words suited to form hexameters being alone previously calculated, the harmonious combination of which will be found to be practically interminable.

THE EUREKA.

"*The Eureka,*" *poetry machine, 1845*

could be read through a window at its center—to the tune of "God Save the King." In this age of mechanical improvements and calculating engines, here, inevitably, was a poetry-writing machine.

Its inventor insisted on the machine's philosophical implications. In the midst of the *Vestiges* controversy, John Clark saw his invention as "a practical illustration of the law of evolution." It arranged and rearranged a limited number of elements—words and letters—according to fixed mechanical laws to form poetic verses, from a "practically interminable" set of possible "harmonious combinations," like so many novel species evolved according to a general organic form. Clark called its principle "Kaleidoscopic evolution," after David Brewster's optical toy, which also combined finite elements into an infinite variety of arrangements. At the front of the machine was "a large Kaleidoscope, which regularly constructs a splendid geometric figure" that appeared, in a new configuration, "at the precise time when the line of verse is conceived, previous to its mechanical composition"—a visual analogy of the poetry machine's process of creation, both random and law-bound.

Did Poe see this article before writing "The Philosophy of Composition"? It appeared in a popular magazine, on the same page as an illustrated story about a fire that destroyed the Pennsylvania Academy of the Fine Arts—a tragedy in his old neighborhood that would have caught his eye. Even more suggestively, the name of this poetic machine and illustration of "the law of evolution" was the Eureka—the title Poe would give his own evolutionary cosmology, three years later.

With "The Philosophy of Composition," Poe was already touching on the cosmological debates around the *Bridgewater Treatises* and *Vestiges*. "It is my *design*," he wrote, "to render it *manifest* that no one point in its composition is referable either to accident or intuition." Just as the *Vestiges* author said of the universe, and as Charles Babbage argued with reference to his calculating engine's pre-programmed miracles, Poe insisted that *his* artwork was entirely the product of a progressive, uniform law. It unfolded "step by step, to its completion, with the precision and rigid consequence of a mathematical problem."

Readers might take this as an enormous *diminishment* of the role of the poet. It seems to do away with the romantic claim to godlike, creative inspiration—just as critics saw *Vestiges* doing to the cosmos, by promoting the "horrible vision" of a material, mechanical world drained of divinity.

But one might also read "The Philosophy of Composition" in line with *Vestiges*' claim to be offering a *more* reverent view of the creator. *Vestiges*, like Babbage's *Ninth Bridgewater Treatise*, argued that a creator who was able to plot out in advance every step of his work's development would be far more admirable than one who was forced to adjust, alter, and repair it along the way, as mainstream natural theologians held.

At first, "The Philosophy of Composition" appears to be a fierce denial of romantic notions of inspired, spontaneous creativity. If we read it as a hoax, by contrast, it becomes an over-the-top (and recognizably romantic) satire on attempts to mechanize every domain of life. But there is another possibility. When we take into account its cosmological resonances—including its strong resemblance to *Vestiges* and the Eureka machine, a poetic automaton intended to illustrate law-bound evolution—it hints at even more miraculous claims to creative omniscience.

Or perhaps it holds all these contradictory truths at once, which makes it something more. *You decide.*

War of the Literati

In Fordham with Muddy, Poe lavished attentions on Virginia. He also looked after the tortoiseshell Catterina, cherry trees, sunflowers, dahlias, and a pet bird. Poe himself was often seriously ill; Dr. John Francis diagnosed him with heart disease. But he continued to write.

He published the revenge fantasy "The Cask of Amontillado." With the promise of a taste of rare sherry, the insulted Montresor leads his enemy, Fortunato, to an agonizing live burial. Poe also advanced directly against his literary opponents. He had been planning a book on American poetry, *American Parnassus*, but he now concentrated on "The Literati of New York City," a series he published in the Philadelphia magazine *Godey's* from May to October 1846.

He praised his friends—recognizing, for instance, Margaret Fuller's "high genius," though he worried that her book *Woman in the Nineteenth Century* failed to consider "the intention of the Deity as regards sexual differences." He also ravaged his foes, torching his former partner Charles Briggs as derivative, vain, and "grossly uneducated," describing Hiram Fuller's *Mirror* as "a desert of stupidity," rapping Lewis Gaylord Clark as pointless and indistinct: "he is as smooth as oil," and his "forehead is, phrenologically, bad—round and what is termed 'bullety.'"

Toward Thomas Dunn English he directed lethal condescension: "No spectacle can be more pitiable than that of a man without the commonest school education busying himself in attempts to instruct mankind on topics of polite literature." He thoughtfully advised, "Mr. E. is yet young—certainly not more than thirty-five—and might, with his talents, readily improve himself at points where he is most defective. No one of any generosity would think the worse of him for getting private instruction." He infuriatingly sealed his judgment as impartial: "I do not personally know Mr. English."

Briggs replied in the *Mirror* with a caricature of Poe in an upstate asylum; he lopped five inches from his height and described his eyes as "gray, watery, and always dull." English took to *The Morning Telegraph* to detail Poe's crimes: the fight over Elizabeth Ellet's letters, an unreturned loan for *The Broadway Journal*, and the previous year's (debunked) forgery accusation. Poe was "an assassin in morals" and "a quack in literature." In the *Mirror*, Hiram Fuller sketched Poe "in a condition of sad, wretched imbecility, bearing in his feeble body the evidences of evil living," and caricatured Maria Clemm as "an aged female relative, who was going a weary round in the hot streets, following his steps to prevent his indulging in a love of drink." Poe "was evidently committing a suicide upon his body, as he had already done upon his character. Unhappy man!"

In a printed reply to these attacks, in *The Spirit of the Times*, Poe acknowledged his "weakness" for drink but cited Dr. Francis's verdict that his irregularities were "the *effect* of a terrible evil rather than its cause." He would have ignored English's allegations were it not for the "criminal" accusations of taking money on false pretense and forgery; for these slanders, he would exonerate himself "in a court of justice." He filed a declaration of grievances against Hiram Fuller in July. The trial was repeatedly postponed, draining his funds, though he eventually won two hundred dollars and legal fees.

Virginia's health was growing worse; Poe was exhausted and ill. "The Literati of New York City" continued to pay, but it made him more enemies than friends. By summer, cruel whispers made their way across the country. In St. Louis, the *Daily Reveille* heard that "Mr. Edgar A. Poe, the poet and author, has become deranged, and his friends are about to place him under the charge of Dr. Brigham, of the Insane Retreat at Utica. We sincerely hope that this is not true." Even his old friend Snodgrass in Baltimore repeated the rumor.

Thomas Dunn English's literary satire "1844," published in Fuller's *Mirror*, depicted the author of "The Humbug and Other Tales" in an asylum, begging for coins and reading out in "a sing-song voice" an article that used "the calculus of probabilities" to prove "that Mr. Carlyle is an ass." The blows kept coming: Clark's *Knickerbocker* published an "Epitaph on a Modern 'Critic,'" presenting Poe lying in a grave, "cold as his muse, and stiffer than his style," killed by either "excessive genius or excessive gin!"

William Gilmore Simms, a member of the Young America set, warned Poe, "You are now perhaps in the most perilous period of your career—just in that position—just at that time of life—when a false step becomes a capital error—when a single leading mistake is fatal in its consequences." Literary squabbles "vex your temper, destroy your peace of mind, and hurt your reputation." He urged him, "Change your tactics & begin a new series."

Poe's reputation was battered, but it might recover. In spite of Fordham's trees, its river and healthy air, and the care of her mother and husband, Virginia's health would not.

PART V

To the Plutonian Shore

And this was the reason that, long ago,
In this kingdom by the sea,
A wind blew out of a cloud, chilling
My beautiful Annabel Lee;
So that her highborn kinsmen came
And bore her away from me,
To shut her up in a sepulchre
In this kingdom by the sea.

—Poe, "Annabel Lee"

Hekla Volcano, Iceland, hand-colored magic lantern slide,
nineteenth century

15

A Spectacle for Angels

In a Bed of Misery During the deadly cholera epidemic of the 1830s, a man with a superstitious turn of mind retreats to a relative's home north of New York City to escape the disease. Through a window he sees a terrifying sight: on the far bank of the Hudson stands a "monster of hideous conformation, larger than a ship, with huge wings covered with metal scales and a mouth at the end of a sixty foot long proboscis, with hair at its root, and a crystal prism on either side of this nose."

Horror strikes: "I doubted my own sanity—or at least the evidence of my own eyes; and many minutes passed before I succeeded in convincing myself that I was neither mad nor in a dream." He tremulously points out the monster to his host, convinced it is an omen of imminent death.

His companion smiles, casually musing that "the principal source of error in all human investigations [lies] in the liability of the understanding to underrate or to overvalue the importance of an object through mere misadmeasurement of its propinquity." Picking up a book of natural history, he reads out "a school-boy account of the genus *Sphinx*, of the family *Crepuscularia*, of the order *Lepidoptera*, of the class of *Insecta*." Sitting where the narrator had sat when he saw "the monster," the host points to a small winged insect: the death's-head sphinx moth described in his book. The horrific vision is a simple distortion, a hallucination produced by misconstruing distance and proportion, intensified by fear.

Fear stretches the senses—feeds itself—creates monsters out of mites. It can be dispelled by listening to facts, adjusting perspectives to their proper scale. Poe offered this reassuring Enlightenment myth, with its echo of the "animalcules" magnified by the oxyhydrogen microscope in scientific magic lantern shows, in "The Sphinx," first published in 1846.

Another cholera epidemic would strike in 1848, but in the year that Poe published "The Sphinx"—this tale of contagious disease, terror, and omens hovering over the Hudson—he was living through real events just as fearsome. In Fordham, monstrous threats were bearing down on him and Virginia.

In November 1846, Maria Clemm told the health reformer Mary Gove that Virginia was "very ill" and "dying of want." Gove took the train north and found a pitiful scene: everything was "scant and poverty-stricken," with "no clothing on the bed."

Virginia had "the dreadful chills that accompany the hectic fever of consumption"; her "pale face, her brilliant eyes, and her raven hair gave her an unearthly look." When she coughed, it was obvious "that she was rapidly passing away." She lay on a straw bed, "wrapped in her husband's great-coat, with a large tortoise-shell cat on her bosom." These were "the sufferer's only means of warmth, except as her husband held her hands, and her mother her feet."

In New York, Gove contacted Marie Louise Shew, a kindhearted nurse and medical writer who started up a subscription for donations. She brought the Poes sixty dollars in gifts a few days later. Over the next weeks, Shew visited regularly "and ministered to the comfort of the dying and the living."

By December, word of the Poes' miserable state had spread. The *Morning Express* included a notice, ILLNESS OF EDGAR A. POE, describing both Poe and his wife as "dangerously ill with the consumption" and "barely able to obtain the necessaries of life." Even Hiram Fuller said, "His friends ought not to wait for publishers to start a movement in his behalf, and if they do not, we, whom he has quarrelled with, will take the lead." The young Walt Whitman (whose essay "Art-Singing and Heart-Singing" Poe published in *The Broadway Journal*) wrote in the *Brooklyn Daily Eagle* of Poe's "brain fever" and his condition, "without money and without friends, actually suffering from disease and destitution."

Nathaniel Willis took charge: he published a request for aid for "one of the most original men of genius, and one of the most industrious of the literary profession of our country." He would forward any "generous gift" to Poe and went so far as to urge that the reforming zeal of the age be directed to founding a "hospital for disabled labourers with the brain." Hiram Fuller suggested balancing it with "an asylum for those who have been ruined by the diddlers of the quill." Jane Locke, from Lowell, Massachusetts, sent Willis "An Invocation for Suffering Genius," inspired by Poe's state, while Poe's Bostonian opponent Cornelia Wells Walter reported that "Poe and his wife are both down upon a bed of misery . . . In a christian land, where *millions!* are wasted in a heathenish war, in rum, in toasting and feasting swindlers." She piously observed that "no object of humanity should be permitted to die of hunger," urging Poe to a "*reformation of habits*" and helpfully reminding him "how much of his pecuniary distress he has brought on through the indulgence of his own *weaknesses.*"

Moved by his friends' generosity, but mortified to be placed on "a level with the common objects of public charity," Poe published a statement "of what is true and what erroneous": "That my wife is ill, then, is true"; he acknowledged having been "long and dangerously ill," but he planned to answer "the innumerable paragraphs of personal and literary abuse with which I have been latterly assailed" once he returned to health. His illness left him without money, but never beyond "the extent of my capacity for suffering." Nor was he "without friends"—a "gross calumny" that "a thousand noble-hearted men would have good right never to forgive me for permitting to pass unnoticed and undenied."

He reassured his friends, "The truth is, I have a great deal to do; and I have made up my mind not to die till it is done."

An Intermediate or Secondary Nature

As Virginia lay dying, Poe revised and expanded his earlier article "The Landscape Garden" into a haunting aesthetic testament. He named it "The Domain of Arnheim."

The protagonist, simply called Ellison, is an advocate of philosophies of improvement and progress and heir to a gigantic fortune. He tries his hand at many kinds of art before realizing that "the august purposes for which the Deity had implanted the poetic sentiment in man" as well as "his own destiny as a poet" would be best fulfilled by becoming a landscape artist—a gardener and artificer working with stones, soil, trees, flowers, and water.

To define his aesthetic, he reflects on the irregularities we see in nature: "No position can be attained on the wide surface of the *natural* earth, from which an artistical eye, looking steadily, will not find matter of offence in what is termed the 'composition' of the landscape." If, as natural theologians argued, the earth's surface was arranged to fulfill "at all points man's sense of perfection in the beautiful, the sublime, or the picturesque," why do these "geological disturbances," these inharmonious patches of "form and color-grouping" exist, marring our aesthetic pleasure?

Ellison suspects these apparent flaws in the weave of nature are divinely designed, but for eyes other than our own. Natural landscapes that *our* eyes perceive as "unpicturesque" might be perfectly adapted to the eyes of angels, to whom "our disorder may seem order." He imagines that God arrayed the earth's vast surface, "the wide landscape-gardens of the hemispheres," for the angels' "death-refined appreciation of the beautiful." Apparent irregularities are adapted for the more perfect, more expansive senses of beings not limited by bodies.

This notion fires his artistic ambition. He abandons the thought of designing a landscape according to *merely human* ideas, such as "nice relations of size, proportion, and color"—vague terms that debase divine creation to the standard of humans. Nor will he simply re-create a wilderness, a state too dispersed, vast, and irregular to move us.

Avoiding both the *too orderly* order that human art typically imposes on nature and the hidden, imperceptible order of wilderness, Ellison imagines *an order in between*, an art that allows "the Almighty design to be *one step depressed*—to be brought into something like harmony or consistency with the sense of human art." This "intermedium" between the plainly human and the incomprehensibly divine would be a material landscape that allows humans to feel and sense themselves moving from the bounds of ordinary bodily experience toward the ether-like perception enjoyed by angels.

Those who experienced such a landscape would feel a design at work but just beyond reach, without "the harshness and technicality of Art." The rules and reasons behind the composition would be sensed and yet somehow transcendent, giving an impression of "spiritual interference"; material forms would provide a bridge leading to experience beyond bodily limits. Such a landscape's "united beauty, magnificence, and *strangeness*" would create the overwhelming impression of "an intermediate or secondary nature," appearing as "the handiwork of the angels that hover between man and God."

Ellison searches for a place to realize his vision—the Pacific Islands are too remote, while a site comparable to the view from the volcano Mount Etna is too exposed—before finding the Domain of Arnheim. The place now enjoys "a species of secret and subdued if not solemn celebrity."

Ellison's name recalls the magnetizer John Elliotson. Like a mesmeric lecture, Poe's tale began with an exposition of principles, then led its audience into a vivid dream—a hypnotic, immersive landscape crafted of words. The tale's second half, describing a visit to the Domain of Arnheim, resembles a magic lantern show with a sequence of dissolving views; it depicts the realization of creative spirit in matter, elevating the interplay of the human senses to the subtlety and splendor of an angel's mind.

Weird Symmetry   Arnheim is a few hours from the nearest city. The visit starts early in the morning, with the solitary traveler boarding a small boat on a river, pulled by an invisible force past green meadows dotted with sheep, conveying a sense of "merely pastoral care." Undergoing "a thousand turns," the vessel seems "imprisoned within an enchanted circle"; entering a steep gorge with "an air of funereal gloom," the passenger is "enwrapt" in "an exquisite sense of the strange." The "thought of nature still remained, but her character seemed to have undergone modification." The landscape again shifts, assuming odd proportions of "thrilling uniformity" and "weird symmetry."

The boat enters a basin surrounded by flowery hills, a "sea of odorous and fluctuating color" exuding "a miraculous extremeness of culture." The flowers seem to be cared for by "a new race of fairies, laborious, tasteful, magnificent, and fastidious." The hills covered in flowers resemble a jeweled waterfall, "a panoramic cataract of rubies, sapphires, opals and golden onyxes, rolling silently out of the sky."

As the setting sun appears between two hills, the passenger changes boats, entering a crescent-shaped ivory canoe "stained with arabesque devices," setting off to "soothing but melancholy music." Velvety green plateaus alternate with chasms and forests. After more windings "with gentle but gradually accelerated velocity," the vessel approaches a gigantic gate of burnished gold, "reflecting the direct rays of the now fast-sinking sun," wreathing the forest in flames. The gates slowly open, letting the boat into a "vast amphitheater entirely begirt with purple mountains." The story's tense switches from past to present and ends with two sentences, one short, one long:

> Meantime the whole Paradise of Arnheim bursts upon the view. There is a gush of entrancing melody; there is an oppressive sense of strange sweet odor;—there is a dream-like intermingling to the eye of tall slender Eastern trees—bosky shrubberies—flocks of golden and crimson birds—lily-fringed lakes—meadows of violets, tulips, poppies, hyacinths and tuberoses—long intertangled lines of silver streamlets—and, upspringing confusedly from amid all, a mass of semi-Gothic, semi-Saracenic architecture, sustaining itself by miracle in mid-air; glittering in the red sunlight with a hundred oriels, minarets, and pinnacles; and seeming the phantom handiwork, conjointly, of the Sylphs, of the Fairies, of the Genii, and of the Gnomes.

The tale ends with this panorama of heightened nature, seemingly cared for by fairies and sylphs—those nature-divinities whose disappearance Poe mourned in "Sonnet—To Science."

Poe was describing what a lovable, wise science, guided by a true poetic sentiment, might accomplish: a super-cultivated nature, the brightly colored scenes of a magic lantern blooming in four dimensions, a more-than-earthly paradise tasted by human senses. More than any of Poe's other

works, "The Domain of Arnheim" embodies his aesthetics of technolog-
ical transcendence: his view that art, at its finest, is nature continued by
other means. The notion of art as a mirror of nature and even a heightening or perfection of it flourished in the Renaissance, but by the eighteenth
century "nature" and "art" (especially "the mechanical arts") were often
seen as diametrically opposed. Entering into the flow of nature's creativity
and carrying it further, Poe wove nature and art together once more. Yet
in the age of mechanical industry, with the sense of an opposition between
nature and art still lingering, the result was beautiful if uncanny, both uplifting and disorienting.

Poe's reverie also resonated with the ideal of "second nature" at the heart
of American visions of western settlement, a "middle landscape" cultivated
and improved by pastoral care, industry, and labor. This was often someone
else's labor, whether of hired hands or slaves, and Poe's allusion to the work
of "a new race of fairies," like contemporary evocations of "Providential" expansion, seems a wishful dodge of the often brutal arrangements that built
the country's canals, railroads, and neoclassical edifices.

Yet in Poe's vision, a deliberate wildness remains: the earth's excess and
caprice have been not tamed or contained but concentrated and heightened
with knowledge and care. As in "Landor's Cottage," a tale he wrote later as
an accompaniment to "The Domain of Arnheim," tremendous effort and
knowledge are required for an art that appears truly artless and graceful;
the waterworks of "Arnheim" are more ethereal, more individualized, more
gratuitous than the utilitarian canals of the American System, and far more
difficult to realize. "The Domain of Arnheim" is less a vision of power than
an ode to the power of vision—the fantasy of a highly personal aesthetic
wedded to technical virtuosity, of nature transmuted into grace, prolonged
into a deliberate, exceptional *strangeness*.

In its gradual approach to an otherworldly realm, the tale echoes the
revelatory canoe journey at the end of *Pym*; it also resembles Thomas Cole's
celebrated series of paintings displayed at New York's Society Library, *The
Voyage of Life*, which depicts the movement of life from childhood to old age
as a journey on a small boat. The voyage to Arnheim takes its passenger
backward through stages of human development—from a neighboring city,

Thomas Cole, The Voyage of Life: Youth, *1842*

to farmed and pastoral landscapes, through dark, funereal passages, to a "semi-Gothic, semi-Saracenic" building, a floating "miracle in mid-air" akin to the pleasure dome of Coleridge's trippy "Kubla Khan."

Poe's phantasmagoria has inspired artists from Charles Baudelaire, J. K. Huysmans, and Oscar Wilde to René Magritte and John Lennon, who rewrote it as "Lucy in the Sky with Diamonds." Taken literally, it anticipates other technically amplified, sensuously immersive dreamworlds—the Great Exhibitions of the nineteenth century, the theme parks of the twentieth, and the digital fantasias of augmented and virtual reality in the twenty-first. Read allegorically, it depicts the journey taken by the thinker, investigator, seer, or artist who returns with treasures from beyond the veil of ordinary experience; it presents artworks as technologies for transforming reality by repatterning matter, perception, and thought.

It may also present a warning. As in much of Poe's most memorable work, the tale's beauties carry a palpable creepiness. It draws the reader along in a hypnotic, seductive, possibly malevolent movement; the passenger is led by "invisible guardians" into an all-encompassing, synthetic realm—as if lured by eldritch powers into an alien penitentiary.

As the critic Joan (Colin) Dayan has argued, Poe's language in the final vista echoes Milton's in *Paradise Lost*—describing the approach to hell. Arnheim's enticing heaven may be a diabolical fabrication, perhaps a trap of artistic pride: the self-set snare of the artist and the philosophers of progress who inspired him, who ignore human limits and take on the monstrous ambition to supplant God.

The vision's ambivalence—wavering between salvation and damnation, between creation, death, and rebirth, between nature and artifice—is intrinsic to its uncanny effect. As a crystallization of Poe's "mystical philosophy of nature," "Arnheim" opened the route his cosmology, *Eureka*, would trace to the universe's end.

A Wind from a Cloud

In mundane reality, Virginia's condition spiraled down. Poe wrote to Marie Louise Shew in late January 1847, "My poor Virginia still lives, although failing fast and now suffering much pain." He implored her, "Come—oh come to-morrow!" On January 30, Shew was in Fordham, looking after Virginia. "She called me to her bedside, took a picture of her husband from under her pillow kissed it and gave it to me." Soon she spoke no more.

The *New-York Daily Tribune* and *Herald* carried the notice:

On Saturday, the 30th ult., of pulmonary consumption, in the 25th year of her age, VIRGINIA ELIZA, wife of EDGAR A. POE. Her friends are invited to attend her funeral at Fordham, Westchester county, on Tuesday next, (to-morrow,) at 2 PM. The cars leave New-York for Fordham, from the City Hall, at 12.

She was buried in a small ceremony before a handful of friends including Willis, Duyckinck, and the women who had lately helped care for her, Estelle Lewis, Mary Gove, and Marie Louise Shew.

In March, Poe responded to Jane Locke's "Invocation for Suffering Genius," to assure her "of the sweet emotion made up of respect and gratitude alone with which my heart was overflowing" on reading it; he had been "overwhelmed by a sorrow so poignant" that it deprived him "for several weeks of all power of thought or action." On a manuscript copy of his poem "Eulalie," he added two lines:

> Deep in earth my love is lying
> And I must weep alone.

Floored by grief, Poe set aside his petty literary battles. Virginia's death—and his urgent attempt to realize, in "Arnheim," a physical voyage into mysteries beyond ordinary existence—pushed him onward, to face his boldest challenge. He now prepared himself to defy the emerging establishment of science and to confront his only truly worthy adversaries: the universe and its creator, the greatest enigmas of all.

16

The Plots of God

Dream Within a Dream   After Virginia died, Marie Louise Shew diagnosed Poe with a "brain fever, brought on by extreme suffering of mind and body" and by his exertions "to supply food, medicine, and comforts to his dying wife." Throughout 1847 he recuperated under the care of Shew and Maria Clemm. He wrote "Ulalume," an eerie, complex poem of mourning, following a sleepwalker's journey through haunted woods lit by celestial machinery and a newborn star:

> And now, as the night was senescent
> And star-dials pointed to morn—
> As the star-dials hinted of morn—
> At the end of our path a liquescent
> And nebulous lustre was born.

The newborn nebulous star guides the semiconscious speaker to the tomb of Ulalume—the name of his love, the ululating light, dead a year before.

Later in 1847, friends found Poe in Fordham in determined spirits. Mary Bronson, who visited with her father, a professor of elocution, expected to find him "grave and melancholy" but instead met Mr. Poe in the garden, "a very handsome and elegant-appearing gentleman, who welcomed us with a quiet, cordial, and graceful politeness that ill accorded with my imaginary somber poet. I dare say I looked the surprise I felt, for I saw an amused look on his face as I raised my eyes." Conversation gravitated toward Poe's obsessions: "After dinner, we all walked along the banks of the Bronx . . . Mr. Poe spoke much and well of the science of composition."

Beneath social graces, Poe was grieving and desolate. He walked through a landscape alternating between stark, stony cliffs and picturesque meadows only recently pierced by the railroad, suffused with a ghostly vividness. A reeling, world-weary sense of sorrow and panic pulsed through his new poems:

> I stand amid the roar
> Of a surf-tormented shore,
> And I hold within my hand
> Grains of the golden sand— . . .
> While I weep—while I weep!
> O God! can I not grasp
> Them with a tighter clasp?
> O God! can I not save
> One from the pitiless wave?
> Is all that we see or seem
> But a dream within a dream?

What survives the heaving storms of this brief life? If its succession of disasters is due to Providence, how could the Deity who designed it be benevolent, his universe harmonious? Drawing on years of thought and reading, Poe began work on his cosmological lecture, "The Universe," which would become the book *Eureka*.

Cosmic Clashes

Poe was stepping into an agitated arena. The previous three years had seen political and regional tensions threatening to explode. The opposition between "the slave power" and slavery's opponents hardened, exacerbated by the nation's deepening attachment to King Cotton. The 1844 election, putting Polk into power, had ratified further military expansion into the West and opened Texas for plantations. Polk provoked a war with Mexico—a landgrab carried out under false premises. The nation now stretched from the Atlantic to the Pacific, whatever the cost in public decency and lives.

The war rode on providential prophecies of "Manifest Destiny." Religious enthusiasm also fueled fears—and hopes—for the end of the world. The increasingly militant wing of abolitionism led by William Lloyd Garrison claimed an authority higher than society's law, ardently seeking to end a social system dependent on sin and brutality. Henry David Thoreau, imprisoned in 1846 for refusing to pay taxes for the Mexican War, declared the right to disobey an unjust government.

Amid this tumult, novel cosmologies vied for public attention. Thoreau's mentor, Emerson, had combined scientific facts with a sense of nature as living and conscious, inspired by Alexander von Humboldt and the spiritualism and pantheism of *Naturphilosophie*. Visionary philosophers such as the phreno-magnetists, Swedenborgians, and Fourierists buttressed their arguments for a world to come with facts from physiology, meteorology, and physics. Many, like Andrew Jackson Davis, based their authority on personal knowledge obtained beyond ordinary experience.

The "horrible vision" of *Vestiges*—whether it was understood as deist, materialist, or pantheist—still scandalized the devout with its compelling portrait of the universe as an autonomously evolving machine. (Only in 1884 would its author be confirmed as the Edinburgh publisher Robert Chambers, thirteen years after his death.) The controversy roiling about the book made vividly clear to Bache, Henry, and Peirce the urgent need to draw lines in scientific questions between "charlatans" and "men of proven reputation" and secure the latter's authority as experts on nature.

Some help came from abroad. The first volume of Alexander von Humboldt's magnum opus, *Cosmos*, was published in 1845, followed by the second in 1847. With sparkling prose and profound learning, the elder scientific statesman made an irresistible case for the intellectual, empirical, and aesthetic unity of the natural world that science revealed. Humboldt presented the nebular hypothesis only tentatively; the universe was marvelous enough without adding wild speculation. Joseph Henry made one of his students purchase *Cosmos*; the author of *Vestiges* "understands the mere Literature of Science," he told him, while "Baron HUMBOLDT comprehends the Science of Science." Poe published a translation of a German review of *Cosmos* in *The Broadway Journal*; its author hoped that enthusiasm for the work would "advance the taste for the sublime in natural sciences" while avoiding theological conflict.

The Swiss naturalist Louis Agassiz, a protégé of Humboldt and Cuvier, was brought to Boston in 1846 to deliver lectures on geology and natural history. Agassiz continued Cuvier's fight against the impious doctrine of the transmutation of species. Speaking to large audiences, first in Cambridge and soon in other cities, he pointedly dismissed *Vestiges* as a book "very much spoken of, but which I consider unworthy of notice by any serious scientific man." Asa Gray, *Vestiges*' fiercest American opponent, praised Agassiz's lectures as "the most original and fundamental refutation of materialism I ever heard."

Barrel-chested and energetic, with a charming French accent, Agassiz was invited to stay on as an instructor at Harvard's newly opened Lawrence Scientific School. He took a tour to meet prominent men of science in his adopted homeland. In Philadelphia he formed a bond with Samuel Morton, entirely persuaded by the skull collector's theory that the human races were distinct species with white Europeans ranked above all others. In Washington, Bache warmly welcomed Agassiz into his inner circle of elite science reformers alongside Henry and Peirce.

In September 1847, Agassiz was the star lecturer at the Association of American Geologists and Naturalists' eighth annual meeting, held that year in Boston. The geologists were in an expansive mood; the *Daily Journal* reported the group's "large number of new members" recently added "in view of its re-organization." Agassiz gave three rousing lectures and argued— despite the missteps of the National Institute in Washington—for the advantages of a national scientific organization.

That year, for the first time, Joseph Henry attended the AAGN. No longer a mere professor, he now held extraordinary scientific authority. In 1846, a congressional committee (which included Bache) determined that James Smithson's five-hundred-thousand-dollar bequest would be used to found a "Smithsonian Institution"—and appointed Joseph Henry its director. Though Henry feared the politics involved, he welcomed the opportunity; he would try to focus the institution on research rather than diffusion and steer it away from crowd-pleasing ventures like the National Institute or a museum. Henry soon brought in James Espy to make the Smithsonian the center of a telegraphic network, collecting meteorological reports and monitoring the nation's weather—the precursor to the National Weather Service.

Bache, who had tirelessly maneuvered to get Henry appointed, exulted to Peirce, "Science is triumphant in the election of Joseph Henry." The reformers' vision of a federal framework for American science—to build exclusive institutions to collect facts, using standard measures, procedures, and instruments, eventually joining facts into general laws—was taking firm root within the state, paid by public funds. They were also advancing their ambitions to compete on equal terms with European scientists.

Throughout 1847, Peirce had been working on calculations of the observed positions of Neptune, the remote planet first detected the previous year. The discovery was celebrated as a thrilling demonstration of the power of mathematical theory: "Its magnitude and exact place in the heavens were assigned, from considerations purely theoretical." A dispute over priority arose between French and English scientists: Urbain Le Verrier, an astronomer at the Paris Observatory, was the first to claim visual confirmation (with the aid of a colleague in Berlin, Johann Galle), but a young British mathematician, John Couch Adams, had independently predicted Neptune's location in nearly the same place.

Peirce intrepidly declared that the controversy was moot, because the discovery was a fluke. By his reckoning (supported by the calculations of Sears Cook Walker, a protégé of Bache's at the Naval Observatory), both Le Verrier and Adams had significantly miscalculated Neptune's orbit. That the planet happened to be passing through the region they predicted was just a "happy accident." When Peirce held his ground against Le Verrier's angry rebuttal—printed in Ormsby Mitchel's *Sidereal Messenger*—Asa Gray congratulated him: "As one zealous for the highest interests and character of American Science and American *Savans*, I thank you most sincerely."

Yet Peirce, Bache, and Henry still saw their greatest enemies not in Paris or Greenwich but at home. They saw charlatans and humbugs of popular science maintaining their eager audiences, while the unorthodox "scientific romance" of *Vestiges* remained a dangerous challenge. As reported in Boston newspapers, at the 1847 AAGN meeting Peirce read out a letter by W. C. Bond, director of Harvard's new astronomical observatory. Lord Rosse's "Leviathan" had not definitively resolved the Orion nebula, but on September

22, 1847, Bond pointed Harvard's new telescope, the largest in America, toward the constellation: the nebula's central cluster (or "trapezium") was "resolved into bright points of light." With this "ocular evidence" that the nebula was made of stars, not clouds of gas, Peirce declared the nebular hypothesis "totally incapable of advancing Astronomy." Peirce was using the platform of the AAGN—the united voice of "true men of science," conveyed by the press—to unravel *Vestiges* by pulling at its central thread.

Now standing in an authoritative position over the nation's scientific research at the Smithsonian, Joseph Henry finally conceded, "It is time that we should have among us a scientific *esprit de corps.*" The country was ready for—and scientists needed—a "tribunal" to "put down" the quackery of unlicensed speculation and fend off blows to their unity and authority. The AAGN's members resolved that at their next meeting they would adopt a new, more august and embracing name: the American Association for the Advancement of Science.

That autumn, at his retreat in Fordham, Poe was composing his own most ambitious scientific work. "My health is better—best," he told a friend. By January 1848 he was preparing a tour to gather subscribers for *The Stylus*, starting with a lecture in New York.

He asked the lawyer Henry Chapin for aid in this "endeavour to re-establish myself." With a loan of fifteen dollars he booked the lecture room of the Society Library; "I think that, without being too sanguine, I may count upon an audience of some 3 or 4 hundreds." On January 22 he announced the title: "The Universe." Notices of Poe's comeback performance appeared in Willis's *Home Journal* and Bryant's *Evening Post* and as far away as Philadelphia.

The night of February 3 was cold and wet; a storm kept most people at home. Yet the sixty-odd attendees "listened with enchained attention throughout" as "Mr. Poe fastened the attention of his audience" on his subject.

Introductory Leaps   He opened big. He would speak of the "*Physical, Metaphysical and Mathematical—of the Material and Spiritual Universe:—of its Essence, its Origin, its Creation, its Present Condition and its Destiny.*" To mark his destination, he announced his general theme: "*In the Original Unity of the First Thing lies the Secondary Cause of All Things, with the Germ of their Inevitable Annihilation.*"

He next invited his audience to join him in a thought experiment.

Humboldt's recent book, *Cosmos*, he said, showed the connections among the laws of each domain of nature. But Humboldt presented the universe "*not* in its individuality but in its generality." The "mere multiplicity" of details made it impossible to grasp the universe in its "individuality of impression"—as a unitary *effect*.

Poe proposed a more effective conceptual technology. He took his audience to the edge of an ancient volcano: "He who from the top of Ætna casts his eyes leisurely around, is affected chiefly by the *extent* and *diversity* of the scene. Only by a rapid whirling on his heel could he hope to comprehend the panorama in the sublimity of its *oneness*." To grasp "the full uniqueness of the prospect"—to take it all "into his brain"—he imagined a "mental gyration on the heel": to spin like a top (or to whirl like a kaleidoscope, or a phenakistoscope, an optical toy producing an illusion of motion), with eyes open. In this rapid "revolution of all things about the central point of sight," details "vanish altogether," while larger objects "become blended into one."

Before his listeners could regain their mental balance, Poe leaped again— into the future. He pulled out a letter from the year 2848 that he said he found floating on the *Mare Tenebrarum*—the sea of darkness, the place beyond the known continents on ancient maps.

The letter's author is an antiquarian; she takes time during a high-speed hot-air balloon cruise across the continental empire of "Kanadaw" to share her recent discoveries about the bizarre civilizations of the past.

Those strange people of the nineteenth century were convinced there were "*only two practicable roads to truth*": one was the "a priori" or deductive road that started with general laws or axioms, and from there discovered consequences; the other was the inductive road, "observing, analyzing, and classifying facts" and "arranging them into general laws." Both these "narrow and crooked paths" slowed mental traffic to a crawl. Real scientific advance, everyone in 2848 knew, was achieved through "seemingly intuitive *leaps*." The soul "loves nothing so well as to soar" in "regions of illimitable intuition."

In that benighted past, the followers of "Hog" (or Bacon) had lauded "merely perceptive men," the timid and narrow "diggers and peddlers of minute *facts*." The other school, followers of the "Ram," "Aries Tottle" (or Aristotle), the *a priorists*, failed to realize that "*no such things as axioms ever existed or can exist at all.*" They were foolish enough to believe, for instance, in what an ancient mill horse (J. S. Mill) had called the axiomatic "principle of non-contradiction"—that, for example, "a tree must be either a tree or not a tree."

Poe's letter-writer knew that depending on one's perspective, two contradictory views might both be true: "That a tree can be both a tree and not a tree, is an idea which the angels, or the devils, *may* entertain."

The antiquarian instead praised "the majestic highway of the *Consistent*," traveled by mutually reinforcing facts, speculations, and intuitions. This, she said, is the method of "the *only* true thinkers"—the "generally-educated men of ardent imagination." It was the method of Johannes Kepler, the seventeenth-century Platonic astronomer, whose laws of planetary motion established the fixed relations among planetary orbits. Kepler discovered his laws in part through a dream, followed by intuitive and imaginative leaps. She admired thinkers like Kepler because they "speculate—theorize—and their theories are merely corrected—reduced—sifted—cleared, little by little, of their chaff of inconsistency—until at length there stands apparent an unencumbered *Consistency* . . . an absolute and unquestionable *Truth*."

Kepler's intuitions were well-informed and logical, but fast: the result of "*deductions* or *inductions* of which the processes are so shadowy as to have escaped his consciousness." This is the method of cryptologists, the decoders of "the golden secret of the Egyptians." If asked about his method, Kepler would have replied, "I know nothing about *routes*—but I *do* know the machinery of the Universe. Here it is."

Poe's starting leaps took his listeners to distant imaginary lookouts, from which New York in 1848 began to look strange. Dashing past contemporary scientists—"one idea'd, one-sided and lame of a leg"—he was drawing a new timeline connecting ancient Egypt, Kepler, and the future, planting his own work as a landmark on the way.

The Design   Poe moved now to his "general design": *the Universe*.

He laid out two basic concepts: Infinity, and God. Both conjure up a "shadowy and fluctuating domain, now shrinking, now swelling, with the vacillating energies of the imagination."

Within infinite space lies "the Universe of Stars"—a bounded sphere populated by matter, created by pure spirit, or by the Godhead, "out of Nothing, by dint of his Volition."

But in its original state, matter took the form of a perfectly unified, un-differentiated entity: a "primordial Particle."

Under a divine impulsion, this particle exploded outward: in a flash, minute atoms of matter were "radiated spherically," distributed through space to form the universe of stars.

The principle and aim of this rapid emanation of matter was "multi-plicity out of unity—diversity out of sameness—heterogeneity out of homogeneity—complexity out of simplicity." This diffused state of matter is "abnormal"; the particles long to return to their "lost parent," that first particle to which they still belong, their "normal" state.

Their "appetite for unity" takes a material form. This is the force of at-traction, known to Newton as gravity, the longing of matter for matter.

But the particles' desire cannot be satisfied all at once. It is opposed by "a separative *something*," a resisting force. This is the "spiritual Ether"—understood "now as heat, now as magnetism"—electricity for short. This is also the determining factor in "phaenomena of vitality, consciousness, and *Thought*."

These, then, are the two essential constituents of the universe: attraction (identified with gravity and matter), and repulsion (identified with ether, electricity, life, mind, and spirit). "*No other principles exist.*"

After the first diffusion, each atom rushes toward the others. Here and there they swerve together into masses, growing in rotational velocity as more are added, forming swirling clouds of diffused matter—or nebulae.

As in Laplace's nebular hypothesis, as each mass condenses, it throws off its outer layer, forming a ring of matter; these rings condense into rotating planets. The nebula's core concentrates into a burning sun.

But this process of material condensation is constantly opposed by the resisting electrical force; without it, all bodies would have collapsed at once, returning to the original particle. In its current state, the universe's agglom-erations of matter—bodies, planets, solar systems—are drawing toward their own centers and toward each other, toward "absolute coalition," while the counterforce of repulsion pushes in the opposite direction, toward relation and difference.

All bodies, Poe explained, are "mere assemblages of more or fewer differences," and bringing together "any two differences" results in "a development of electricity." As attraction brings the diffused parts of matter together, they form new relations and new differences; as a result, more resistance arises, more electricity.

All existence faces this struggle between attraction and repulsion, matter and spirit. This is "the great *Now*—the awful Present."

Ahead lies a "still more awful Future."

Eventually, the balance between the two great forces will tip. Electrical repulsion will be overwhelmed by gravitational attraction: moons will crash into planets, planets into suns. The "myriad now-existing stars of the firmament" will rush together into "almost infinitely superior spheres." At this point, "amid unfathomable abysses" will appear "glaring unimaginable suns"—a "climacic magnificence" and a "new genesis" that forebodes "the great End."

These mega-stars race together; "and now, with a million-fold electric velocity" proportionate to their growing size and their "passion for oneness," "the majestic remnants of the tribe of Stars flash, at length, into a common embrace."

In this "inevitable catastrophe," only "one material globe of globes" now remains. This immeasurable, absolutely dense particle of matter will have attained its object, thus becoming "objectless." Where there are no parts, only "absolute Unity," there can be no differences.

> Matter, finally, expelling the Ether, shall have returned into absolute Unity,—it will then (to speak paradoxically for the moment) be Matter without Attraction and without Repulsion—in other words, Matter without Matter—in other words, again, *Matter no more*.

The complete, restored primary globe will "sink at once into that Nothingness which, to all finite perception, Unity must be." A perfect realization of unity is the goal of all difference and relation—a culmination of being, indistinguishable from emptiness.

The metaphysical curtain fell.

Encore   But another act remained. Poe reminded his listeners of "the great law of periodicity" and indulged a further hope—"that the processes we have here ventured to contemplate will be renewed forever, and forever, and forever; a novel Universe swelling into existence, and then subsiding into nothingness, at every throb of the Heart Divine." He imagined the universe expanding and contracting again and again—the heartbeat of the creation, in a millennia-long repetition.

If matter and spirit were separated only in order to reveal their identity—if every material element of the universe is but a fragment of a spiritual whole, driven by the appetite to return to its original unity—can these terms, "matter" and "spirit," "creation" and "Creator," be truly distinguished? Must there not be at every point a perfect *mutuality of adaptation* between the cause and its effect, pulsing through the cosmos?

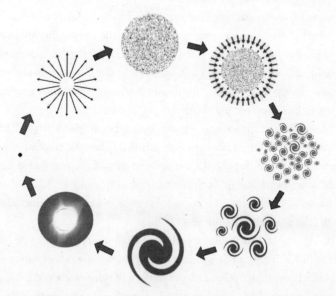

Eureka's cyclical plot (from center left, clockwise): *the original particle; irradiation; diffused matter;*
confrontation of gravity (attraction) with electricity (repulsion); agglomerations forming nebulae,
then galaxies; collapse into mega-stars; globe of globes; matter no more; original particle

The hidden inner chamber of the work now bursts into view: "And now—this Heart Divine—what is it? It is our own."

He invited his audience into a future where "the sense of individual identity will be gradually merged in the general consciousness."

We all knew this truth as children; memories whisper that our own souls are infinite, eternal; that for every person, nothing "exists *greater than his own soul*."

The limitlessness of every soul affirms not a hierarchy but the thorough-going *equality* of souls and essences: "The utter impossibility of any one's soul feeling itself inferior to another; the intense, overwhelming dissatisfaction and rebellion at the thought;—these, with the omniprevalent aspirations at perfection, are but the spiritual, coincident with the material, struggles to-wards the original Unity." This certainty and striving is proof "that no one soul *is* inferior to another," that "each soul is, in part, its own God—its own Creator," that each being is the individualization of divine spirit, while "the regathering of this diffused Matter and Spirit will be but the re-constitution of the *purely Spiritual* and Individual God."

Shattering and scattering God into "the diffused Matter and Spirit of the Universe," restoring the fragments to unity, then exploding them out-ward again, Poe invested every element of matter with the same sacred aura and vitality as living things and as God himself. For Poe this metaphysical truth, the identity of God, individual souls, and nature, was also a narrative necessity—the reason behind the design.

The painful incompleteness we experience as earthlings is an illusion; our intimations of immortality are fully justified. The total amount of plea-sure enjoyed by all diffused beings is equal to that of the regathered divine particle. Our awareness of both our individuality and our "identity with God" grows as "the bright stars become blended"—merging distinct identi-ties into "the general consciousness."

The striving and suffering of each individual life turns out to be only a dream within a dream, a tale within a tale in which there is "beginning eter-nally behind beginning." There may even be "a limitless succession" of other universes "more or less similar to ours," growing and shrinking but beyond our ability to observe them, each existing "apart and independently, *in the bosom of its proper and particular God.*" Poe's closing words retraced the spiraling orbits and staggered architecture of his design. "In the meantime bear in mind that all is Life—Life—Life within Life—the less within the greater, and all within the *Spirit Divine.*"

The same words had appeared in "The Island of the Fay," the phantasmagoric reverie that he offered in 1841 as an answer to the question of his "Sonnet—To Science"—how a poet should love science.

The secret at the center of *Eureka's* labyrinth is that the heart is the whole; an identity is shared between the core and the shell, the inner and the outer—between soul and body, spirit and matter, creator and creation, feeling and reason, beauty and truth, self and other, beholder and beheld, poetry and science. They form an infinitely resonant structure, made of weirdly fractal symmetries repeated at every scale of the composition.

Reviewers were struck by Poe's delivery, his "captivating and energetic style of expression." Some witnesses said the lecture held its audience much like a mesmeric séance: throughout a "rhapsody of the most intense brilliancy," the speaker "appeared inspired, and inspiration affected the scant audience almost painfully."

The Reason for the Darkness

For a work insisting on symmetry and unity of effect, *Eureka* is punishingly digressive and lopsided. Its tone veers from hushed awe at the universe's majesty to broad humor and painful puns, from nitpicking technical analysis to polemic with hints of mockery and deliberate mystification, from soaring ecstasies to touching sincerity. Poe's lecture, as well as *Eureka*, the one-hundred-page book that came from it, was a mess: a serious mess, a glorious mess, but a mess. It is impossible to decide how much of its disorienting, maddening effect was deliberate—perhaps as an illustration of the complexity, instability, and ultimately unspeakable mystery of the universe—or simply the result of the impossible task Poe had set himself.

Poe wanted to make the scale and complexity of the universe graspable. To convey the distance between our sun and the nearby Alpha Lyrae "we should need the tongue of an archangel." Lacking that organ, he proposed analogies to forge the "chain of *graduated impression* by which alone the intellect of Man can expect to encompass the grandeurs" of the "majestic totality." A cannonball traveling at normal velocity would take six centuries to reach Neptune, the newly discovered planet Poe considered the outer limit of our solar system; some of the light that reaches us from nebulae is three million years old, while Lord Rosse's "magical tube" is "whispering in our ears the secrets of *a million of ages* by-gone." Such examples confirm that "*Space and Duration are one.*"

Poe framed his lecture as a risky, technologically supported performance, akin to those assisted by magic lanterns, orreries, or dioramas—or the cabinet-sized poetry-making machine, "The Eureka," the "practical illustration" of "Kaleidoscopic evolution." The long dashes and italics with which Poe stippled his text functioned as the bolts—the pivots—the adjusting screws—of his rhetorical apparatus; they *focus* and *redirect* the readers' attention to notions with which he most sought to *impress*.

Like a magic lanternist, Poe spun the reader's perceptions, blending images and arguments into an undivided, "individual" effect—a vision of the universe as a whole. He wanted the reader to fly with the lightning thought of a cryptographer, with the untethered limbs of an angel—or join with the atomist Empedocles, who died by leaping into the volcano at Etna, in the molten zone between matter and thought.

Yet for all its dizzying movement and play across scales, *Eureka* took as its touchstone the facts and arguments of modern science—Newton, Bacon, Humboldt, the astronomers Encke and Laplace. His view of matter as the interplay of attraction and repulsion drew from both Kant and the chemist Boscovich. His long discussions of Neptune—which he saw as the first planet condensed out of our solar system's nebula—incorporated the recent debates over the planet's discovery and orbit in which Peirce, along with Herschel and Arago, was engaged.

Poe also answered challenges to the nebular hypothesis—defended just days before his lecture by the Scottish astronomer John Pringle Nichol in speeches to the Mercantile Library Association of New York. These were printed immediately in Greeley's *Tribune* and soon gathered into a pamphlet. Poe confronted the astronomical reports, by Rosse in Ireland, Peirce in Cambridge, and Mitchel in Cincinnati, that the nebula in Orion had been "resolved" by powerful telescopes. Nichol ducked the controversy in his lectures—admiring the reach of Rosse's telescope while presenting the nebular hypothesis as solid confirmation of his guiding conviction: the universal law of progress, of "evolution, ceaseless and irresistible—advancing from the imperfect to the perfect." Poe boldly declared that merely "ocular evidence" was unlikely to be conclusive and that the nebular hypothesis was *"beautifully true,"* "far too beautiful, indeed, *not* to possess Truth." Objections and unexpected evidence would lead to the theory's being "corrected—reduced—sifted—cleared," until all that remains would be an "unencumbered *Consistency.*"

This was a judgment appropriate to a very general theory or to a work of art. *Eureka* included a brief preface that aligned his book closely with his own critical theories and with the "scientific romance" of *Vestiges.* He offered *Eureka* as "an Art-Product alone:—let us say as a Romance; or, if I be not urging too lofty a claim, as a Poem," Poe staked the work's reception on "the Beauty that abounds in its Truth."

The thread connecting aesthetics and science for Poe was "design." The term was a staple of natural theology, but Poe was not propping up mainstream Christianity. For years he had been building a case against the *Bridgewater Treatises*—not because he doubted the existence of a divine plan, but because he saw their view of a harmonious, purely benevolent, human-centered creation contradicted on all sides. Protestant natural theologians, he thought, failed to reckon with the ubiquity of destruction and chaos, the perversity of human motivation, the beauty of the imperfect and strange.

He also rejected their linear view of causality—that all things could be explained by reference to a single divine intention or act. In *Eureka* he returned to his argument against the *Bridgewater* authors about the reversibility of cause and effect: "In Divine constructions . . . we may take at any time a cause for an effect, or the converse—so that we can never absolutely decide which is which." For example, he argued, in polar regions humans need highly energetic food, which is precisely what is available in the form of seal and whale oil. Is the food there to feed the humans, or are the humans there to consume the food?

Such "mutuality of adaptation" was built into the universe's symmetrical design: Poe found "the end of all things metaphysically involved in the thought of a beginning." The final collapse was a consequence of the opening eruption, just as his two driving principles were bound up in mutual dependence: "*Attraction* and *Repulsion*—the Material and the Spiritual—accompany each other, in the strictest fellowship, forever." Poe argued that this design was even more fundamental than gravity. The wish of the diffused particles to return to each other, expressed as gravity, is simply a consequence of their original unity. Each element of his plot springs "out of the bosom of the thesis—out of the heart of the ruling idea."

Poe was applying his theory of literary composition to God's creation. Both were focused on *unity of effect*. "In fictitious literature," he argued,

> we should aim at so arranging the incidents that we shall not be able to determine, of any one of them, whether it depends from any one other or upholds it. In this sense, of course, *perfection* of plot is really, or practically, unattainable—but only because it is a finite intelligence that constructs. The plots of God are perfect. The Universe is a plot of God.

Every true artist—or scientist—seeks to follow, emulate, and anticipate the perfect symmetries of God's design. True critics know it: "the pleasure which we derive from any display of human ingenuity" grows as a work approaches an "absolute reciprocity of adaptation."

Poe found further confirmation of this design—what Whewell called "consilience"—in solutions it offered to fundamental questions: biological, astronomical, and moral.

Charles Darwin, born the same year as Poe, similarly tried to make sense of novelty and conflict in nature, wrestling with natural theology, *Vestiges*, and Charles Babbage's arguments for uniform natural law. They both explained the origin of species (though Darwin's publication came in 1859) as the result of uniform law and universal struggle. For Darwin, the law of natural selection drew limits on the variations that survived from each generation, though his theory shed no light on the cause of these variations.

For Poe, the central cosmic struggle was between the pressure of gravity and the reciprocal increase of electricity, a strife that spurred increases in "vitality, consciousness, and thought." This conflict kept pace, "very closely, with the heterogeneity, or complexity, of the animal species," producing more vigorous, more capable, "superior and still superior" creatures. As an illustration he cited the extravagant *"ultra-tropical vegetation"* found on South Seas islands—the same evidence of luxuriant and varied life reported by Humboldt and other voyagers that drew Darwin to the Galápagos.

Poe's offhand solution to another puzzle earned him a place in the history of astronomy. The Prussian astronomer Olbers—who explained the existence of asteroids by the nebular hypothesis—had pointed out a paradox. If the universe is infinite, as Newton and his followers believed, and space is filled with infinite numbers of celestial bodies, the light from those stars should fill the night sky.

But the night sky is dark. Poe stated what has been called Olbers's paradox thus: "Were the succession of stars endless, then the background of the sky would present us an uniform luminosity, . . . since *there could be absolutely no point, in all that background, at which would not exist a star.*"

The explanation Poe provided is the one that twenty-first-century physicists accept. The universe is finite; it had a beginning in time and, though bewilderingly vast, may well be bounded in space. Because the quantity of stars is limited, they leave dark gaps between them. Poe's solution did away with any notion of an eternal or static universe.

More important, Poe thought, was the moral insight his cosmology provided. The unfolding of the universe's plot creates a maximum of heterogeneity, differences, and relations: "Right is positive; wrong is negative—is merely the negation of right; as cold is the negation of heat." As in "Mesmeric Revelation," pain exists that we may know pleasure. *Eureka* argued that any two opposed terms are intrinsically, essentially bound to each other, part of the divine substance, and only superficially distinct. We experience limitations, losses, conflicts, and disappointments because of the oppositions and separations of matter.

Yet every atom of every being, material and spiritual, beats with the divine heart. In Poe's narration, "the existence of Evil becomes intelligible; but in this view it becomes more—it becomes endurable. Our souls no longer rebel at a *Sorrow* which we ourselves have imposed upon ourselves." Our separation from others, and from the original source that both grounds and contains us, is a temporary illusion. Knowledge of our eventual and complete return, Poe says, is our consolation.

Swerves and Crashes   Poe's cosmology drew from wildly eclectic sources. From evangelical millenarianism he borrowed apocalyptic fire and Almighty God, though few Christians would recognize his vision of eternal life as an impersonal disintegration into a divine substance, and Christ was nowhere in sight. He siphoned the mesmerists' quasi-material ethers and echoed metaphysical schemes of participation in a living, dynamic nature from Swedenborg, transcendentalism, neo-Platonism, and *Naturphilosophie*. His dying and reborn cosmos, underwritten by a fundamental law, harked back to the Stoics. He might even have been channeling recently translated Hindu and Buddhist cosmologies making the rounds in philological circles.

Eureka owed a particularly large debt to the Epicurean view that life and thought arose from the interactions of minute particles of matter. As in Lucretius's poem, *The Nature of Things*, Poe described vortices and maelstrom-like motions as the source of life and order from the smallest to the largest scale, from the first agglomerations of matter following the primordial swerve of atoms to the whorls of the spiral nebula. This cosmic vision of material growth echoed Erasmus Darwin's epic poem of materialist creation, *The Botanic Garden*, which also predicted that the cosmos would collapse and be reborn. Darwin wrote,

> *Star after star from Heaven's high arch shall rush,*
> *Suns sink on suns, and systems systems crush,*
> *Headlong, extinct, to one dark center fall,*
> *And Death and Night and Chaos mingle all!*
> *—Till o'er the wreck, emerging from the storm,*
> *Immortal Nature lifts her changeful form,*
> *Mounts from her funeral pyre on wings of flame,*
> *And soars and shines, another and the same.*

Poe took chance seriously. But like Lucretius's nineteenth-century translator and apologist, John Mason Good—and unlike Erasmus Darwin—in *Eureka* he sought to reconcile chaos with divine providence and design.

Poe was carrying on the project of natural theology, though with radical twists. In his design, conflict and disharmony were everywhere on display, while species and individuals lacked intrinsic borders. Life emerged from matter, and matter thrummed with life. Symmetry and unity were realized in fiery destruction. His paradoxical conviction of a built-in drive toward perverseness was the moral equivalent of a preordained Epicurean swerve.

Poe was certainly influenced by what he had read of *Vestiges*, with its law-based, ongoing generation of solar systems and living things. For the *Bridgewater* authors, God was always active and present, governing and preserving inert matter; in contrast, *Vestiges* and Charles Babbage held that God did not need to maintain and adjust his laws. He could have programmed them in the beginning and left them to run their course. Poe wrote in "The Mystery of Marie Rogêt," "It is not that the Deity cannot modify his laws, but that we insult him in imagining a possible necessity for modification."

Despite great similarities with *Vestiges*, however, *Eureka* thoroughly diverted the linear optimism of its "law of progress." Poe's scientific romance began as a rising line, but at its climax it swerved to form a loop. It interrupted "progress" with a collapse, followed by a sublime conflagration—before beginning all over again.

Poe's cosmology was no stalking horse for atheism either, as some considered *Vestiges*. *Eureka* was an intensely theological text. It presented God not just as the universe's Creator, Governor, and Preserver but as its *actual and entire substance*. If everything can be understood as either cause or effect, there can no longer be a distinct "First Cause"; everything is, or is a part of, the prime mover. Poe embedded the universe and the thoughts that think it in a fluid pantheism and unstable materialism: matter shaded in and out of states resembling spirit and mind, all nested "within the Life divine." Earlier he had sharpened his analytic skills by empathically entering the thoughts of a chess-playing machine, inhuman murderers, a wily minister, and a host of poets, novelists, scientists, and philosophers. In *Eureka* he adapted his thoughts to the level of God and his creation. Sharing divine thoughts, thinking of and with the cosmos, Poe realized that he, the universe, and the Deity must be one. *Eureka*'s epiphanies took readers through repeated processes of separation, judgment, and opposition—at once material and mental—only to reunite them with what they behold.

Eureka was one of the most creative, audacious, and idiosyncratic syntheses of science and aesthetics in nineteenth-century America. Its capitalized phrase the "Universe of Stars" may suggest a parallel with the "United States." The book's effort to establish a balance between individuality and unity, between equality and difference—its *declaration of interdependence*—could be read as a restatement of his nation's enduring tensions. But if this was an allegory of America, the road Poe saw ahead would oscillate between paradise and inferno while somehow keeping both in view—"an idea which the angels, or the devils, *may* entertain."

Eureka saved its harshest criticisms for contemporary science. Though Poe drew competently and respectfully on physics, astronomy, and natural history, he opened with a scorched-earth attack on the narrowness of professional "men of science." Swimming against the tide of scientific specialization, empiricism, and professionalization, *Eureka* was a belated contribution to the eclipsed project of natural philosophy, the search for a unified system of causes for all of nature. Hence the book's dedication, "With very profound respect," to Alexander von Humboldt—whose multivolume *Cosmos* dared to offer a comprehensive and untimely view of the increasingly dispersed fields of science.

Poe knew he was doing something different, something doomed: he accepted that his book would be appreciated only by "the few who love me and whom I love," by "those who feel rather than those who think." Despite his conviction that his "Book of Truths" would "revolutionize the world of Physical & Metaphysical Science," he stated that "it is as a Poem only that I wish this work to be judged after I am dead." He suspected his literary renown would prevent envious critics from recognizing his scientific ideas: "A poet who has achieved a great (by which I mean an effective) poem, should be cautious not to distinguish himself in any other walk of Letters. In especial—let him make no effort in Science—unless anonymously, or with the view of waiting patiently the judgment of posterity." *Eureka* was explicitly addressed to readers who would find it after his death.

He urged his editor to print a first edition of fifty thousand copies; Putnam agreed to five hundred. The publication met both praise and condemnation. *Hunt's Merchants' Magazine* called *Eureka* a "startling work, properly located in the province of poetry or romance," seeing "much of the *true* in the grand Utopia of the universe thus imaged forth." Walt Whitman's *Brooklyn Daily Eagle* noted its "new and startling" thoughts, while Nathaniel Willis set Poe alongside the *Bridgewater* author Thomas Chalmers and the NYU scientist John W. Draper in the modern "tendency to elucidate scientific truth by addressing the sense of beauty." Greeley's *Tribune* acknowledged "the daring" with which Poe "throws aside all previous systems of philosophers and theologians."

Theologians were less thrilled. *The New Church Repository*, a Swedenborgian journal edited by George Bush, offered a reserved response: "He calls his work a poem, perhaps because, with Madame De Staël, he regards the Universe itself as more like a poem than a machine." The review recommended Swedenborg's *Outlines of a Philosophical Argument on the Infinite* as a correction to Poe's pantheist tendencies.

The theology student John H. Hopkins, a friend of Marie Louise Shew's, met with Poe to try to talk him down from his heresies. In his review in *The Literary World* he worried for the soul of anyone who heeded *Eureka*'s claims. He protested against the work's vacillations among Christianity, deism, and polytheism. Most damningly, he observed, "the system of Pantheism" was "more or less in-woven into the texture." The book was "extraordinary nonsense, if not blasphemy; and it may very possibly be *both*."

Under the title GREAT LITERARY CRASH, the dependably snide Thomas Dunn English reported the collapse of a book rack at Putnam's publishing house: "A new porter, not yet acquainted with the specific gravity of the various American authors, had imprudently piled the entire edition of POE'S new poem 'Eureka,' upon these shelves." Its "immense ponderosity" brought them crashing to the ground.

Circling the Scientific Wagons

In *Eureka*, Poe accepted scientific facts as far as they went. But mere facts never went far enough: taken alone, they limited knowledge to a narrow, disjointed surface. In the tool kit of official science, he said, intuition and imagination had been replaced by the cautious accountancy of deduction and induction; general views and leaping analogies were banned.

He was also challenging the emerging institutional infrastructure of science and the authority it granted to a small, self-defined group of experts.

The *Boston Journal*, excerpting the report of the lecture in the *Morning Courier*, sarcastically commented, "Mr. Poe is already a great man. If he establishes this theory to the satisfaction of learned and philosophical astronomers, his greatness will be greater than ever."

But many of the learned astronomers—and other professional men of American science—were at that very moment, very deliberately, turning away from the public. After years of plotting, they were creating institutions in which they could conduct their business at a safe distance from the interjections, speculations, and deceptions of the crowd.

As they'd planned the year before, in September 1848 the former AAGN met in Philadelphia. They now called themselves the American Association for the Advancement of Science—the AAAS. Henry Darwin Rogers, on a committee with Peirce and Agassiz, drew up a constitution: the association would meet each year in a different city, under an elected president and officers. Membership was open to anyone connected with the sciences—as long as they had been nominated by a member and were approved by the standing committee. The AAAS would include some women, including the comet spotter Maria Mitchell, but membership depended on being already known to, and approved by, those whom "true men of science" had determined to be "true men of science."

At the 1848 meeting, the association's members declared their common intent. With "periodical and migratory" meetings, they would connect disparate regions, provide a "general" and "more systematic direction to scientific research," and secure financial backing and "a wider usefulness" to their labors.

The AAAS's first president was William Redfield, a New York steamboat engineer and meteorologist who had earned international kudos for his whirlwind theory of storms. He commissioned the AAAS's first collective, publicly funded project: they would request funds from the secretary of the navy for Matthew Maury, director of the Naval Observatory, to gather information from ships' logs to draw up improved oceanographic charts.

Bache did not attend the first meeting of the AAAS, which helps explain why it conferred such important roles on Redfield—a fierce rival of Bache and Henry's associate the "Storm King" James Espy—and on Maury. As the head of the Naval Observatory, Maury competed against Bache's

Coast Survey for government support. Maury's florid, theologically tinged, crowd-pleasing communications (and, according to Bache, weak grasp of mathematics) infuriated "the chief."

Bache quickly made up for his oversight. Henry was elected as the AAAS's second president; third came Bache, followed by Agassiz and Peirce; after the first year, the association's steps were entirely guided by Bache's inner circle. With a handful of others, the clique soon took to calling themselves by the conspiratorial name "the Lazzaroni," after Neapolitan beggars—a shadowy brotherhood prepared to go to great lengths to secure government handouts and reach their ends.

During his presidency, Bache oversaw a bureaucratic reform to make the AAAS function not as a gentlemen's club but "as a *system*," as he had for the Bureau of Weights and Measures. All its aspects would assume a standard, methodical regularity. Now that America's "real men of science" had a nationwide organization, Bache and his allies would keep a firm grip on it—as they were doing with the Smithsonian and the Coast Survey.

Henry's presidential address summarized the AAAS's mission, to vindicate "the claims of science to public respect" and promote "the nature and dignity of the pursuit." Religion had a place; the association could point out those "things which have a bearing through science on the material and spiritual improvement of man." But most important, it would be the nation's scientific authority, the tribunal for separating real science from quackery, always ready to "expose the wiles of the pretender." Bache and his allies were establishing a supreme position over science and the previous decades' unregulated debates. They would define and defend the line between legitimate and illegitimate topics, methods, arguments, and authors.

Poe's cosmology in *Eureka*, published the year the AAAS began, was precisely the kind of publicly oriented, freewheeling, generalizing, and unlicensed speculation that the AAAS was created to exclude. Just like *Eureka*, the AAAS appeared in the immediate wake of the *Vestiges* controversy, which had shown the men of science how easily their authority could be shaken by a good story, backed with evidence and argument and well framed by the press. The AAAS would stand above fleeting enthusiasms and destructive passions, whether in science, religion, or politics.

As the 1848 election approached, the campaigns of the Whig candidate, the Mexican War hero Zachary Taylor, and of the Democratic Party's Lewis Cass were as shrill and contentious as any before. The clash over slavery was putting the Union into doubt. Despite early attempts by Agassiz, Morton, and Josiah Nott to give polygenesis and race science the AAAS's stamp of legitimacy—by featuring it, for example, at the Charleston meeting of 1850, where Nott gave a lecture titled "The Physical History of the Jews, in Its Bearing on the Question of the Unity of the Races," and Agassiz presented arguments to prove that "the Caucasian and the Negro" had always been distinct zoological species—Bache and Henry sidelined such topics in the interest of scientific unity. Yet whatever individual members' views about the Mexican War and the expansion of slavery to the West, the lands this conquest opened to white settlement meant a steady demand for scientific experts: surveyors, geologists, and civil engineers. Members of the AAAS, working along with Bache, Henry, and Peirce in institutional bases in the Coast Survey, Smithsonian, and elite universities, would train and supply them.

American science was being forged. A new profession was staking its claim in public life, dutifully extending the American empire and putting nature under universal law, standard methods, modest personal habits, and central administration. It had a vast future. It would map new territories and advise on telegraphs, railroads, factories, ports, and lighthouses. It would assist in piling up munitions and building defenses for both the Union and the Confederacy—led by Bache's West Point friend Jefferson Davis. After the Civil War, U.S. scientists would go to work for the transcontinental industries of oil, chemistry, steel, and finance, hastening the next century's planet-threatening "great acceleration" of production and consumption.

Poe foretold much of this: he saw method and mechanism extending their domains, growing in power while narrowing reality to questions of utility and profit. With the earth's divinities in hiding, he prophesied a bleak outcome in "The Colloquy of Monos and Una": "Huge smoking cities arose, innumerable. Green leaves shrank before the hot breath of furnaces. The fair face of Nature was deformed as with the ravages of some loathsome disease."

With its serpentine logic, *Eureka* brought forward an alternative vision.

Instead of a dead machine, this cosmos was alive with thought and passion, known through leaps of intuition and sympathy. It would never be fully tamed by analysis, tables, or grids; at best it might be followed and extended along routes suggested by nature itself. Though shot through with the aesthetics of design, it offered no naive return to Edenic harmony. Poe affirmed disorder and destruction at the core of existence, and the fragile enmeshment of human ideas and actions with the world.

Despite *Eureka*'s facts and "undying truths," Poe dedicated it to "dreamers and those who put faith in dreams." Its narrator was humbled and intoxicated by the world's splendor, its giddy intricacies, its endless rebirths. He was awed by our essential, absolute participation in its life and by that life's unresolvable obscurity to our view. Beauty, symmetry, and intuition offered Poe insight into a magnificently designed but uncontrollable cosmos whose wild whirlings between creation and destruction were identical to his own.

A year later, desperate for money, suffering frightful hallucinations, Poe would write to Maria Clemm, "It is no use to reason with me *now*, I must die. I have no desire to live since I have done 'Eureka.' I could accomplish nothing more."

17

Falling Star

A Leap in Providence   With *Eureka* in print in July 1848, Poe summoned his forces for another lecture tour to promote *The Stylus*, seeking support in the "South and West, among my personal and literary friends." In a desperate grasp for stability, he also sought to remarry. That autumn he undertook a feverish pursuit of Sarah Helen Whitman, a poet and widow from Providence with a passion for mesmerism.

The strange affair began when Whitman, a friend of Poe's New York circle, wrote an ode to Poe, whose voice "like distant thunder" charmed her "vagrant fancy"; it was first read aloud at Anne Lynch's Valentine party in 1848. Though Lynch warned her of "a deeply rooted prejudice against him," Whitman published it in March in *The Home Journal* as "The Raven."

In return Poe sent her his early poem "To Helen," torn from his published collection, inscribed with a personal dedication, together with a new poem he had published in *The Home Journal*, also called "To Helen." He praised her eyes, which he believed he had seen from a distance when he visited Providence with Frances Osgood in 1845: "They fill my soul with Beauty (which is Hope)."

Osgood deduced Whitman's intended reader: "Your beautiful invocation has reached 'The Raven' in his eyrie and I suppose ere this he has swooped upon your little *dove-cot* in Providence. May Providence protect you if he has!—He is in truth, 'A glorious devil, with large heart and brain.'"

Whitman waited until late that summer to reply directly, sending the two of them on a tormented, overwrought courtship.

Poe began his planned lecture tour in Richmond. He was welcomed by friends of his youth. Thomas White had died, and his first encounter with the *Messenger*'s new editor, John R. Thompson, was inauspicious: "A man calling himself Poe had been wandering around Rocketts (a rather disreputable suburb of Richmond) in a state of intoxication and apparent destitution." Ten days later, Thompson met him at the Alhambra, a saloon "frequented by gamblers and sporting men," where he was standing on a marble-top table reciting passages from *Eureka* "to a motley crowd, to whom it was as unintelligible as so much Hebrew." He also drunkenly challenged an editor to a duel.

On other occasions, Poe's composure pleased the pallid gentry: Thompson acknowledged him as "unmistakably a gentleman of education and refinement, with the indescribable marks of genius in his face, which was of almost marble whiteness. He was dressed with perfect neatness; but one could see signs of poverty." He visited his sister, Rosalie, at the home of her adoptive parents, the Mackenzies, and recited "The Raven" when asked.

Upon receiving a forwarded letter from Sarah Helen Whitman, he hurried back north. From Fordham he wrote to her under an assumed name asking for her autograph, an unnecessary ruse, because a mutual friend supplied an introduction.

Poe appeared at Whitman's door in Providence on September 21. He was ready to make an impression, prepared to be impressed. Laying eyes on her, he felt "the existence of spiritual influences altogether out of the reach of the reason."

In the language of the mesmeric poetics they shared, mingling physiology, reverie, and mystic transport, he later told her, "My brain reeled beneath the intoxicating spell of your presence, and it was with no merely human senses that I either saw or heard you. It was my soul only that distinguished you there."

They spoke of poetry. Whitman asked Poe if he knew the recently published, anonymous "Ulalume."

"I wrote it," he said.

As a fittingly extreme conclusion to their heady first date, Poe asked Whitman to marry him. A week later, after Poe had returned to New York, she declined, but with spiritual fervor: "Had I youth and health and beauty I would live for you and die with you. Now were I to allow myself to love you, I could only enjoy a bright brief hour of rapture and die." Her frail health and age—forty-four, to his thirty-nine—doomed their hopes. Poe promised in reply, "I would comfort you—soothe you—tranquilize you. My love—my faith—should instill in your bosom a praeternatural calm. You would rest from care—from all worldly agitation. You would get better, and finally well." Other outcomes were possible, he added with macabre reassurance: "And if *not*, Helen,—if not—if you *died*—then at least would I clasp your dear hand in death, and willingly—*oh, joyfully—joyfully—joyfully*—go down *with* you into the night of the Grave." She mentioned rumors that he had "no principle—no moral sense"; he indignantly fended them off.

In late October he returned north, staying first a few days in Lowell, Massachusetts, at the home of Nancy Richmond, who would call herself "Annie." According to copied letters she claimed had been sent by Poe, professing his passionate love for her, he reluctantly left Lowell for Providence in a wretched state. After "a long, long, hideous night of despair" in a hotel, he awoke. "I endeavored to quiet my mind by a rapid walk in the cold, keen air—but all *would* not do—the demon tormented me still." He tried to end his life: "I procured two ounces of laudnum & without returning, to my Hotel, took the cars back to Boston." He wrote a deathbed letter, but before he arrived at the post office, "my reason was completely gone," and his body rejected the laudanum. Horribly ill, he returned to Providence.

Whatever had occurred the previous day, Whitman found him waiting at the Providence Athenaeum in a horrid condition. She did little to assuage him, confronting him with letters from "friends" in New York who questioned his character. That night he sent her a farewell note in a hand shaky with emotion. He returned to her house in the morning in "a state of wild & delirious excitement," raving about "some terrible impending doom." Whitman said, "The tones of his voice were appalling & rang through the house. Never have I heard anything so awful." A doctor was called and found "symptoms of cerebral congestion."

Whitman understandably feared a marriage "which, under any circumstances, seemed to all my friends full of evil portents." Yet having seen him in such a pitiful state, she was moved in a way she had not been.

Whitman's family detested the match. "My mother did say more than once in his presence that my death would not be regarded by her so great an evil as my marriage under circumstances of such ominous import." Yet

Whitman was willing to take up the "burden of responsibility" to save Poe from himself: "No person could be long near him in his healthier moods, without loving him & putting faith in the sweetness & goodness of his nature."

In Providence, Whitman encouraged him to have a daguerreotype portrait taken. The result was the unforgettable image of a haunted man staring into grim destiny, struggling to maintain self-possession. In Whitman's state of excitement, everything "seemed a portent or an omen"; bidding him farewell on November 13, she looked to the horizon "& saw there *Arcturus* shining resplendently through a rift in the clouds." That night, "under a strange accession of prophetic exaltation," she wrote a poem evoking Poe's astronomical mystique: "Hast thou not stooped from heaven, fair star! to be / So near me in this hour of agony?— / So near—so bright—so glorious, that I seem / To lie entranced as in some wondrous dream."

Back in New York, Poe reassured her, "You say that all depends on my own firmness. If this be so, all is safe—for the terrible agony—known only to my God and to myself—seems to have passed my soul through fire and purified it from all that is weak." Her mother, still "inflexibly opposed"

Sarah Helen Whitman, daguerreotype by J. White, 1856; Poe, "Ultima Thule" daguerreotype, taken in Providence, 1848

to their union, drew up legal documents to take control of Whitman's inheritance.

Poe returned to Providence to give a lecture on December 20, organized through Whitman's considerable local influence. He wrote, "Keep up heart—*for all will go well*. My mother sends her dearest love and says she will return good for evil and treat you *much* better than *your* mother has treated me." Yet according to Mary Hewitt, who ran into him before his departure, he had doubts. "That marriage may never take place," he said. The lecture was held in Providence's Franklin Lyceum before a crowd of two thousand. He read "The Poetic Principle," a new declaration of literary faith, addressing passages to Whitman in the front row.

After the lecture, she accepted his proposal. Marriage banns were drawn up the next day.

Yet at a gathering on December 22, 1848, Poe—though quiet and composed—was intoxicated. He made new promises that night, and the next day the two went riding in a carriage. Stopping in at the Athenaeum's library, Whitman recalled, "a communication was handed me cautioning me against this imprudent marriage & informing me of many things in Mr. Poe's recent career with which I was previously unacquainted. I was at the same time informed that he had *already* violated the solemn promises that he had made to me & to my friends on the preceding evening" by drinking a glass of wine with his breakfast. "I felt utterly helpless of being able to exercise any permanent influence over his life." As she later concluded, "If I had never seen Poe intoxicated, I should never have consented to marry him; had he kept his promise never again to taste wine, I should never have broken the engagement."

They returned to her house. "While he was endeavouring to win from me an assurance that our parting should not be a final one, my mother saved me from a response by insisting upon the immediate termination of the interview." She never saw him again.

Last Jests   In early 1849, Poe published a poem, "For Annie," shuddered up from the depths of torment, where death is sweet relief:

Thank Heaven! the crisis—
The danger is past,
And the lingering illness
Is over at last—
And the fever called "Living"
Is conquered at last.

Nathaniel Willis ran it in *The Home Journal*, under the heading ODD POEM. Willis kept his readers' attention on this suffering national treasure: "Money (to tell a useless truth) could not be better laid out for the honor of this period of American literature—neither by the government, by a society, nor by an individual—than in giving EDGAR POE a competent annuity."

Poe was reorienting, seeking his fortunes elsewhere. His failed mesmeric romance with Whitman deepened his resentment against literary New York: "They are a heartless, unnatural, venomous, dishonorable *set*." Yet at times, despite "present anxieties and embarrassments," he could still feel "in my inmost soul a *divine joy*—a happiness inexpressible—that nothing seems to disturb." His poem "Eldorado" registered weary admiration for the heroic quest, in this case spurred by the discovery of gold in California. A shadow tells an explorer,

"Over the Mountains
Of the Moon,
Down the Valley of the Shadow,
Ride, boldly ride,"
The shade replied,—
"If you seek for Eldorado!"

He had new prospects. In Richmond, the *Southern Literary Messenger* accepted a new set of his "Marginalia"; in Boston, *The Flag of Our Union*, a family broadsheet with a wide distribution "independent of party or sect," struck up a deal for poems and tales. It "pays well as times go—but unquestionably it ought to pay ten prices; for whatever I send it I feel I am consigning to the tomb."

In the tales he wrote in 1849, Poe reworked long-standing obsessions. The most lighthearted, "X-ing a Paragrab," mocked self-important magazine editors, depicting a literary battle foiled by a printing error. When a compositor replaces *o*'s with *x*'s, a polemical article is transformed into what appears to be a "mystical and cabalistical" cryptogram.

"Mellonta Tauta" (things of the future) was set in 2848—reworking the letter Poe quoted at the start of *Eureka*. Its narrator scoffs at the nineteenth century's strange beliefs: that individuals matter, that plagues and wars are bad, that universal suffrage leads to good government. Most of that ancient civilization's record was lost in the *Mare Tenebrarum*, the dark sea of forgetting. The tale sought to undercut his age's faith in progress. As Poe had told Lowell, "I think that human exertion will have no appreciable effect upon humanity. Man is now only more active—not more happy—nor more wise, than he was 6000 years ago."

That year, the restless activity of his countrymen was leading them to California. The gold rush inspired his tale "Von Kempelen and His Discovery," a scientific hoax whose protagonist shared a name with the man who invented the chess-playing automaton. Von Kempelen has made a major scientific discovery, but the article doesn't reveal it until the end. Instead, it wades through *responses* to the discovery, referring to the "very minute and elaborate paper by Arago, to say nothing of the summary in 'Silliman's Journal,' with the detailed statement just published by Lieutenant Maury"; it quotes passages from Humphry Davy's notebook (which turn out to be about the effects of laughing gas). In the final pages we learn that Von Kempelen has achieved "the old chimera of the philosopher's stone": the transmutation of lead into gold. Poe told Duyckinck that he saw this fictional experiment as "a *check* to the gold-fever," aiming to "create a *stir* to some purpose." If the discovery of a cheap way of making gold were true—or simply widely believed— the metal would cease to be precious, and its value would plunge.

The most horrifying of Poe's late tales, "Hop-Frog; or, The Eight Chained Ourang-Outangs," told of a jester—a crippled dwarf from "some barbarous region"—exacting revenge on his tormentors. The enslaved artist, Hop-Frog, plans a masquerade; he persuades the cruel king and his ministers to dress up as "chained ourang-outangs," disguised with tar and flax. "Your majesty cannot conceive the *effect*" it will produce on the other partygoers, he boasts. As they run through the crowd, Hop-Frog snags the king and his ministers in the ring of an iron chandelier and hoists them to the ceiling, setting them ablaze; they "swung in their chains, a fetid, blackened, hideous, and indistinguishable mass." Hop-Frog announces, "I am simply Hop-Frog, the jester— and *this is my last jest*."

Since his upbringing in Allan's house, Poe had traveled far. Toiling without relief in the "magazine prison-house" at the whim of readers and publishers had taught him harsh lessons; during one unrelenting season he wrote, "I never knew what it was to be a slave before." In "Hop-Frog," depicting a brutalized servant who takes revenge on his inhumanly abusive masters, Poe identified with a slave in revolt. The tale depicted no gentle reform, no condescending half measures, only the violent overthrow of an outrageously unjust social order. After years of deprivation, Poe now allowed himself such identifications and dreams of vengeance—at least behind the mask of a tale.

In 1849, Poe also completed "Annabel Lee," a heartfelt lyric that worked through Virginia's death (and perhaps other losses). Two childish lovers share a happiness so great that jealous angels, "not half so happy in heaven," send "a wind from a cloud" that chills and kills her:

> But our love it was stronger by far than the love
> Of those who were older than we—
> Of many far wiser than we—
> And neither the angels in Heaven above
> Nor the demons down under the sea
> Can ever dissever my soul from the soul
> Of the beautiful Annabel Lee:—

The narrator now lies down by her side, "in her tomb by the sounding sea."

Poe pinned his mortal hopes once again on the great chimera of his life: the launch of a "five dollar journal." A young man named Edward Patterson wrote Poe to offer up his inheritance to launch *The Stylus* from his home in Oquawka, Illinois. Poe thought New York, or even St. Louis, might be a better base but approved the idea of printing simultaneously in the East and the West. Once more he would start a lecture tour to drum up support, beginning in Richmond.

On June 30, 1849, he bade Maria Clemm farewell in Fordham. "God bless you, my own darling Mother," she recalled him saying, "do not fear for Eddy! See how good I will be while I am away." Ten days later, she wrote to a friend, "If Eddy gets to Richmond and can succeed in what he intends doing, we will be relieved of part of our difficulties; but if he comes home in trouble and sick, I know not what is to become of us."

She kept silent about another possibility: that he would not make it home at all.

Indian Summer   Poe took the steamboat from Brooklyn and turned up in Philadelphia on July 2. The city was nearly deserted. The epidemic of the previous decade had returned; there were "cholera bulletins upon every newspaper door." Poe appeared at the home of his friend John Sartain in extreme distress. He said he had overheard two men talking about him, planning to kill him. Poe asked for Sartain's help in cutting off his mustache as a disguise and, because his shoes were worn down, borrowed a pair of slippers. Sartain gave him a cup of tea and went with him for a walk and an omnibus ride to the reservoir where they over-looked the Schuylkill River.

Poe told him he had been in prison in South Philadelphia, arrested on suspicion of "trying to pass a fifty-dollar counterfeit note," but was released on the word of one of the court officers who recognized him. While in prison he had a vision: a young female figure who stood upon the prison's stone tower across from his cell, "brightly radiant, like silver dipped in light," questioned him in a frightful "ordeal."

Poe wrote to Maria Clemm, "I have been *so* ill—have had the cholera, or spasms quite as bad, and can now hardly hold the pen." He now said he was in prison "for being drunk; but *then* I was not. It was about Virginia." Sartain bought his poems "The Bells" and "Annabel Lee," for his *Union Magazine*; the kindly novelist George Lippard also managed to collect five dollars to put shoes on his feet and help him on his way.

He left Philadelphia on Friday, July 13, and arrived in Richmond without his suitcase. He wrote to Clemm, ascribing the hallucinations he had suffered in Philadelphia to "mania à potu," delirium tremens. Alcohol was corroding his brain. From the Swan Tavern, he wrote thanking Lippard for his help, and Patterson in Oquakwa, who had sent him a lifeline of fifty dollars—an advance on *The Stylus*—for "the tour." Patterson was well pleased with the cover Poe had mocked up and sent him, showing a disembodied hand engraving the word "Truth" and the slogan *aliquando ferreus, aureus aliquando*—sometimes a pen of iron, sometimes of gold.

Outfitted with new clothes and hat, Poe visited his sister, Rosalie, and the Mackenzies. Passing his former home, the Moldavia mansion, he saw a new addition on its side, built by the second Mrs. Allan to accommodate her three children—Allan's recognized heirs.

Despite rumors of Poe's irregularities, his childhood friends welcomed him back. He confirmed his sober intentions by joining the Richmond chapter of the Sons of Temperance, Shockoe Hill Division No. 54. A future poet, Susan Archer Talley, then a young girl, recalled meeting him at her family's home, impressed by this "refined, high-bred, and chivalrous gentleman . . . So dignified was his manner, so reserved his expression, that I experienced an involuntary recoil, until I turned to him and saw his eyes suddenly

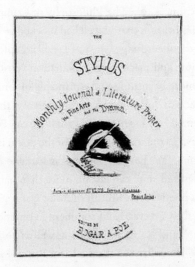

Cover design for The Stylus, *drawn by Poe, 1849*

brighten as I offered my hand; a barrier seemed to melt between us"; such was the "indescribable charm, I might almost say magnetism, which his eyes possessed."

He called on his teenage love, his neighbor Sarah Elmira Royster, now a wealthy widow. "I went down and was amazed to see him—but knew him instantly—He came up to me in the most enthusiastic manner and said: 'Oh! Elmira, is this you?'" Poe visited her often, "very pleasant and cheerful," and asked her to marry him. Her answer, though not definite, involved an engraved ring and gave him hope for a new life in Richmond.

He also dropped in at the *Southern Literary Messenger*. Though Thompson had doubts about the "Germanic" tint of Poe's tales, and had seen Poe in a sodden, befuddled state the previous year, Poe's poetry and critical writings—including his new, highly technical analysis of meter and rhyme, "The Rationale of Verse"—brought the journal prestige and readers.

In August, to "quite a full, and very fashionable audience" in the Exchange Concert Rooms, Poe recited poems and delivered his lecture "The Poetic Principle." "We were never more delighted in our lives," reported the *Richmond Whig*. Arrangements were made for further lectures in Richmond and Norfolk. His return was going as splendidly as Poe could have hoped.

In "The Poetic Principle" Poe testified to his highest poetic ideals—with neither the "pokerishness" of "The Philosophy of Composition" nor those technicalities of versification that Lowell had mocked as "three-fifths . . . genius" and "two-fifths sheer fudge." Just as *Eureka* accepted the value of calculation, observation, and mechanical explanation yet sought a truth more fundamental and more beautiful in "the principle of Unity" so "The Poetic Principle" identified not only the rules of verse but also the underlying hopes and ideals that formed "the essentiality of what we call Poetry."

True poetry elevates the soul; "the value of the poem is in the ratio of this elevating excitement." Yet because the soul is variable and states of mind change, such an elevated state can last no more than half an hour. Thus no long poem is truly poetic; even Milton's *Paradise Lost* is merely several short poems strung together. Poe read and commented on poems by Shelley, Byron, Thomas Moore, and Nathaniel Willis, as well as Longfellow's prefatory poem for *The Waif*—the collection that had sent Poe on his disastrous "Longfellow War."

Such poems respond to the "immortal instinct, deep within the spirit of man," for the beautiful, in "manifold forms, and sounds, and odours, and sentiments." Yet true poetry also goes beyond merely echoing and repeating these forms. "We have still a thirst unquenchable" belonging to "the immortality of Man. It is at once a consequence and an indication of his perennial existence." He took a famous phrase of Percy Shelley's, "the desire of the moth for the star," and made it his definition of poetry.

This image—a moth fluttering toward a distant star—captures the astronomical distance between the ideal of beauty and the doomed efforts of the minuscule creature who strives toward it: "Inspired by an ecstatic prescience of the glories beyond the grave, we struggle, by multiform combinations among the things and thoughts of Time, to attain a portion of that Loveliness whose very elements, perhaps, appertain to eternity alone." In *Eureka*, as the struggle between gravity and the vital ether or electricity increased, more wild and varied forms of life and art emerged. Here "the struggle to apprehend the supernal Loveliness" gives the world all it can "*feel as poetic*."

In a poem, he argued, the intellect and truth may be engaged but only to serve beauty. Conversely, in a work of reason and intellect, such as *Eureka*, beauty may be employed to serve truth. These are complementary routes toward the same remote ideal.

Though addressing an audience in his hometown, Poe strikingly identified himself with the city of his birth, Boston, just as he had at times identified himself as a Virginian while in the North. The "didactic heresy," he argued, is particularly pronounced in the North: "We Americans, especially, have patronised this happy idea; and we Bostonians, very especially, have developed it in full."

Poe ended with a poem by William Motherwell, asking his audience to identify, "in fancy, with the soul of the old cavalier" fighting a doomed battle:

Our business is like men to fight,
And hero-like to die!

Returning to the troubadour traditions and cavalier fantasies of his childhood home, Poe cast poetry as struggle, striving, aspiration, and experiment—the moth's desire for the star, the expression of its instinct for death and rebirth.

A viewer described him as he spoke: "his eyes dark and restless—in the mouth firmness mingled with an element of scorn and discontent. Firm and erect gait, but nervous and emphatic manner." Though he was a "man of fine address and cordial," one could see an inner battle across his features: "the great struggle for self control in which he seemed to be constantly engaged."

He was making plans. He would go to St. Louis to meet Patterson in October, and *The Stylus* would launch the following year. He would marry Elmira Royster and bring Maria Clemm to live with them in Richmond. He had an offer of a hundred dollars from a rich man in Philadelphia to "edit" his wife's poems.

Before setting off, he visited a deserted manor that had belonged to old friends of the Allans'. There he appeared to a companion, Susan Archer Talley, "unusually silent and preoccupied" with "memories associated with the place." In the empty house "he passed from room to room with a grave, abstracted look, and removed his hat, as if involuntarily, on entering the saloon, where in old times many a brilliant company had assembled. Seated in one of the deep windows, over which now grew masses of ivy, his memory must have borne him back to former scenes, for he repeated the familiar lines of Moore:

'I feel like one who treads alone,
Some banquet hall deserted'—

and paused, with the first expression of real sadness that I had ever seen on his face."

To Clemm he wrote, "I *never* was received with so much enthusiasm. The papers have done nothing but praise me before the lecture & since . . . I have been invited out a great deal—but could seldom go, on account of not having a dress coat. To-night Rose [his sister] & I are to spend the evening at Elmira's . . . Since the report of my intended marriage, the McKenzies have overwhelmed me with attentions."

He declared that "the last few weeks in the society of his old and new friends had been the happiest that he had known for many years, and that when he again left New York he should there leave behind all the trouble and vexations of his past life."

Nevertheless, the sense of foreboding, of a fate that exceeded his will, hovered around him. After one party his host recalled, "We were standing on the portico, and after going a few steps he paused, turned, and again lifted his hat, in a last adieu. At the moment, a brilliant meteor appeared in the sky directly over his head, and vanished in the east. We commented laughingly on the incident, but I remembered it sadly afterward."

Journey to Arnheim   Poe prepared his departure from Richmond, paying a visit to Sarah Elmira Royster on the evening of September 26: "He was very sad, and complained of being quite sick. I felt his pulse, and found he had considerable fever." At midnight he set out for the steamboat and docked in Baltimore on the twenty-seventh.

There he met people who persuaded him to have a drink, or several. He might have traveled to Philadelphia, where one witness claims he was hosted by the musician James P. Moss, who found him ailing and unfit to travel.

Six days later, on October 3, he was in Baltimore, the day of a congressional election. In a typical ruse of campaigns, "the candidates saw that every voter had all the whiskey he wanted."

Late that afternoon, from Gunner's Hall—a combined tavern and polling place—a typesetter sent a note to Poe's old friend Joseph Snodgrass: "There is a gentleman, rather the worse for wear, at Ryan's 4th ward polls, who goes under the cognomen of Edgar A. Poe, and who appears in great distress, & he says he is acquainted with you, and I assure you, he is in need of immediate assistance." Snodgrass went to the bar and "instantly recognized the face of one whom I had often seen and knew well, although it wore an aspect of vacant stupidity which made me shudder. The intellectual flash of his eye had vanished, or rather had been quenched."

Poe had been robbed of his clothes, or had exchanged them, and was now wearing a soiled palm-leaf hat, a dingy coat, and ill-fitting trousers "with neither vest nor neckcloth." He was "so utterly stupefied with liquor that I thought it best not to seek recognition or conversation." Snodgrass contacted Henry Herring, one of Mrs. Clemm's relations, who came to the tavern but "declined to take private care of him" due to Poe's irascibility on a previous occasion. They carried him in a state of insensibility to a carriage;

his "muscles of articulation seemed paralyzed to speechlessness, and mere incoherent mutterings were all that were heard."

The carriage delivered him to Washington College Hospital, north of Fells Point; Neilson Poe brought linens and clothing. The doctor John Moran spoke with him, "but his answers were incoherent & unsatisfactory." His trunk was missing; he said he had a wife in Richmond.

When the doctor asked what he could do to bring him comfort, Poe replied that "the best thing his best friend could do would be to blow out his brains with a pistol." He dozed; the doctor returned to find him in "a violent delirium, resisting the efforts of two nurses to keep him in bed."

He remained in this fluctuating state until 3:00 a.m. on Sunday, October 7, 1849, when "enfeebled from exertion he became quiet and seemed to rest for a short time." One of the doctors reported him saying, "*Lord help my poor Soul.*" One account had him crying out, "Reynolds!"—the name of the visionary who inspired the expedition beyond the limits of the known world.

He died. Neilson and Herring organized a funeral "utterly without ostentation": a single hack cab followed the procession carrying Poe's "neat mahogany coffin" along Fayette Street. In attendance were Snodgrass and a local lawyer and classmate from the University of Virginia, as well as Poe's former schoolteacher from Richmond Joseph Clarke and two cousins.

Because the service took place the day after his death, it was a quiet affair. One onlooker felt that "the burial ceremony, which did not occupy more than three minutes, was so cold-blooded and unchristianlike as to provoke on my part a sense of anger." At the Presbyterian cemetery on Fayette and Greene, words were spoken by the minister William Clemm, Poe's cousin. His defeated corpse was lowered into the plot of his ancestor, the Revolutionary hero General Poe.

Infamous Memorial   The day of Poe's burial, Horace Greeley received the news by telegraph from Baltimore. He assigned Rufus Griswold to report it. In his diary, Griswold noted, "Wrote, hastily, two or three columns about Poe, for the Tribune."

Those "two or three columns" would have an outsized influence on Poe's memory. Griswold's resentment toward the critic and poet who so greatly outshone him spilled onto the page:

> Edgar Allan Poe is dead. He died in Baltimore the day before yester-day. This announcement will startle many, but few will be grieved by it. The poet was well known personally or by reputation, in all this country; he had readers in England, and in several of the states of Continental Europe; but he had few or no friends.

That Poe lacked friends was an outright lie. And Griswold's dissection of Poe's character, as "shrewd and naturally unamiable," would have been a sur-prise to those many who, though aware of his foibles, held deep affection not only for the author but also for the admittedly imperfect man.

Griswold compared Poe to a character in Bulwer's *Caxtons*, quoting the novel directly: "Irascible, envious [changed from "arrogant"]—bad enough, but not the worst, for these salient angles were all varnished over with a cold repellent cynicism, his passions vented themselves in sneers." Griswold con-tinued, "There seemed to him no moral susceptibility; and what was more remarkable in a proud nature, little or nothing of the true point of honour."

Such an assault would have led a living Poe to confront or even sue him. But Griswold's takedown of a man already lowered by circumstance and freshly buried faced no retribution. It was quickly reprinted in New York, Richmond, and Philadelphia. While those who knew Poe might recognize in the hasty obituary the spite of a rival and the many points at which a weakness was amplified into an abomination, Griswold's account cast a long shadow.

More generous and fond obituaries appeared in the next days, though they were forced to answer Griswold's attack. The *New York Herald* on Octo-ber 9 conceded that Poe "was a most eccentric genius, with many friends and many foes, but all, I feel satisfied, will view with regret the sad fate of the poet and critic," while for the *Richmond Whig* the news of his death "will be read with profound regret by all who appreciate generous quali-ties, or admire genius." Cornelia Wells Walter's paper, the Boston *Evening Transcript*, moralized once more: "He had talents, with which he might have done great things, had he united to them stable principles, earnest purposes and self-denying habits." In Baltimore, John Pendleton Kennedy, who had started Poe on his literary life, now lamented its end: "Poor Poe! He was an original and exquisite poet, and one of the best prose writers in this country . . . His taste was replete with classical flavor, and he wrote in the spirit of an old Greek philosopher."

Lambert Wilmer, under the heading EDGAR A. POE AND HIS CALUMNIATORS, lambasted "the slanderous and malicious miscreant" behind the first obituary. Even Longfellow, despite Poe's petty attacks, mourned the "melancholy death" of "a man so richly endowed with genius," for whose "vigorous, direct, and yet affluent" prose and poetry he "always entertained a high appreciation." He magnanimously dismissed Poe's harsh criticism as nothing but "the irritation of a sensitive nature, chafed by some indefinite sense of wrong." Poe's unshakable friend Nathaniel Willis remembered him as "a quiet, patient, industrious, and most gentlemanly person, commanding the utmost respect and good feeling by his unvarying deportment and ability"; Willis conveyed the "*goodness in Edgar Poe*" by describing his profound devotion to his mother-in-law, his "ministering angel."

Maria Clemm heard of his death on the morning of October 9. She wrote to Neilson, who confirmed the worst: "I assure you, My dear Madam, that, if I had known where a letter would reach you, I would have communicated the melancholy tidings in time to enable you to attend his funeral . . . Edgar had seen so much of sorrow—had so little reason to be satisfied with life—that, to him, the change can scarcely be said to be a misfortune."

Clemm's disjointed note to Annie Richmond conveyed her extreme distress: "Annie my Eddy is dead he died in Baltimore yesterday—Annie my Annie pray for me your desolate friend. My senses *will leave me*—I will write the moment I hear the particulars, I have written to Baltimore—write and advise me what to do." Richmond invited her to her home in Lowell, with a cool eye on Poe's posterity: "Oh my mother, my darling darling mother oh what shall I say to you—how *can* I comfort you . . . stay with us long as you please—Do dear mother, gather up *all his papers and books*, and take them and come."

Rufus Griswold also put his mind to gathering Poe's papers. Bizarrely, he had become Poe's literary executor. Maria Clemm asked for his help in preparing Poe's collected works and granted him "full power of attorney" over his manuscripts in October 1849, and Griswold claimed that Poe had expressed this as his own wish that summer. If true, Poe's choice of one of his most bitter literary enemies as the guardian of his reputation was an act of either sheer perversity or astounding prescience. Perhaps he anticipated the flood of corrections that would follow Griswold's defamation.

With Clemm's help, Griswold tracked down Poe's suitcase, papers, and letters. Though Griswold claimed not to be receiving any remuneration, less than a month after Poe's death six people were "employed in setting up the copy," and four others would join the project, while Maria Clemm would receive no money from the publication, only copies of the finished collection. Within three months, in January 1850, Griswold published two volumes of Poe's works, prompting further rounds of attacks and defenses of Poe's life and character.

Griswold's most damaging accusations appeared a few months later in his third volume; it featured a twenty-seven-page "Memoir of the Author" containing selected facts of Poe's life, adorned with anecdotes and interpretations with no basis in reality. Griswold forged a portrait of Poe as ingenious but morally bankrupt, perpetually drunk, ruthless, coarse, and generally low of character. He furnished quotations from nonexistent letters, as well as from several documents that he altered—heightening Griswold's reputation and diminishing Poe's and lending force to his damning conclusion that Poe exhibited "scarcely any virtue in either his life or his writings."

Griswold's falsifications took nearly a century to unravel, helping turn Poe's survivors—even those who knew him as generous, kind, and idealistic—against him. Poe's character became a subject of enduring debate, as did the value and significance of his works. Frances Osgood tried to set the record straight before her own death to consumption a few months later, and Sarah Helen Whitman became his enduring defender, while spiritualists of the next decades contacted Poe directly in séances and transcribed new poems.

Poe died in Baltimore, on his way from Richmond (where he called himself a Bostonian) to Philadelphia and New York. Each of these cities has tried to claim him as its own. Even the cause of his death—was it alcohol poisoning, heart disease, a brain tumor, rabies?—has been hotly debated, with medical expertise and speculation called in to solve the mystery.

His death, at the age of forty, brought new puzzles: a final exclamation point, followed by a question mark. Poe mastered the art of posing a claim in such a way that questions quickly followed, until the very grounds for an answer began to decompose. The facts in the case grow more vivid and more uncertain; strange and contradictory possibilities crawl forward, causing tremors, anxious laughter, and more resonant, unshakable doubts.

The lantern goes out; there is a moment of darkness and silence. Then the house lights are lit, casting a warm glow and deep shadows, and the chattering starts up again.

From a Lighthouse

The Hall of Fantasy   The portrait of Poe as an immoral reprobate took root. His most perverse, dissolute characters became stand-ins for himself. Griswold's "Memoir" strongly influenced later presentations of Poe's works, including one in 1856 from *The North American Review* which ended with a prayer for a "potent chemistry to blot out from our brain-roll forever, beyond the power of future resurrection, the greater part of what has been inscribed upon it by the ghastly and charnel-hued pen of Edgar Allan Poe." Griswold accomplished his aim: to bury Poe in such a way that he would be shrouded in infamy. But bodies don't always stay buried.

For a long time, moralistic condemnations—along with suspicion about his tales' sensational, crowd-pleasing thrills—kept Poe from a high place in the canon of American authors. His movements among popular genres, poetry, philosophy, and such experiments as *Eureka* also made him difficult to place within literary histories. The influential historian of American intellectual life F. O. Matthiessen struggled to incorporate him into *American Renaissance*, his landmark 1941 study of antebellum ideas and literature. With a strong focus on Massachusetts, Matthiessen baptized the mid-nineteenth century as "the age of Emerson and Whitman" and devoted chapters to Thoreau, Melville, and Hawthorne while relegating Poe to footnotes and a later, halfhearted review.

Many of Poe's admirers, however, became influential innovators. From his blueprints they built the modern genres of detective fiction (Arthur

Conan Doyle), science fiction (Jules Verne), and horror, particularly of a weird and psychological cast (Robert Louis Stevenson, H. P. Lovecraft, and eventually Stephen King). While these advances in genre fiction have had a remarkable, long-lasting popular success, for a long time they were not treated as "high literature" worthy of serious critical consideration.

Poe's gradual recognition as a significant literary figure came as a result of the insightful introductions and gorgeous, unified French translations of Charles Baudelaire. Baudelaire's shocking and sickly vision of the modern soul, expressed in classical poetic form in *The Flowers of Evil*, made him one of the founding voices of symbolism and a prophet of modernism.

Baudelaire first encountered Poe in a translation of "The Black Cat" in a Fourierist journal. He quickly began to translate his tales, starting in 1846 with "Mesmeric Revelation," which he presented to French readers as an unprecedented, utterly modern fusion of science, mysticism, imagination, and literary rigor, perfectly aligned with the utopian cosmologies flourishing before the workers' Revolution of 1848. After the reactionary coup d'état of Napoleon III in 1851, Baudelaire downplayed the scientific and technical aspects he had earlier celebrated in Poe's work (translating "The Philosophy of Composition," for example, under the theologically inflected title "La Genèse d'un poème"). He depicted Poe as the exemplary "accursed poet"—a transgressive artist pursuing an otherworldly ideal of beauty in a hostile, utilitarian world who paid for his opposition to modern banality and hypocrisy with a tragic, untimely death.

The icon Baudelaire painted of Poe—as a self-conscious craftsman of "pure poetry"—deeply inspired Stéphane Mallarmé and Paul Valéry. The stellar reputation of these poetic innovators, and their immense esteem for Poe, led English and American authors and critics to reevaluate his work. T. S. Eliot, however, disdained the Europeans' enthusiasm, while D. H. Lawrence tellingly considered him "almost more a scientist than a poet." Poe was recruited as a precursor for surrealism and other avant-gardes; his intense and off-kilter introspection seeded the fiction of Dostoevsky, Conrad, Kafka, and myriad later writers who have experimented with genre conventions, unreliable narration, and point of view.

In Latin America, Poe was a model for Jorge Luis Borges and Julio Cortázar, while more recently Roberto Bolaño confessed that "with Edgar

Allan Poe, we would all have more than enough good material to read" and urged aspiring writers to reflect on this advice "on bended knees." Poe's place has been secured as one of the world's most widely enjoyed and celebrated authors. His impact on popular culture remains nearly biblical; his techniques reverberate through detective novels, horror films, and sci-fi series, adapted to changing historical circumstances and to histories that remain painfully unchanged. In recent years, the filmmaker Jordan Peele has cast Black protagonists in Poe-like scenarios—*Get Out* uncovers a conspiracy of mesmeric enslavement, while in *Us* vengeful doppelgängers torment an affluent family named Wilson—to retell the ongoing horror of forced exclusion by race and class.

Cosmological Premonitions

But Poe's influence is not limited to literary and artistic genealogies. Alfred Russel Wallace—the explorer and naturalist who independently came up with a theory of natural selection at the same time as Charles Darwin—was a great admirer, as was Charles Sanders Peirce, the son of Benjamin Peirce and the brilliantly eccentric founder of philosophical pragmatism. Peirce tested his own theories about the energies of words by studying the physiological effect of "The Raven." His semiotics and "abduction" (rapid reasoning not reducible to the "creeping and crawling" of induction or deduction) and his cosmological writings suggest the influence of Poe's criticism, tales of detection, and *Eureka*.

In the early twentieth century a greater interest in Poe as a scientific thinker arose. By then, the Newtonian view of a stable, mechanical universe had been stretched and transformed by evolutionary accounts of life, by geometric systems that went beyond Euclid, and by the circumscription of Newton's clockwork by relativity and quantum mechanics. Poe's rewriting of the story of the universe—including *Eureka*'s declaration that "Space and Duration are one"—suddenly seemed prophetic.

The author of a landmark 1941 biography of Poe, Arthur Hobson Quinn, devoted considerable energy to grappling with *Eureka*'s anticipations of the new physics. Quinn corresponded with Arthur Eddington, the English physicist celebrated for establishing empirical proof of relativity. Eddington was a Quaker, attuned to the possibility of immediate contact with the divine. He saw *Eureka* as "the work of a man trying to reconcile the science of his time with the more philosophical and spiritual cravings of the mind," while Poe's "idea of 'unity in diversity and diversity in unity'" was "now becoming actually realised in scientific theory." Yet Eddington wanted to make sure any leaps of intuition were retraced in cautious empirical and mathematical footsteps: even if Poe had stated "an ultimate truth," science "must be left gradually to find it by its own pedestrian progress." Eddington was enforcing the values of his profession, hard won in the century since *Eureka*: "Poe's more definite suggestions (in the contemporary state of science) were not unintelligent but amateurish." He placed Poe on the wrong side of a line separating scientific professionals from amateurs—a line that was only beginning to be sharply drawn in Poe's lifetime.

Quinn also sought the opinion of the founder of relativity himself, Albert Einstein. Intriguingly, just after taking up his position at the Institute for Advanced Study in Princeton, in 1933 Einstein had replied (in German) to an inquiry about Poe's cosmology from Richard Gimbel—a book collector, Yale aeronautics professor, and department store heir; after a quick read, Einstein described *Eureka* as "a very beautiful achievement of an unusually independent mind." But in 1940, answering Quinn, Einstein was far less complimentary. Though he found the first half of the work "very witty and remarkable" for recognizing "that true science is possible only through combination of systematic experimentation and logical construction," in the second part Poe lost "every sense of that critical mood"; the overall presentation showed "a striking resemblance to the scientific crank-letters I receive every day." As he wrote, Einstein was years into his battle against aspects of quantum physics that conflicted with his views: the role it gave to chance (or "playing dice" with the universe) and its suggestion that perception impacts physical reality (Heisenberg and Bohr's "observer effect"). As we have seen, in *Eureka* Poe argued for versions of these ideas, as well as an expanding universe—a view Einstein once rejected.

Later in the twentieth century, scientists gave *Eureka* a fresh look. An Italian astronomer, Alberto Cappi, awarded the work landmark status. By extending the nebular hypothesis to the universe as a whole—not just a solar system—and by offering a physical explanation for how the nebular clouds first appeared, Poe did what no one in the eighteenth or nineteenth century—astronomer, physicist, or poet—had done. With *Eureka*, Poe was the first person "to imagine an evolving Universe in a Newtonian frame."

The more modest yet more specific honor of solving "Olbers's paradox"—the reason for the darkness of the night—was confirmed by Edward Harrison, an astrophysicist. Poe's answer to the puzzle left by Newton's infinite universe—if there are indeed stars throughout infinite space, why are the gaps we see between them dark and unlit?—involved seeing the universe as an object with a history whose present state holds clues to its past. The universe must have a plot; it must have unfolded *in time*. Though staggeringly immense today, it had a beginning.

Even more tantalizing possibilities of Poe's scientific influence were suggested by Cappi and by a Dutch chemist, René van Slooten, who holds that Poe's anticipations of contemporary physical cosmology are more substantial than has been acknowledged. *Eureka*'s depiction of the universe, beginning with a single particle and exploding outward in a rapid flash, is spookily reminiscent of the "big bang" theory advanced in the early twentieth century—the mathematically grounded cosmogony that took into account the constraints of relativity.

Credit for the earliest formulations of the big bang theory goes to two physicist-mathematicians: the Russian Alexander Friedmann and the Belgian Georges Lemaître. Friedmann published in 1924 a mathematical exposition of the curvature of space that allowed for an expanding universe; his favorite authors were Dostoevsky and Poe. Lemaître, a Catholic priest who became an adviser to Popes Pius XII and John XXIII, proposed in 1927 that the universe was expanding; in 1931 he published a version of his "hypothesis of the primeval atom," addressing developments in quantum mechanics, with a strong resemblance to *Eureka*'s "primordial particle." Lemaître studied with Eddington in Cambridge; though most of his papers were destroyed in a fire during World War II, he was a serious student of literature who published work on Molière. As a member of Belgian intellectual circles in the early twentieth century, he would also have known the work of the leading figure in French poetry, Paul Valéry.

Valéry was obsessed with the thought processes and methods shared by scientists and artists; he studied Leonardo da Vinci's notebooks and exchanged ideas with Einstein and the physicists Henri Poincaré and Louis de

Broglie. In 1921, Valéry published "Au sujet d'*Eureka*," a passionate appreciation of Poe's cosmogony, insisting on its resonances with relativity: "The essential characteristic of Einstein's universe is, in fact, its formal symmetry. Therein lies its beauty." At the same time—writing in the international heyday of the philosopher of creative evolution, Henri Bergson—Valéry noted the vitalist heart of Poe's universe: "in matter an eternal fever rages," and "everything is stirred by deeper and deeper agitations, rotations, exchanges, radiations."

Though we have as yet no outright confirmation from Friedmann or Lemaître, Poe's *Eureka* might well have contributed to modern science's leading cosmological narrative, the big bang. If you're so inclined, other anticipations can be found: some physicists advocate a "cyclical universe" that passes endlessly from big bang to "big crunch" and back again, as in *Eureka*. Even Einstein came to argue, as Poe had before him, that gravity is not a fundamental cosmic force, but a mere consequence of geometry; and Poe's suggestion of "a limitless succession of other Universes" alongside ours resembles theories of a multiverse. Though *Eureka*'s preface offered the work "to the dreamers and those who put faith in dreams as in the only realities," there's a strange symmetry to the thought that even those who believe only the truths of science may now be living in a cosmos Poe helped imagine.

How Deem Thee Wise?

As we have seen, in antebellum America the institutional markers separating professional scientists from amateurs or cranks were only just beginning to take shape, and Poe made the most of this ambiguity to put forward his own analyses in aeronautics, conchology, and psychology and put his stamp on cryptology, information theory, and cosmology. With these neglected achievements we might be inclined to add him to the pantheon of contributors to what his contemporary David Brewster called the "one vast miracle" of modern science.

No doubt, modern society as we know it has inherited much from Poe's age—including the efforts of Bache, Henry, and their collaborators to establish national and international institutions. Recognized authorities over what counts as science—such as the AAAS and the even more exclusive National Academy of Sciences, which Bache, Peirce, and Agassiz ushered into existence in 1863—provide forums for researchers to agree upon standards, evaluate claims, and communicate with a unified voice. They offer a counterweight to the endless multiplication of reckless speculations and dangerous frauds; they air disagreements, validate consensus, and provide informed advice on matters of common concern—such as public health, industrial safety, contagious diseases, and climate change. Poe often lent his voice to these projects, championing science publications and a national research institute. His efforts on behalf of American literature had similar civic goals—to encourage the arts and recognize their achievements.

But Poe also drew attention to science's tendency to overreach. As he showed with the hapless prefect in the Dupin tales, those who fixated on formulaic procedures often overlooked the obvious. He saw an unquestioning faith in established methods and "ocular evidence" leading to a static and narrow image of knowledge, while method, mechanism, efficiency, and profitability were forcing their way into domains in which they had no necessary place (one reading of "The Philosophy of Composition" is as a satire of just such intrusions). Foreclosing the world's essential mystery, science was crystallizing a view of nature as a passive, lifeless storehouse of matter and facts to be observed, mastered, and exploited to humans' chosen ends.

The ideal of impersonal "objectivity" becoming a guiding norm in Poe's lifetime—characterized by detachment, discipline, and value-free observation—set the conditions for an explosive growth of technology and information. These became the hallmarks of "progress." But as Poe saw— and as we see even more starkly now—the stance of "neutrality" allowed science to be complicit in destructive and unjustifiable designs. For some, including Henry, Bache, and Poe himself, objectivity meant avoiding the conflict over slavery and a tacit endorsement of the status quo. For others, including Samuel Morton, Louis Agassiz, and later eugenicists, it meant using science's prestigious "universality" to justify oppression; as Frederick Douglass observed in 1854, "It is somewhat remarkable, that, at a time when knowledge is so generally diffused . . . there should arise a phalanx of learned men—speaking in the name of *science*—to forbid the magnificent reunion of mankind in one brotherhood." Scientists took a leading role in mapping and claiming Native Americans' lands in the bloody process of western settlement and assisted imperial ventures worldwide. Their successors built weapons with the ability to end life on earth, while others promoted "rational choice" theories in which global politics is played out on a chessboard of mutually assured destruction.

The idea that matters of fact must be kept utterly separate from matters of value gave science and its associated technologies a free hand. Along with undeniable attainments, "objective science" helped develop the industries of production and extraction that now ravage the planet—the "huge smoking furnaces" Poe prophesied blasting the "fair face of Nature." Over the nineteenth and twentieth centuries, natural theologians' faith in the wisdom and goodness of the Creator was transformed into a widespread faith in the wisdom and goodness of Science. Too often, the material successes and moral prestige of science have allowed it to disown its deadly by-products, granting an undeserved authority to manipulative frauds such as social Darwinism, free-market fundamentalism, and rule by algorithm. Such unjustified faith leads some to believe that all we need to clean up the messes made by science and technology is more science and technology.

Poe pointed to the threads of destruction, disharmony, and perverseness running through existence. These insights deepened his skepticism toward claims of rational mastery or linear progress. They showed him the frailty of human constructions and our fragile dependence on our surroundings. Though tangled up with his fits of grandiosity, Poe's recognition of human limitations came with a strong sense of shame—a revulsion at his own faults and at what humans have done and have let each other do to the earth and one another.

Yet in spite of his warnings about science's excesses and blind spots—and in spite of his love of hoaxes and ambiguities—Poe was no nihilist of knowledge. He recognized the value of precise observation, encompassing theory, intuitive discovery, and pragmatic, well-tested consensus. While some of his moves resembled those of exploitative charlatans such as Barnum (whose line of descent can be traced directly to the White House in the election of 2016), even Poe's mystifications served high ideals.

How could he love science, how deem it wise? From that early question in "Sonnet—To Science" through to *Eureka*, Poe immersed himself in the standard models of science being forged and expanded in his day. He also placed himself within the sites of popular science and embraced the challenges that mesmerism, vitalism, and *Naturphilosophie* posed to ordinary mechanism, theology, and lifeless materialism. He gave a hearing to alternative approaches to knowledge: shattering insights that race past stepwise logic, leaps of intuition that follow the longing of the moth for the star.

Though he expected that *Eureka* would be misunderstood, his cosmology gave voice to the intelligence of feeling and to humans' profound involvement in what they study. *Eureka* proposed a vision of science in which beauty and sympathy are valued as much as efficiency and accumulation and where the ends of knowledge—its guiding values, uses, and consequences—are as important as short-term results and eye-catching innovations.

Poe offered a bewitching allegory in "The Domain of Arnheim" of a natural world we don't simply know and inhabit but have a responsibility for making and choosing. Showing art—in its widest sense, including technology aided by science—as nature continued by other means, he cast humans and their inventions as actors in a cosmic drama with both good and evil effects. To make a livable world requires imagination and aesthetic judgment, along with a wide-open empathy, and the recognition that while the earth may yield to our suggestions, it always escapes our grasp.

If we survey Poe's works with a rapid "mental gyration on the heel," an underlying "philosophy of composition" comes into focus. Poe rendered the world as a "manifold" and "multiform" reality, revealing truth after truth, "beginning eternally behind beginning." His writings—taken individually and together—often present incompatible truths, one after another or all at once. In his tales and essays, detached objectivity is mirrored by a charged, participatory empathy; vitalized holism stands with and against mechanical reduction; matter becomes spirit becomes matter.

In this kaleidoscopic realism, we see empirical observation, mystical revelation, and irrevocable skepticism continually trading places: as he said in *Eureka*, "That a tree can be both a tree and not a tree, is an idea which the angels, or the devils, *may* entertain." New configurations of reality arise according to the domains on which we focus and the tools that shape our answers (sense organs, concepts, language, technologies of perception and communication)—with all sublimely embedded, "the less within the greater, and all within the Spirit Divine."

This is not a philosophy likely to be endorsed by builders of solemn institutions or defenders of a monolithic rationality. But it may appeal to those who suspect that what passes for consensual reality is never as consensual or as real as it's cracked up to be, and who believe that other worlds—more beautiful and more strange—are always possible.

Into the Sea of Darkness   Poe's plots masterfully conveyed the sense of being encircled and inhabited by weird forces we neither master nor fully understand. To hold back the destruction that seems always just ahead, his characters design schemes, devices, and rationales to bring order to chaos, to send light into the shadows.

Fear drives us to barricade ourselves into habits for protection and reassurance. We build cocoons, burrows, and walls to avoid being distracted or injured, diminished or betrayed. But these structures of isolation, aiming to defend our individuality and identity, grow brittle. They, too, are battered by the tide.

In a little-known manuscript that can fairly be called Poe's *most obscure tale*, we hear Poe's ever-renewed striving against chaos and destruction and its inevitable failure. Discovered after Poe's death, the story was never completed.

"The Light-House"—a title added later—was likely written in 1849, but it echoes Poe's tales from the early 1840s, when he was in Philadelphia at the center of the nation's technical, scientific, and literary activity. A few years earlier, John Quincy Adams's inauguration promised "light-houses of the skies"—astronomical observatories to anchor the communications networks of an American science of stars and storms; as part of the Coast Survey, Bache and Henry later launched a federal lighthouse commission.

With its fretful, unreliable narrator, Poe's manuscript resembles "The Tell-Tale Heart." Its detailing of a paranoid enclosure echoes "Ligeia," "The Pit and the Pendulum," and "The Cask of Amontillado," while its remote maritime setting recalls "A Descent into the Maelström" and *Pym*.

But the fact that the tale reaches us as a fragment, undated and fugitive, makes it a true "MS. Found in a Bottle," sender and destination unknown.

As we have ended with Poe going silent, buried—if only to be brought back to life in the works he continually influences—"The Light-House" lets us hear his unmistakably nervous, observant voice again.

Below is the tale in its broken entirety.

Jan 1—1796. This day—my first on the light-house—I make this entry in my Diary, as agreed on with De Grät. As regularly as I *can* keep the journal, I will—but there is no telling what may happen to a man all alone as I am—I may get sick, or worse . . . So far well! The cutter had a narrow escape—but why dwell on that, since I am *here*, all safe? My spirits are beginning to revive already, at the mere thought of being—for once in my life at least—thoroughly *alone*; for, of course, Neptune, large as he is, is not to be taken into consideration as "society." Would to Heaven I had ever found in "society" one half as much *faith* as in this poor dog:—in such case I and "society" might never have parted—even for the year . . . What most surprises me, is the difficulty De Grät had in getting me the appointment—and I a noble of the realm! It could not be that the Consistory had any doubt of my ability to manage the light. *One* man had attended it before now—and got on quite as well as the three that are usually put in. The duty is a mere nothing; and the printed instructions are as plain as possible. It never would have done to let Orndoff accompany me. I never should have made any way with my book as long as he was within reach of me, with his intolerable gossip—not to mention that everlasting mëerschaum. Besides, I wish to be *alone* . . . It is strange that I never observed, until this moment, how dreary a sound that word has—"alone"! I could half fancy there was some peculiarity in the echo of these cylindrical walls—but oh, no!—this is all nonsense. I do believe I am going to get nervous about my insulation. *That* will never do. I have not forgotten De Grät's prophecy. Now for a scramble to the lanterns and a good look around to "see what I can see" . . . To see what I can see indeed!—not very much. The swell is subsiding a little, I think—but the cutter will have a rough passage home, nevertheless. She will hardly get within sight of the Norland before noon to-morrow—and yet it can hardly be more than 190 or 200 miles.

Jan. 2. I have passed this day in a species of ecstasy that I find it impossible to describe. My passion for solitude could scarcely have been more thoroughly gratified. I do not say *satisfied*; for I believe I should never be satiated with such delight as I have experienced to-day . . . The wind lulled about day-break, and by the afternoon the sea had gone down materially . . . Nothing to be seen, with the telescope even, but ocean and sky, with an occasional gull.

Jan. 3. A dead calm all day. Towards evening, the sea looked very much like glass. A few sea-weeds came in sight; but besides them absolutely *nothing* all day—not even the slightest speck of cloud . . . Occupied myself in exploring the light-house . . . It is a very lofty one—as I find to my cost when I have to ascend its interminable stairs—not quite 160 feet, I should say, from the low-water mark to the top of the lantern. From the bottom *inside* the shaft, however, the distance to the summit is 180 feet at least. thus the floor is 20 feet below the surface of the sea, even at low-tide . . . It seems to me that the hollow interior at the bottom should have been filled in with solid masonry. Undoubtedly the whole would have been thus rendered more *safe*:—but what am I thinking about? A structure such as this is safe enough under any circumstances. I should feel myself secure in it during the fiercest hurricane that ever raged—and yet I have heard seamen say occasionally, with a wind at South-West, the sea has been known to run higher here than any where with the single exception of the Western opening of the Straits of Magellan. No mere sea, though, could accomplish anything with this solid iron-riveted wall—which, at 50 feet from high-water mark, is four feet thick, if one inch . . . The basis on which the structure rests seems to me to be chalk . . .

Jan 4 . . .

Here Poe's fragment ends, but we might guess what comes next. Gradually, inevitably, catastrophically, the tower will come crashing down along with the narrator's fragile defenses. His quest for solitude and safe enclosure—

observing, measuring, recording—will, paradoxically, expose him ever more fully to forces of destruction both within him and without.

One way or another, we will all sink back into the sea from which we came. Meanwhile, we might rework the magic lantern shows and landscape gardens of our arts, our sciences, our thoughts—to recompose our worlds, to bring things briefly together in beautiful and revealing forms, with an eye to their inevitable collapse.

Poe described the novel certainties and undying anxieties of his place and time. Beyond the facts of his own case, he telegraphed the fascinations and the terrors of simply being alive. Sending signals to readers who would discover him long after he was gone, Poe sharpened the piercing light of reason—and deepened the darkness in its wake.

Lighthouse, hand-colored magic lantern slide, nineteenth century

Notes

ABBREVIATIONS

AHQ Arthur Hobson Quinn, *Edgar Allan Poe: A Critical Biography* (1941; Baltimore: Johns Hopkins University Press, 1997).

ATQ American Transcendental Quarterly

eapoe.org The Edgar Allan Poe Society of Baltimore website

EAPR Edgar Allan Poe Review

LOA *Edgar Allan Poe: Poetry and Tales*, ed. Patrick F. Quinn (New York: Library of America, 1984).

Ostrom *The Collected Letters of Edgar Allan Poe*, 2 vols., ed. John Ward Ostrom, Burton R. Pollin, and Jeffrey A. Savoye, 3rd ed. (New York: Gordian Press, 2008).

PS Poe Studies/Dark Romanticism

SLM Southern Literary Messenger

TPL Dwight Thomas and David Kelly Jackson, *The Poe Log: A Documentary Life of Edgar Allan Poe, 1809–1849* (Boston: G. K. Hall, 1987).

A NOTE ABOUT SOURCES

The website of the Edgar Allan Poe Society of Baltimore (eapoe.org) provides an extraordinarily valuable service to those who work on Poe or wish to learn more, with the largest store anywhere of documentation on Poe's life, works (with crucial bibliographical information), commentary, and criticism. I am extremely grateful to Jeffrey Savoye, who created, maintains, and updates the site. Along with Thomas and Jackson's monumental *Poe Log*, it offers researchers more than enough material to get lost in. To find my way, I have benefited from the constantly growing scholarship on Poe, not all of which is noted in my endnotes, including biographies and editions by Arthur Quinn, Kenneth Silverman, Jeffrey Meyers, Paul Collins, George Woodberry, Thomas Ollive Mabbott, Hervey Allen, and James Harrison. When I began working on Poe's connections to science (in an essay in *The British Journal for the History of Science* in 1996), the most prominent works on the topic were Valéry's 1921 essay, Margaret Alterton's 1925 book on his critical theory, Carroll Dee Laverty's 1951 PhD dissertation, and Harold Beaver's tantalizing endnotes in 1976 to *The Science Fiction of Edgar Allan Poe*; since then, I have benefited from work by Barbara Cantalupo, Paul Grimstad, Maurice Lee, John Limon, Dana Medoro, Naomi Miyazawa, Sean Moreland, Cristina Pérez, Laura Saltz, Robert Scholnick, David Stamos, Susan Sweeney, René van Slooten, and many others. Poe's complicated relations with his complicated era have been increasingly untangled thanks to examinations of Poe and the press by Jonathan Elmer, Kevin Hayes, Leon Jackson, Meredith McGill, and Terrence Whalen, and landmark collections by editors and authors including Shawn Rosenheim and Stephen Rachman, Eric Carlson, Benjamin Franklin Fisher IV, Richard Kopley, Kevin J. Hayes (both the *Cambridge Companion* and *Poe in Context*), J. Gerald Kennedy and Liliane Weissberg, and the recent omnibus from Kennedy and Scott Peeples, *Oxford Handbook of Edgar Allan Poe Studies*, in which detailed history increasingly informs holistic interpretation. On nineteenth-century American science, see the endnotes to the introduction.

Introduction: Subject: The Universe

3   *"Edgar A. Poe will lecture"*: New York Daily Tribune, Feb. 2, 1848, 3.

3   *"There is but one thing certain"*: Nathaniel P. Willis, "Mr. Poe's Eureka," *New York Home Journal*, Feb. 5, 1848, 2, in *TPL*, 720.

3   *"Mr. Poe is not merely a man of science"*: *Weekly Universe*, Jan. 1848, quoted by George W. Eveleth to Poe, July 9, 1848, in *TPL*, 719.

4   *"It is written in a Stanza"*: Charles Eames, New York New World, Feb. 15, 1845, in *TPL*, 503.

4   *"his proud beautiful head"*: Frances Osgood, quoted in Hervey Allen, *Israfel: The Life and Times of Edgar Allan Poe*, 2 vols. (New York: George H. Doran, 1926), 2:643.

4   *"People seem to think"*: Sarah Helen Whitman, quoting a letter from 1846, in *The Complete Works of Edgar Allan Poe*, ed. James A. Harrison, 17 vols. (New York: Thomas Y. Crowell, 1902), 1:246.

4   *"I became insane"*: Poe to George W. Eveleth, Jan. 4, 1848, in Ostrom, 2:354–57, and *TPL*, 716.

4   *"Where is Mr. Poe"*: Eveleth to Duyckinck, Nov. 23, 1847, in *TPL*, 709.

5   *"We understand"*: John Donkey (Philadelphia), Jan. 1, 1848, *TPL*, 710. Walt Whitman excerpted the forthcoming inaugural issue of the *John Donkey* in the Brooklyn Daily Eagle on Dec. 17, 1847, and praised this joke as "the real coarse, but deep, true stuff," in *TPL*, 711. English's novel, first released in 1843, was republished in late 1847 as *Walter Woolfe; or, The Doom of the Drinker*; *TPL*, 711.

5   *"to re-establish myself"*: Poe to H. D. Chapin, Jan. 17, 1847, in Ostrom, 2:644–45, and *TPL*, 717.

6   *"I must get a list"*: Poe to Nathaniel P. Willis, Jan. 22, 1848, in Ostrom, 2:647, and *TPL*, 718.

6   John Pringle Nichol: Nichol's lectures were transcribed and printed in the *New York Daily Tribune*, Feb. 3, 1848, and collected afterward as a pamphlet: J. P. Nichol, *Views of Astronomy: Seven Lectures Delivered Before the Mercantile Library Association of New York in the Months of January and February, 1848* (New York: Greeley & McElrath, 1848); J. P. Nichol, *Views of the Architecture of the Heavens, in Letters to a Lady* (Edinburgh: Tait, 1837).

7   The theological consequences: Poe once denied having read *Vestiges* (see his letter to George E. Isbell, Feb. 29, 1848, in Ostrom, 2:658–61), but he was well aware of the book's arguments and implications (see chapter 14 below).

7   *"I used to sit up with him"*: Maria Clemm, in George E. Woodberry, *The Life of Edgar Allan Poe, Personal and Literary, with His Chief Correspondence with Men of Letters*, 2 vols. (Boston: Houghton Mifflin, 1909), 2:236.

8   *"select, but highly appreciative"*: Albion, Feb. 5, 1848, in *TPL*, 722.

8   *"I have seen no portrait"*: Maunsell B. Field, *Memories of Many Men and of Some Women: Being Personal Recollections of Emperors, Kings, Queens, Princes, Presidents, Statesmen, Authors, and Artists, at Home and Abroad, During the Last Thirty Years* (New York: Harper & Brothers, 1874), 224, in *TPL*, 720.

8   *"the Universe of Stars"*: Poe, *Eureka*, 8; this and subsequent citations are from the critical edition of *Eureka*, ed. Stuart Levine and Susan F. Levine (Urbana: University of Illinois Press, 2004).

8   *"phaenomena of vitality"*: Ibid., 28.

8   *"with a million-fold velocity"*: Ibid., 100.

8   *"absolute unity"*: Ibid., 23.

8   *"the most elaborate and profound"*: John Henry Hopkins, New York Morning Express, Feb. 4, 1848, in *TPL*, 721.

8   *"revolutionize the world"*: Poe to Eveleth, Feb. 29, 1848, in Ostrom, 2:650.

8   *"I have no desire"*: Poe to Maria Clemm, July 7, 1849, in Ostrom, 2:452, and *TPL*, 814.

9   *"the second scientific revolution"*: Thomas Kuhn, *The Essential Tension* (Chicago: University of Chicago Press, 1977), chap. 3. For Kuhn, science was transformed from 1800 to 1850 through the rise of precision measurement and experiment in numerous fields; see M. Norton Wise, ed., *The Values of Precision* (Princeton, N.J.: Princeton University Press, 1995); David Cahan, ed., *From Natural Philosophy to the Sciences: Writing the History of Nineteenth-Century Science* (Chicago: University of Chicago Press, 2003); John Pickstone, *Ways of Knowing: A New History of Science, Technology, and Medicine* (Chicago: University of Chicago Press, 2001).

10 The scientific situation in the United States: Key sources on antebellum science include Sally Gregory Kohlstedt, *The Formation of the American Scientific Community* (Urbana: University of Illinois Press, 1976); Sally Gregory Kohlstedt, "Parlors, Primers, and Public Schooling: Education for Science in Nineteenth-Century America," *Isis* 81, no. 3 (1990): 424–45; Nathan Reingold, *Science, American Style* (New Brunswick, N.J.: Rutgers University Press, 1991); Hunter Dupree, *Science in the Federal Government: A History of Policies and Activities* (Baltimore: Johns Hopkins University Press, 1986); Robert V.

Bruce, *The Launching of American Science, 1846–1876* (New York: Knopf, 1987); George H. Daniels, *American Science in the Age of Jackson* (Tuscaloosa: University of Alabama Press, 1994); Britt Rusert, *Fugitive Science: Empiricism and Freedom in Early African American Culture* (New York: New York University Press, 2017); Katherine Pandora, "Popular Science in National and Transnational Perspective: Suggestions from the American Context," *Isis* 100, no. 2 (2009): 346–58; Conevery Bolton Valenčius et al., "Science in Early America: Print Culture and the Sciences of Territoriality," *Journal of the Early Republic* 36, no. 1 (2016): 73–123; Marc Rothenberg, ed., *The History of Science in the United States: An Encyclopedia* (New York: Garland, 2001); Joshua Nall, *News from Mars: Mass Media and the Forging of a New Astronomy, 1860–1910* (University of Pittsburgh Press, 2019). American science in this period saw the emergence of professional, elite institutions (Reingold, Bruce, Daniel, Dupree) as well as an expansion of popular and everyday sites for science (Pandora, Valenčius, literature on Barnum and "pseudoscience"). Though often studied separately, Poe's life and work make it possible to see how these two tendencies intersected. On previous studies of Poe, see "A Note About Sources" above.

11 *Freud's student Marie Bonaparte*: Marie Bonaparte, *The Life and Works of Edgar Allan Poe: A Psycho-analytic Interpretation*, trans. John Rodker (London: Imago, 1949).

12 *the building of the United States*: General histories include Daniel Walker Howe, *What Hath God Wrought: The Transformation of America, 1815–1848* (New York: Oxford University Press, 2007); David S. Reynolds, *Waking Giant: America in the Age of Jackson* (New York: Harper, 2008); Charles Sellers, *The Market Revolution: Jacksonian America, 1815–1846* (New York: Oxford University Press, 1991); Sean Wilentz, *The Rise of American Democracy: Jefferson to Lincoln* (New York: W. W. Norton, 2006); Jill Lepore, *These Truths: A History of the United States* (New York: W. W. Norton, 2018); Howard Zinn, *A People's History of the United States: 1492–Present* (New York: Routledge, 2015); Eric Foner, *The Story of American Freedom* (New York: W. W. Norton, 1999).

Part I: From Allan to Poe

13 *"A dark unfathom'd tide"*: Edgar Allan Poe, "Imitation," in *Tamerlane and Other Poems* (Boston: Calvin Thomas, 1827), 29–30, and LOA, 34; an early "imitation" of Byron.

1. The Young Astronomer

15 *instrument of glass, wood, and brass*: Allen, *Israfel*, 1:130; the instrument is held at the Edgar Allan Poe House and Museum, Amity Street, Baltimore.

15 *"Astronomy, Conic Sections"*: Advertisement for Richmond Academy, *Richmond Compiler*, Oct. 2 and 4, 1821, in *TPL*, 48.

15 *"a chaos of deep passion"*: Poe, "Dreams," in *Tamerlane and Other Poems*, 26–27, and LOA, 31–32.

17 *"with the slaves, stocks, and properties"*: William Galt's will, probated March 29, 1824, MS, Deed Book 117-B, 99, Circuit Court of the City of Richmond, Va., Division I, in *TPL*, 63–64.

17 *"one of the handsomest women"*: *Norfolk Herald*, July 26, 1811; AHQ, 41–42, in *TPL*, 11.

17 *compound interest on a loan*: The page is reproduced in Allen, *Israfel*, 76; on Poe in Allan's firm, see Terence Whalen, *Edgar Allan Poe and the Masses: The Political Economy of Literature in Antebellum America* (Princeton, N.J.: Princeton University Press, 1999), 23.

17 *"Evening Star"*: Poe, "Evening Star," in *Tamerlane and Other Poems*, 28–29, and LOA, 33–34.

19 *"A singular fashion"*: Samuel Mordecai (a Richmond merchant and historian) to his sister Rachel, Nov. 2, 1811, Jacob Mordecai Papers, Duke University Library, in *TPL*, 13.

19 *"Mrs. Poe, lingering"*: *Richmond Enquirer*, Nov. 29, 1811, in *TPL*, 13.

19 *Plantation owners depended on traders*: David Leverenz, "Poe and Gentry Virginia," in *The American Face of Edgar Allan Poe*, ed. Shawn Rosenheim and Stephen Rachman (Baltimore: Johns Hopkins University Press, 1995), 210–36. Leverenz also shows the changing views toward slavery in Virginia. In the 1820s, debates in the state legislature flared over representation: which census to use, whether slaves should be counted (increasing the voting power of their owners), and various schemes for abolishing slavery—including recolonization of Liberia, established in 1822.

19 *White Sulphur Springs*: "Medicinal Springs of Virginia in the 19th Century," Claude Moore Health Sciences Library, University of Virginia, exhibits.hsl.virginia.edu/springs/introessay/.

19 *"charming everyone"*: Eugene L. Didier, summarizing the recollections of "several persons now living in Richmond, who remember seeing him" in 1812–15 at the springs, in *Life of Edgar A. Poe* (New York: W. J. Widdleton, 1877), 28, in *TPL*, 18.

20 *sexual liberties*: David S. Reynolds, *Beneath the American Renaissance: The Subversive Imagination in the Age*

of Emerson and Melville (New York: Oxford University Press, 2011); Teresa A. Goddu, *Gothic America: Narrative, History, and Nation* (New York: Columbia University Press, 1997).

20 *more than one third of the population*: Midori Takagi, *Rearing Wolves to Our Own Destruction: Slavery in Richmond Virginia, 1782–1865* (Charlottesville: University of Virginia Press, 2000).

20 *Dabney Dandridge*: James H. Whitty, "Memoir," in *The Complete Poems of Edgar Allan Poe*, ed. James H. Whitty (Boston: Houghton Mifflin, 1911), xxii, n2; Allen, *Israfel*, 1:196–97.

20 *African tales*: Susan A. T. Weiss, "The Allan Home," in *Home Life of Poe* (New York: Broadway, 1907), 18. On Poe and cultures of the African diaspora, see Thomas Bledsoe and Thomas Mabbott, "Poe and Armistead Gordon," *Phylon* 7, no. 4 (1946): 355–57; Joan Dayan, "Amorous Bondage: Poe, Ladies, and Slaves," *American Literature* 66, no. 2 (1994): 239–73; Colin Dayan, "Legal Terrors," *Representations* 92, no. 1 (2005): 42–80; Yvonne P. Chireau, *Black Magic: Religion and the African American Conjuring Tradition* (Berkeley: University of California Press, 2006).

20 *"terror is not of Germany"*: Poe, *Tales of the Grotesque and Arabesque* (Philadelphia: Lea and Blanchard, 1840), 5.

21 *Common sense*: Sophia A. Rosenfeld, *Common Sense: A Political History* (Cambridge, Mass.: Harvard University Press, 2011).

21 *Along with Benjamin Franklin*: See I. Bernard Cohen, *Science and the Founding Fathers: Science in the Political Thought of Jefferson, Franklin, Adams, and Madison* (New York: W. W. Norton, 1995); Tom Shachtman, *Gentlemen Scientists and Revolutionaries: The Founding Fathers in the Age of Enlightenment* (New York: Palgrave Macmillan, 2014).

22 *enlightened natural science*: See James Delbourgo, *A Most Amazing Scene of Wonders: Electricity and Enlightenment in Early America* (Cambridge, Mass.: Harvard University Press, 2006); Susan Scott Parrish, *American Curiosity: Cultures of Natural History in the Colonial British Atlantic World* (Chapel Hill: University of North Carolina Press, 2006); Andrew J. Lewis, *A Democracy of Facts: Natural History in the Early Republic* (Philadelphia: University of Pennsylvania Press, 2011).

22 *"bold, sublime experiment"*: Jefferson in Russell Blaine Nye, *The Cultural Life of the New Nation, 1776–1830* (New York: Harper & Row, 1960), 100.

22 *"a glorious task"*: Franklin to Samuel Cooper, May 1, 1777, in *The Writings of Benjamin Franklin*, ed. Albert Henry Smyth, 10 vols. (New York: Macmillan, 1905–7), 7:56.

22 *"presented in politics"*: Thomas Paine, *The Rights of Man* (London: J. S. Jordan, 1791), quoted in Nye, *Cultural Life*, 66.

22 *"trinity of the three greatest men"*: I. Bernard Cohen, "Science and the Growth of the American Republic," *Review of Politics* 38, no. 3 (July 1976): 365.

22 *referred to Newtonian laws*: John F. Kasson, *Civilizing the Machine: Technology and Republican Values in America, 1776–1900* (New York: Grossman, 1976).

22 *Attempts to present*: see Brooke Hindle, *The Pursuit of Science in Revolutionary America, 1735–1789* (Chapel Hill: University of North Carolina Press, 1956).

22 *five editions of Lucretius's*: James Gilreath and Douglas L. Wilson, eds., *Thomas Jefferson's Library: A Catalogue with Entries in His Own Order* (Washington, D.C.: Library of Congress, 1989), 119, cited in Robert J. Scholnick, *Poe's "Eureka," Erasmus Darwin, and Discourses of Radical Science in Britain and America, 1770–1850* (Lewiston, N.Y.: Edwin Mellen Press, 2018), David N. Stamos, *Edgar Allan Poe, "Eureka," and Scientific Imagination* (Albany: State University of New York Press, 2017), 131–32.

22 *"Botany I rank with the most valuable sciences"*: Jefferson to Thomas Cooper, Oct. 7, 1814, in *The Writings of Thomas Jefferson*, ed. H. A. Washington, 9 vols., new ed. (Washington, D.C.: Taylor & Maury, 1853–54; Cambridge, U.K.: Cambridge University Press, 2011), 6:390.

23 *"shriveled and diminished"*: Georges-Louis Leclerc, Comte de Buffon, *Histoire naturelle*, 36 vols. (Paris: Royal, 1749–88), 11:103–4, quoted in Lee Alan Dugatkin, "Buffon, Jefferson, and the Theory of New World Degeneracy," *Evolution: Education and Outreach* 12, no. 1 (2019): 2; J. C. Greene, *American Science in the Age of Jefferson* (Ames: Iowa State University Press, 1984).

23 *"the real distinctions"*: Jefferson, *Notes on the State of Virginia* (Richmond: J.W. Randolph, 1853), 149.

23 *"to wean yourselves"*: Benjamin Banneker to Thomas Jefferson, 19 Aug. 1791, Founders Online, National Archives, https://founders.archives.gov/documents/Jefferson/01-22-02-0049.

23 *"hideous blot"*: Jefferson to William Short, Sept. 8, 1823, Jefferson Papers, Founders Online, National Archives, founders.archives.gov/documents/Jefferson/98-01-02-3750; on Jefferson and colonization, see Peter S. Onuf, *Jefferson's Empire: The Language of American Nationhood* (Charlottesville: University of Virginia Press, 2000).

23 *a British warship had fired*: This incident became known as the *Chesapeake-Leopard* affair, after the two ships involved, the USS *Chesapeake* and the HMS *Leopard*; see James Fenimore Cooper, *History of the Navy of the United States of America* (New York: Stringer & Townsend, 1839).

24 *its merchants anxiously watched*: Letters of Dr. Thomas Massie of Richmond, in "Richmond During the War of 1812 (Continued)," *Virginia Magazine of History and Biography*, April 1900, 406–18.

24 *"Edgar says Pa say something"*: Allan to Ellis, Sept. 21, 1815, Ellis-Allan Papers, Library of Congress, in *TPL*, 26.

24 *The city that indelibly marked*: See J. Gerald Kennedy, "The Realm of Dream and Memory: Poe's England," in *Poe and Place*, ed. Philip Edward Phillips (Cham, Switzerland: Palgrave Macmillan, 2018), 71–96.

24 *"sitting by a snug fire"*: Allan to Ellis, Oct. 30, 1815, Ellis-Allan Papers, in *TPL*, 26.

25 *"I expect Edgar does not know"*: Catherine Poitiaux, May 18, 1816, in Killis Campbell, "New Notes on Poe's Early Years," *Dial*, Feb. 17, 1916, 144, in *TPL*, 30.

25 *"Taxes heavy, debt large"*: Allan to General John H. Cocke, Feb. 3, 1817, E. V. Valentine Collection, Valentine Museum, Richmond; AHQ, 70–71.

25 *Workers were increasingly desperate*: For a panorama of the "age of 'the spirit of the age,'" see James Chandler, *England in 1819: The Politics of Literary Culture and the Case of Romantic Historicism* (Chicago: University of Chicago Press, 1999).

25 *"natural Beauties"*: Allan to William Galt Jr., Nov. 12, 1818, William Galt Jr. Papers, William R. Perkins Library, Duke University, in *TPL*, 39.

25 *"I only wish"*: Frances K. Allan to John Allan, Oct. 15, 1818, Valentine Collection; AHQ, 78.

26 *visible and scandalous figure*: Peter J. Manning, "Childe Harold in the Marketplace: From Romaunt to Handbook," *Modern Language Quarterly* 52, no. 2 (1991): 170–90.

26 *"he was able to speak"*: William Elijah Hunter, "Poe and His English Schoolmaster," *Athenaeum*, Oct. 19, 1878, 496–97; AHQ, 71.

26 *"a palace of enchantment"*: Poe, "William Wilson," *Gift for 1840*, issued about Oct. 1839, 229–53.

26 *"quick and clever boy"*: Hunter, "Poe and His English Schoolmaster"; AHQ, 71.

26 *"Edgar is a fine boy"*: Allan to William Erwin [Ewing], March 21, 1818, Ellis-Allan Papers, in *TPL*, 36.

27 *"remarkable for self-respect"*: Clarke to E. L. Didier, April 16, 1876, Autograph MS, Harvard College Library; AHQ, 83.

27 *"so unwell"*: John Allan to Charles Ellis, 27 July 1820, Library of Congress, Ellis-Allan Papers, *TPL*, 45.

27 *"a leader among boys"*: Thomas H. Ellis, "Edgar Allan Poe," *Richmond Standard*, May 7, 1881, 2; AHQ, 82.

27 *Henry, living with Poe relatives*: See Kenneth Silverman, *Edgar A. Poe: Mournful and Never-Ending Remembrance* (New York: HarperCollins, 1991), 125–26. On Henry's life and relations with Poe, see Hervey Allen and Thomas Mabbott, eds., *Poe's Brother: The Poems of William Henry Leonard Poe* (New York: George H. Doran, 1926), 21; Richard Kopley, introduction to *The Narrative of Arthur Gordon Pym of Nantucket*, ed. Richard Kopley (New York: Penguin Classics, 1999), ix–xxix.

27 *"Of Edgar Poe it was known"*: Colonel John T. L. Preston, "Some Reminiscences of Edgar A. Poe as a Schoolboy," in *Edgar Allan Poe: A Memorial Volume*, ed. Sara S. Rice (Baltimore: Turnbull Brothers, 1877), 40–41; AHQ, 85.

2. In Jefferson's Experiment

29 *"He does nothing"*: Allan to Henry Poe, Nov. 1, 1824, Ellis-Allan Papers, in *TPL*, 61–62.

30 *the German idealists*: Albert J. Lubell, "Poe and A. W. Schlegel," *Journal of English and Germanic Philology* 52, no. 1 (1953): 1–12; Sean Moreland and Devin Zane Shaw, "'As Urged by Schelling': Coleridge, Poe, and the Schellingian Refrain," *EAPR* 13, no. 2 (2012): 50–80.

30 *"wild lake, with black rock"*: Poe, "The Lake," in *LOA*, 37.

30 *"was unhappy at home"*: Maria Clemm to Sarah Helen Whitman, April 14, 1859, Lilly Collection, Indiana University; AHQ, 86.

30 *"was pleasant but"*: Sarah Elmira Royster, in Edward V. Valentine, "Conversation with Mrs. Shelton at Mr. Smith's corner 8th and Leigh Streets, Nov. 19, 1875"; AHQ, 91.

31 *The old soldier and statesman*: Virginius Dabney, *Richmond: The Story of a City*, rev. ed. (Charlottesville: University of Virginia Press, 1990), 104–6.

31 *following the War of 1812*: Howe, *What Hath God Wrought*, 91–124.

31 *"the immense improvements"*: Auguste Levasseur, *Lafayette in America in 1824 and 1825; or, Journal of Travels in the United States*, 2 vols. (New York: Clayton and Van Norden, 1829), 2:14.

31 *"the phenomena of the heavens"*: John Quincy Adams, "First Annual Message (Dec. 6, 1825)," in *A Compilation of the Messages and Papers of the Presidents*, ed. James D. Richardson (Washington, D.C.: Government Printing Office, 1897), vol. 2, pt. 2, 299–317; Marlana Portolano, "John Quincy Adams's Rhetorical Crusade for Astronomy," *Isis* 91, no. 3 (2000): 480–503.

32 *Some showed off inventions*: Poe speaks of the kaleidoscope's effect in "The Philosophy of Furniture," a humorous article published in *Burton's Gentleman's Magazine*, May 1840, 243–45. On magic lantern shows, see X. Theodore Barber, "Phantasmagorical Wonders: The Magic Lantern Ghost Show in Nineteenth-Century America," *Film History* 3, no. 2 (1989): 73–86; Jeremy Brooker, *The Temple of Minerva. Magic and the Magic Lantern at the Royal Polytechnic Institution, London 1837–1901* (The Magic Lantern Society, London, 2013); works by Morus and Wells cited below.

32 *"that science is confined to no favored spot"*: *American Lyceum, with the Proceedings of the Conference Held in N.Y., May 4, 1831, to Organize the National Department of the Institution* (Boston: Hiram Tupper, 1831), 4. On the lyceum movement, see Carl Bode, *The American Lyceum: Town Meeting of the Mind* (New York: Oxford University Press, 1956); Kohlstedt, *Formation of the American Scientific Community*, 8; Kohlstedt, "Parlors, Primers, and Public Schooling"; Donald Zochert, "Science and the Common Man in Ante-bellum America," *Isis* 65, no. 4 (1974): 448–73.

33 *"the land of our birth"*: William H. Allen, *An Address Before the Cuvierian Society of the Wesleyan University, Middletown, Connecticut, July 31, 1838* (New York: Cuvierian Society, 1838), 13, quoted in Kohlstedt, *Formation of the American Scientific Community*, 17; Kent P. Ljungquist, "Lectures and the Lyceum Movement," in *The Oxford Handbook of Transcendentalism*, ed. Joel Myerson, Sandra Harbert Petrulionis, and Laura Dassow Walls (New York: Oxford University Press, 2010), 330–47.

33 *the exuberant Prussian polymath*: See Laura Dassow Walls, *The Passage to Cosmos: Alexander von Humboldt and the Shaping of America* (Chicago: University of Chicago Press, 2009); Mary Louise Pratt, *Imperial Eyes: Travel Writing and Transculturation* (New York: Routledge, 2007); Andrea Wulf, *The Invention of Nature. Alexander von Humboldt's New World* (New York: Vintage, 2016); Ingo Schwarz, "Alexander von Humboldt's Visit to Washington and Philadelphia," *Northeastern Naturalist* 8 (2001): 43–56; Michael Dettelbach, "The Face of Nature: Precise Measurement, Mapping, and Sensibility in the Work of Alexander von Humboldt," *Studies in History and Philosophy of Science Part C* 30, no. 4 (1999): 473–504.

34 *improving the state of learning*: Robert M. Healey, *Jefferson on Religion in Public Education* (New Haven, Conn.: Yale University Press, 1962), 183.

34 *"higher branches of numerical arithmetic"*: Jefferson in Charles Maurice Wiltse, *The Jeffersonian Tradition in American Democracy* (Chapel Hill: University of North Carolina Press, 1935), 141.

34 *"Academical Village"*: Quoted in Nathan Reingold, *Science in Nineteenth-Century America, a Documentary History* (New York: Hill & Wang, 1964), 22.

34 *"natural aristocracy"*: Healey, *Jefferson on Religion in Public Education*, 63.

35 *"destined for learned professions"*: Ibid., 151.

35 *"uncontrolled choice in the lectures"*: Jefferson to George Ticknor, July 16, 1823, in *The Writings of Thomas Jefferson*, ed. Albert Ellery Bergh and Andrew A. Lipscomb, 20 vols. (Washington, D.C.: Jefferson Memorial Association, 1903–4), 15:455.

35 *a French textbook*: For the recruitment of European professors, see Dumas Malone, *The Sage of Monticello*, 6 vols. (Boston: Little, Brown, 1981), 6:397–422; Nicolas Dufief, *Nature Displayed in Her Mode of Teaching Language to Man* (Philadelphia: L. Plowman, 1804).

36 *gambled, cursed, dueled*: AHQ, 107.

36 *part of a genre of literary hoaxes*: Julia Luisa Abramson, *Learning from Lying: Paradoxes of the Literary Mystification* (Newark: University of Delaware Press, 2005).

36 *"was done beyond eating"*: Poe, "Von Jung (Mystification)," in *The Collected Works of Edgar Allan Poe*, 3 vols., ed. Thomas Ollive Mabbott (Cambridge, Mass.: Harvard University Press, 1969–78), 2:296.

36 *"The pillars of the Portico"*: Poe to Allan, Sept. 21, 1826, in Ostrom, 1:6.

36 *"very excitable & restless"*: Miles George to E. V. Valentine, May 18, 1880; AHQ, 108.

36 *"quoting poetic authors"*: AHQ, 104.

37 *"Super Blue Cloth"*: Silverman, *Mournful*, 34.

37 *"To calm & quiet"*: George to Valentine, May 18, 1880; AHQ, 108.

37 *"It was not the taste"*: Thomas Goode Tucker to Douglass Sherley, April 5, 1880, in Woodberry, *Life*, 1:33, and *TPL*, 69–70.

37 *a book of Hogarth prints*: Theodore Pease Stearns, "A Prohibitionist Shakes Dice with Poe," *Out-look*, Sept. 1, 1920, 25–26, in *TPL*, 71–72.

37 *"The uncalculating vanity"*: Poe, "William Wilson," *Gift for 1840*, 245.

38 *"mistaken parsimony"*: Poe to Allan, Jan. 3, 1831, in Ostrom, 1:58–63; AHQ, 110–11.

38 *"inquired into his ways"*: T. H. Ellis, *Richmond Standard*, May 7, 1881, in *TPL*, 75.

38 *"attempting to give him"*: Ellis, "Edgar Allan Poe," 2.

39 *"I saw thee"*: Poe, "Song," in LOA, 31.

39 *"exposing me before those"*: Poe to Allan, March 19, 1827, in Ostrom, 1:10–11.

39 *"I taught you to aspire"*: Allan to Poe, March 20, 1827, quoted in Mary Newton Stanard, *Edgar Allan Poe Letters till Now Unpublished in the Valentine Museum, Richmond, Virginia* (Philadelphia: J. B. Lippincott, 1925), 76, in *TPL*, 78.

40 *"I'm thinking Edgar has gone"*: Allan to his sister, March 27, 1827, Ellis-Allan Papers, in *TPL*, 78.

3. Exile, Artificer, Cadet

41 *"a rather stoop-shouldered clerk"*: Stearns, "Prohibitionist Shakes Dice with Poe," 25–26. On Poe's brief time as a Boston clerk, see Mabbott, *Collected Works*, 1:539.

42 *between forty and at most two hundred*: See Susan Jaffe Tane and Gabriel Mckee, *Evermore: The Persistence of Poe: The Edgar Allan Poe Collection of Susan Jaffe Tane* (New York: Grolier Club, 2014); Leon Jackson, "Poe and Print Culture," *PS/Dark Romanticism* 33, no. 1–2 (2000): 4–9, and "The Italics Are Mine: Edgar Allan Poe and the Semiotics of Print," in *Illuminating Letters: Typography and Literary Interpretation*, ed. Paul C. Gutjahr and Megan L. Benton (Amherst: University of Massachusetts Press, 2001), 53. Only twelve copies of *Tamerlane* are now known; a copy sold at auction in 2009 for a record price. Frank James, "Poe Book Auctioned for $662,500, New Record," NPR, Dec. 4, 2009, www.npr.org/sections/thetwo-way/2009/12/poe_book_auctioned_for_662500.html.

42 *He joined the U.S. Army*: John Samuel Tieman, "Sergeant Major Edgar Allan Poe," *International Journal of Applied Psychoanalytic Studies* 13, no. 4 (2016): 351–66.

42 *Probably half of the army's soldiers*: Ibid.

42 *some biographers pass over*: Quinn (AHQ), Silverman (*Mournful*), and Jeffrey Meyers (*Edgar Allan Poe: His Life and Legacy* [New York: Charles Scribner's Sons, 1992]) give Poe's army years short shrift; in contrast, see William F. Hecker, *Private Perry and Mister Poe: The West Point Poems, 1831* (Baton Rouge: Louisiana State University Press, 2005), and John Thomas Russell, *Edgar Allan Poe: The Army Years* (West Point, N.Y.: U.S. Military Academy, 1972).

42 *promoted to "artificer"*: Michael L. Howard, "Seeds of a Soldier: The True Story of Edgar Allan Poe—the Sergeant Major," *Army Space Journal* (Fall 2003): 60, 56–57.

42 *The work required minute attention*: Hecker (*Private Perry and Mister Poe*) links Poe's "explosive" poetics to the meticulous attention required as artificer; see also Benjamin F. Fisher, *The Cambridge Introduction to Edgar Allan Poe* (Cambridge, U.K.: Cambridge University Press, 2008), 3.

43 *"no longer a boy"*: Poe to Allan, Dec. 1, 1828, Valentine Collection, in Ostrom, 1:14–15, and *TPL*, 86.

43 *"complete master of all exercises"*: William Duane, *A Handbook for Infantry*, 9th ed. (Philadelphia: printed for the author, 1814), quoted in Howard, "Seeds of a Soldier," 60.

43 *"a suit of Black clothes"*: Allan to Powhatan Ellis, March 4, 1829, Ellis-Allan Papers, in *TPL*, 89.

44 *"an experienced soldier"*: Col. James House, Mar. 30, 1829; AHQ, 135.

44 *"From childhood's hour"*: Poe, "Alone," MS copy, Lucy Holmes Balderston Album, 1826–48, MS 1796, Maryland Historical Society; LOA, 60; I. B. Cauthen Jr., "Poe's *Alone*: Its Background, Source, and Manuscript," *Studies in Bibliography* 3 (1950/1951): 284–91.

45 *"habits are good"*: J. Howard (lieutenant, First Artillery), letter of recommendation, April 20, 1829, quoted in Kenneth Walter Cameron, "Young Poe and the Army—Victorian Editing," *ATQ* 20, supplement, pt. 4 (Fall 1973): 158, in *TPL*, 90–91.

45 *"having been born"*: Colonel James P. Preston to Secretary Eaton, May 13, 1829, quoted in Cameron, "Young Poe," 166–77, in *TPL*, 93.

45 *"from the highest"*: Andrew Stevenson to John Eaton, 6 May 1829, in *TPL*, 91.

45 *"a gentleman in whose"*: John Campbell to John Eaton, 6 May 1829, in *TPL*, 91.

45 *politician and lawyer William Wirt*: Wirt was the candidate for president on the Anti-Masonic ticket in 1832. He would argue two Supreme Court cases after being approached by leaders of the Cherokee nation which led to the definition of tribal sovereignty for Indian nations as well as establishing the

federal government and not individual states as the agency with which tribes would negotiate. Andrew Jackson, however, ignored *Worcester v. Georgia* and allowed Georgia to remove Cherokees in what would be known as the Trail of Tears. See Stuart Banner, *How the Indians Lost Their Land: Law and Power on the Frontier* (Cambridge, Mass.: Harvard University Press, 2005).

45 *"a good deal of curious and useful"*: Wirt to Poe, May 11, 1829, in Stanard, *Poe Letters*, 131–32, in *TPL*, 92.

46 *"If the poem"*: Poe to Lea, before May 27, 1829, in Ostrom, 1:28.

46 *"there is much in being"*: Poe to Allan, May 29, 1829, in Ostrom, 1:30.

46 *"men of genius"*: Poe, quoting Allan, to Allan, July 26, 1829, in Ostrom, 1:36.

46 *"might make a beautiful"*: Neal, in the *Yankee, and Boston Literary Gazette*, Sept. 1829, in *TPL*, 98.

46 *"If the remainder"*: "Unpublished Poetry," *Yankee*, Dec. 1829, in *TPL*, 100.

46 *"A part are exceedingly boyish"*: Review of *Al Aaraaf, Tamerlane, and Minor Poems*, by Poe, *American Ladies' Magazine and Literary Gazette* (Boston), Jan. 1830, in *TPL*, 103.

46 *"Next Poe who smil'd"*: *The Musiad, or, Ninead, a Poem, by Diabolus, Edited by Me* (Baltimore, 1830), 8, quoted in Mabbott, *Collected Works*, 1:541, in *TPL*, 103.

47 *"SCIENCE! meet daughter"*: Poe, *Al Aaaraaf, Tamerlane, and Minor Poems* (Baltimore: Hatch and Dunning, 1829), 11.

47 *"Rightly is truth called"*: Francis Bacon, *The New Organon* ed. Lisa Jardine and Michael Silverthorne [Cambridge: Cambridge University Press, 2000 (1620)], 69.

48 *"attack upon Science"*: AHQ, 161.

48 *John Keats's Lamia*: John Keats, *Lamia, Isabella, The Eve of St. Agnes, and Other Poems* (London: Taylor and Hessey, 1820), 1–46.

48 *Poe provided one*: Stephen Rachman, "From 'Al Aaraaf' to the Universe of Stars: Poe, the Arabesque, and Cosmology," *EAPR* 15, no. 1 (2014): 1–19; Jeffrey Einboden, *Islam and Romanticism: Muslim Currents from Goethe to Emerson* (London: Oneworld, 2014) chap 15.

48 *"a medium between Heaven & Hell"*: Poe to Isaac Lea, before May 27, 1829, in Ostrom, 1:26–27.

48 *Tycho Brahe*: See Adam Mosley, *Bearing the Heavens: Tycho Brahe and the Astronomical Community of the Late Sixteenth Century* (Cambridge, U.K.: Cambridge University Press, 2007).

49 *how a poet might love science*: David Van Leer, "Nature's Book: The Language of Science in the American Renaissance," in *Romanticism and the Sciences*, ed. Andrew Cunningham and Nicholas Jardine (Cambridge, U.K.: Cambridge University Press, 1990), 307–21.

50 *"Mr. A is not very often sober"*: Poe to Graves, May 3, 1830, in Ostrom, 1:54.

50 *"a disposition and capacity"*: John H. Eaton, 1830, in Silverman, *Mournful*, 60.

50 *entrance exam*: *General Regulations for the Army*, War Department, 1825, No. 1337, in Karl E. Oelke, "Poe at West Point—a Revaluation," *PS* 6, no. 1 (1973): 5.

50 *"a great many cadets"*: Poe to Allan, June 28, 1830, in Ostrom, 1:56.

51 *"One fellow, on being sent up"*: George Cullum to Alfred Huidekoper, June 21, 1832, quoted in Oelke, "Poe at West Point," 1.

51 *"the only means of opening"*: Albert Church, assistant professor of mathematics at the Military Academy from 1828 to 1831, in "Personal Reminiscences," *Twelfth Annual Reunion of the Association of Graduates of the U.S. Military Academy at West Point, NY, June 9, 1881* (East Saginaw, Mich.: E. W. Lyon, 1881), 151, quoted in Oelke, "Poe at West Point," 2.

51 *"an unauspicious period"*: Thayer in Stephen E. Ambrose, *Duty, Honor, Country: A History of West Point* (Baltimore: Johns Hopkins University Press, 1966), 42.

51 *"rigid in the extreme"*: Poe to Allan, June 28, 1830, in Ostrom, 1:56.

52 *followed a curriculum*: The most reliable account of texts used in the first year of teaching is the 1831 *Board of Visitors Report*, with its "Synopsis of the Course of Studies . . ." including the mathematics texts Lacroix, *Elements of Algebra*, and Legendre, *Elements of Geometry*; see V. Frederick Rickey, David C. Arney, and Joe Albree, eds., *A Station Favorable to the Pursuits of Science: Primary Materials in the History of Mathematics at the United States Military Academy* (Providence: American Mathematical Society, 2000); Oelke, "Poe at West Point," 3.

52 *engineer and science popularizer*: Charles Dupin, *Mathematics Practically Applied to the Useful and Fine Arts*, trans. G. Birkbeck (London: C. Tait, 1827). Poe also likely had in mind Charles's older brother, the even more slippery politician André-Marie-Jean-Jacques Dupin; see John T. Irwin, *The Mystery to a Solution: Poe, Borges, and the Analytic Detective Story* (Baltimore: Johns Hopkins University Press, 1996), 341–45; Richard Kopley, *Edgar Allan Poe and the Dupin Mysteries* (New York: Palgrave Macmillan, 2008),

96n3. See Claudius Berard, "Catalogue of Books Contained in Ten Cases, Lately Received from France, and Sent to the Library of the Military Academy at W-Point by General J. G. Swift," hand-written MS, Sept. 15, 1817, USMA Library, Special Collections.

52 *nebular hypothesis*: See Constance A. Lubbock, *The Herschel Chronicle* (Cambridge, U.K.: Cambridge University Press, 2013), 310; Hervé Faye, *Sur l'origine du monde: Théories cosmogoniques des anciens et des modernes* (Paris: Gauthier-Villars, 1884), 109–11.

52 *"I had no need"*: The source for this immortal quote appears to be Francesco Antommarchi, *Mémoires du docteur F. Antommarchi; ou, Les derniers momens de Napoléon*, 2 vols. (Paris: Barrois L'Aîné, 1825), 1:282.

52 *first American institutions to use blackboards*: Thomas Fleming, *West Point: The Men and Times of the United States Military Academy*, 2nd ed. (New York: Morrow, 1969), 201; Christopher J. Phillips, "An Officer and a Scholar: Nineteenth-Century West Point and the Invention of the Blackboard," *History of Education Quarterly* 55, no. 1 (2015): 82–108; Michael J. Barany and Donald MacKenzie, "Chalk: Materials and Concepts in Mathematics Research," in *Representation in Scientific Practice Revisited*, ed. Catelijne Coopmans et al. (Cambridge, Mass.: MIT Press, 2014), 107–29.

52 *"we have already gone through"*: John C. Pemberton to his parents, Sept. 20, 1822, microfilm, Pemberton Family Papers, USMA Library, Special Collections; original at Historical Society of Pennsylvania. Cited in Oelke, "Poe at West Point," 3n22.

53 *"superiority in scholarship"*: Joseph Henry, "Eulogy on Prof. Alexander Dallas Bache, Late Superintendent of the United States Coast Survey," *Annual Report of the Smithsonian Institution for 1870* (Washington, D.C.: Government Printing Office, 1871), 91.

53 *"how greatly I was gratified"*: Barbour to Sophia Bache, June 10, 1825, series RU 7053, box 3, A. D. Bache Papers, Smithsonian Institution Archives; discussed in Axel Jansen, *Alexander Dallas Bache: Building the American Nation Through Science and Education in the Nineteenth Century* (Frankfurt: Campus, 2011), 59.

54 *"excellent standing"*: Poe to Allan, Nov. 6, 1830, in Ostrom, 1:57.

54 *"worn, weary, discontented"*: Thomas W. Gibson, "Poe at West Point," *Harper's New Monthly Magazine*, Nov. 1867, 754, in TPL, 109.

54 *"the greatest fellow"*: Timothy Pickering Jones interview, *New York Sun*, May 10, 1903; Woodberry, *Life*, 1:369, and TPL, 107.

54 *"a wonderful aptitude"*: Allan B. Magruder, April 23, 1884, quoted in Woodberry, *Life*, 1:70, and TPL, 107.

54 *"forceful and vicious doggerel"*: Gibson, "Poe at West Point," 754, in TPL, 108.

54 *"which does not fit"*: George Cullum to his sister, Feb. 14, 1833, USMA Library, Special Collections, quoted in Oelke, "Poe at West Point," 4.

54 *Other publications available*: Oelke, "Poe at West Point," 4.

54 *a riotous celebration on July 4*: Ibid., 1–2.

54 *"had already acquired"*: Gibson, "Poe at West Point," 755, in TPL, 109.

55 *"he had procured a cadet's appointment"*: Gibson, "Poe at West Point," 754, in TPL, 109.

55 *"I was greatly in hopes"*: Poe to Allan, Nov. 6, 1830, in Ostrom, 1:57–58.

55 *"Did I, when an infant"*: Poe to Allan, Jan. 3, 1831, in Ostrom, 1:58–63.

56 *seventeenth in mathematics and third in French*: "Register of Merit, No. 1, 1817 to 1835," USMA Library, Special Collections, quoted in Oelke, "Poe at West Point," 2.

56 *"will be dismissed"*: John Henry Eaton, "Military Academy Order, No. 7," Washington, Feb. 8, 1831, USMA Library, Special Collections, Post Order Book no. 5, 1827–39.

56 *"fellow of talent"*: David E. Hale to Sarah Josepha Hale, Feb. 10, 1831; AHQ, 171.

56 *"Edgar Poe, at West Point"*: Eugene L. Didier, *The Poe Cult and Other Poe Papers* (New York: Broadway, 1909), 224–25, in TPL, 115.

56 *"Poe's stay at West Point"*: AHQ, 174.

56 *built roads and canals*: William H. Goetzmann, *Exploration and Empire: The Explorer and the Scientist in the Winning of the American West* (New York: Knopf, 1966).

57 *many made fortunes*: Alfred D. Chandler Jr., *The Visible Hand* (Cambridge, Mass.: Harvard University Press, 1993); Captain William V. Judson, Corps of Engineers, U.S. Army, "The Services of Graduates as Explorers, Builders of Railways, Canals, Bridges, Light-Houses, Harbours, and the Like (1886)," in *The Centennial of the United States Military Academy at West Point, New York, 1802–1902* (U.S. Military Academy, 1904), 1:835–74.

Part II: Setting Sail

59   "*I was ambitious*": Poe, "Tamerlane," LOA, 27.

4. A Baltimore Apprenticeship

61   "*I have no money*": Poe to Allan, Feb. 21, 1831, in Ostrom, 1:64, and TPL, 115.

61   "*certificate of standing*": Poe to Thayer, March 10, 1831, in Ostrom, 1:55–56; for provenance of this
      letter, see Ostrom, 2:472n30.

61   *contributed* $1.25: Ledger records, Treasurer of the U.S. Military Academy Thomas J. Leslie,
      cited in Russell, *Army Years*, 29–30, in TPL, 117; on Poe's finances at West Point, see Silverman,
      Mournful, 67.

61   "*a miserable production*": General Allan B. Magruder to George E. Woodberry, July 1, 1884, quoted
      in Woodberry, *Life*, 1:78.

61   *recited in the barracks*: Gibson, "Poe at West Point," 755, in TPL, 118.

61   "*Every thing in the language*": Review by either George P. Morris or Theodore S. Fay, *New-York Mir-
      ror*, May 7, 1831, in TPL, 119.

62   "*a fellow of fine genius*": Elam Bliss, review in the *Morning Courier and New-York Enquirer*, July 8, 1831,
      124, in TPL, 120–21.

62   "*Helen, thy beauty is to me*": Poe, "To Helen," in LOA, 62.

62   "*great barrier in the path*": Poe, "Letter to B——," reprinted in SLM 2, no. 8 (July 1836): 501–3.

62   *Coleridge's verse*: See Floyd Stovall, "Poe's Debt to Coleridge," *Studies in English*, July 8, 1930, 70–127.

62   *description of "indirect" observation*: Roberta Sharp, "Poe's Chapters on 'Natural Magic,'" in *Poe and
      His Times: The Artist and His Milieu*, ed. Benjamin Franklin Fisher IV (Baltimore: Edgar Allan Poe Soci-
      ety, 1990), 154–66; Laura Saltz, "'Eyes Which Behold': Poe's 'Domain of Arnheim' and the Science
      of Vision," EAPR 7, no. 1 (2006): 4–30.

63   "*anxious to remain*": Poe to William Gwynn, May 6, 1831, in Ostrom, 1:66.

63   *The city prided itself on industry*: David Gaylin, *Edgar Allan Poe's Baltimore* (Mount Pleasant, S.C.: Arca-
      dia, 2015); Ruth Schwartz Cowan, *A Social History of American Technology* (New York: Oxford University
      Press, 1997), 94–115.

63   *more than seventy journals*: John C. French, "Poe's Literary Baltimore," *Maryland Historical Magazine*,
      June 1937, 101–12; AHQ, 187n2.

64   *Douglass*: J. Gerald Kennedy, "'Trust No Man': Poe, Douglass, and the Culture of Slavery," in
      Romancing the Shadow: Poe and Race, ed. J. Gerald Kennedy and Liliane Weissberg (New York: Oxford
      University Press, 2001), 225–57. For Douglass's time in Baltimore, see Frederick Douglass, *Narrative of
      the Life of Frederick Douglass, an American Slave, Written by Himself* (Boston: Anti-Slavery Office, 1845), 26–50;
      David Blight, *Frederick Douglass: Prophet of Freedom* (New York: Simon & Schuster, 2018), 48–66.

64   *the preeminent national issue*: See Aziz Rana, *The Two Faces of American Freedom* (Cambridge, Mass.:
      Harvard University Press, 2011); Reynolds, *Waking Giant*.

64   *David Walker*: Rusert, *Fugitive Science*, 40–44.

64   *revival meetings*: Christine Leigh Heyrman, *Southern Cross: The Beginnings of the Bible Belt* (Chapel
      Hill: University of North Carolina Press, 1998).

65   "*A city slave*": Douglass, *Narrative*, 38, quoted in Kennedy, "Trust No Man," 231.

65   "*The idea as to how I might learn*": Douglass, *Narrative*, 43.

65   *a fiercely informed opponent*: Rusert, *Fugitive Science*, 13–14, 124–28.

65   *Baltimore meant living among caring relatives*: Whitty, "Memoir," xix–lxxxvi.

65   *fear of contagion*: Charles Rosenberg, *The Cholera Years: The United States in 1832, 1849, and 1866* (Chicago:
      University of Chicago Press, 1962). Recent work on Poe and medicine has moved beyond a preoccu-
      pation with (his alleged) psychopathology to consider Poe's adaptations of medical writing and his
      reflections on the basis and limitations of its claims. See, for instance, Dana Medoro, "Introduction:
      Edgar Allan Poe and Nineteenth-Century Medicine," PS 50, no. 1 (2017): 2–11; Cristina Pérez, "Edgar
      Allan Poe, MD: Medical Fiction and the Birth of Modern Medicine," *Trespassing Journal* 4 (Fall 2014):
      64–65; Emily Gowen, "A Global Sickness," EAPR 20, no. 2 (2019): 269–88.

65   *teacher and an editorial assistant*: Weiss, *Home Life of Poe*, 62–63, in TPL, 127.

65   *making bricks*: Robert T. P. Allen, a West Point classmate, in "'Edgar Allan Poe,'" *Scribner's Monthly*,
      Nov. 1875, 143, in TPL, 141.

65   "*one of the most hardworking men*": Lambert Wilmer, "Recollections of Edgar A. Poe," *Baltimore Daily*

Commercial, May 23, 1866, reproduced in Lambert Wilmer, *Merlin, Baltimore, 1827; Together with Recollections of Edgar A. Poe*, ed. Thomas Ollive Mabbott (New York: Scholars' Facsimiles & Reprints, 1941); AHQ 197. Wilmer later wrote a play based on Poe's thwarted love affair with Elmira Royster.

65 *general conditions of literary production*: Michael L. Allen, *Poe and the British Magazine Tradition* (New York: Oxford University Press, 1969).

66 *"giving Virginia lessons"*: Wilmer, "Recollections of Edgar A. Poe"; AHQ 198, in *TPL*, 125–26.

66 *"When I think of the long twenty one years"*: Poe to Allan, Oct. 16, 1831, in Ostrom, 1:67–68.

66 *"I am in the greatest distress"*: Poe to Allan, Nov. 18, 1831, in Ostrom 1:47.

66 *"Do not let me perish"*: Poe to Allan, Dec. 15, 1831, in Ostrom 1:48.

66 *inmates of Baltimore's jail*: Lawrence G. Wroth, "Poe's Baltimore," *Johns Hopkins Alumni Magazine*, June 2, 1929, 4; AHQ 190.

67 *"few American authors"*: Editorial in *Baltimore Saturday Visiter*, Aug. 4, 1832; AHQ 195.

67 *"supposed to be read"*: Poe to Joseph T. and Edwin Buckingham, May 4, 1833, in Ostrom, 1:77, in AHQ 199.

67 Tales of the Folio Club: See Alexander Hammond's extensive work, including "Edgar Allan Poe's *Tales of the Folio Club*: The Evolution of a Lost Book," in *Poe at Work: Seven Textual Studies*, ed. Benjamin Franklin Fisher IV (Baltimore: Edgar Allan Poe Society, 1976), 13–43.

67 *The starting point for "A Decided Loss"*: See Michael J. S. Williams, *A World of Words: Language and Displacement in the Fiction of Edgar Allan Poe* (Durham, N.C.: Duke University Press, 1988), 49–53.

67 *"silver fork" novels*: Edward Copeland, *The Silver Fork Novel: Fashionable Fiction in the Age of Reform* (Cambridge, U.K.: Cambridge University Press, 2012); Alexander Hammond, "The Folio Club Collection and the Silver Fork School: Perspectives on Poe's Framestory in Recent Scholarship," *EAPR* 19, no. 2 (2018): 153–76.

68 *"Lionizing"*: Alexander Hammond, "Poe's 'Lionizing' and the Design of *Tales of the Folio Club*," *ESQ* 18, no. 3 (1972): 154–65. On Poe and literary fame, see Leon Jackson, "'The Rage for Lions': Edgar Allan Poe and the Culture of Celebrity," in *Poe and the Remapping of Antebellum Print Culture*, ed. J. Gerald Kennedy and Jerome J. McGann (Baton Rouge: Louisiana State University Press, 2012), 37–61.

68 *"Fine writer!"*: Poe, "Lionizing," LOA, 213.

68 *French precedents*: Ken Alder, *Engineering the Revolution: Arms and Enlightenment in France, 1763–1815* (Chicago: University of Chicago Press, 2010); Merritt Roe Smith, *Harpers Ferry Armory and the New Technology: The Challenge of Change* (Ithaca, N.Y.: Cornell University Press, 1977).

69 *John Latrobe*: John Edward Semmes, *John H. B. Latrobe and His Times, 1803–1891* (Baltimore: Norman, Remington, 1917).

69 *"was so far, so very far superior"*: Latrobe to Charles Chauncey Burr, Dec. 7, 1852, quoted in Jay B. Hubbell, *The South in American Literature, 1607–1900* (Durham, N.C.: Duke University Press, 1954), 837–39, in *TPL*, 132.

70 *"who carried himself erect"*: "Reminiscences of Poe by John H. B. Latrobe," in Rice, *Memorial Volume*, 60, in *TPL*, 133.

70 *"To judge by the outward man"*: Latrobe to Burr, Dec. 7, 1852, in Hubbell, *American Literature*, 837–39, in *TPL*, 132.

70 *"MS. Found in a Bottle"*: *Baltimore Saturday Visiter*, Oct. 19, 1833, 1.

71 *"perishing—absolutely perishing"*: Poe to Allan, April 12, 1832, in Ostrom, 1:73–74.

71 *"Mr. Allan raised his cane"*: Ellis, "Edgar Allan Poe," in *TPL*, 137.

71 *Allan died*: See Silverman, *Mournful*, 97–99.

71 *"has wounded me to the quick"*: Poe to Kennedy, March 15, 1835, in Ostrom, 1:83–84.

71 *"found him in a state of starvation"*: Kennedy, diary entry, Oct. 10, 1849 [shortly after Poe's death], quoted in Woodberry, *Life*, 2:350–51, and *TPL*, 148–49.

72 *"monomania"*: Étienne Esquirol, *Des maladies mentales: Considérées sous les rapports médical, hygiénique et médico-légal*, 2 vols. (Paris: J. B. Baillière, 1838), 2:1–130.

72 *"Would to God"*: Poe, "Berenice," *SLM*, 2, no. 7, March 1835: 333–36.

72 *"far too horrible"*: Poe to White, April 30, 1835, in Ostrom, 1:84–85.

73 *discuss, debate, and reprint it*: Meredith McGill, *American Literature and the Culture of Reprinting, 1834–1853* (Philadelphia: University of Pennsylvania Press, 2007); Jonathan Elmer, *Reading at the Social Limit: Affect, Mass Culture, and Edgar Allan Poe* (Stanford, Calif.: Stanford University Press, 1995); Whalen, *Poe and the Masses*.

73   *the possibility of a job*: AHQ, 209.

73   *"Nothing would give me greater pleasure"*: Poe to White, June 22, 1835, in Ostrom, 1:93.

5. Richmond: The Palpable Obscure

74   *southern intellectual and political circles*: Prominent figures included the novelist and essayist Lucian Minor, the law professor Nathaniel Beverley Tucker, and the president of William & Mary College, Thomas R. Dew. See AHQ, 218; David K. Jackson, *Poe and the "Southern Literary Messenger"* (Richmond: Dietz, 1934); Drew Gilpin Faust, *A Sacred Circle: The Dilemma of the Intellectual in the Old South, 1840–1860* (Baltimore: Johns Hopkins University Press, 1977).

74   *"imparts the same energy"*: Thomas White, "Publisher's Notice," *SLM* 1, no. 1 (Aug. 1834): 2.

74   *"to establish something"*: Thomas White, "Virginia Historical and Philosophical Society," *SLM* 1, no. 3 (Nov. 1834): 123.

74   *"to see the North and South"*: Thomas White, "Prospectus of the *Southern Literary Messenger*," *SLM* 4, no. 1 (Jan. 1838): ii.

75   *rioting and attacking free Blacks*: Howe, *What Hath God Wrought*, 423–31.

75   *"peculiar institution"*: See Kenneth Stampp, *The Peculiar Institution: Slavery in the Ante-Bellum South* (New York: Knopf, 1956).

75   *"positive good"*: John Calhoun, from his 1837 "Speech on the Reception of Abolition Petitions," in *The Works of John C. Calhoun*, ed. Richard K. Crallé, 6 vols. (New York: D. Appleton, 1851–56), 2:625–33.

75   *"average racism"*: Whalen defines this as "a form of racism acceptable to white readers who were otherwise divided over the more precise issue of slavery" (*Poe and the Masses*, 112). From the end of Reconstruction and well into the twentieth century, Poe was often presented as a southern partisan and defender of slavery, in part because the Drayton-Paulding review was attributed to him: his critiques of democracy as "mob rule" and defense of the privilege due to exceptional individuals have been read as expressions of southern aristocratic and slaveholding values. A much more complex and contextualized picture of Poe has emerged in recent years, as marked by the 1995 edited collection by Rosenheim and Rachman, *The American Face of Edgar Allan Poe*, which included Whalen's demonstration that Poe was not the author of the Drayton-Paulding review. Poe's hostility toward the democratic "mob" was directed above all at Jackson's followers, and his stances and identifications have been shown to be mobile, ambiguous, and often subversive. Key texts concerning Poe, slavery, and race include Harry Levin, *The Power of Blackness: Hawthorne, Poe, Melville* (New York: Knopf, 1958); Toni Morrison, *Playing in the Dark: Whiteness and the Literary Imagination* (Cambridge, Mass.: Harvard University Press, 1992); the collection edited by Kennedy and Weissberg, *Romancing the Shadow*; Goddu, *Gothic America*; and Joan (Colin) Dayan's works linking Poe's imagery and metaphysics to the legal status of the enslaved and imprisoned, from *Fables of Mind: An Inquiry into Poe's Fiction* (New York: Oxford University Press, 1987) to *The Law Is a White Dog: How Legal Rituals Make and Unmake Persons* (Princeton, N.J.: Princeton University Press, 2013).

75   *sold a twenty-one-year-old enslaved man*: See Kennedy, "Trust No Man," 235; John Miller, "Did Edgar Allan Poe Really Sell a Slave?," *PS* 9, no. 2 (1976): 52–53; the "letter of conveyance" is reproduced at eapoe.org (www.eapoe.org/works/docs/d2912100.htm).

76   *"No early American author"*: Morrison, *Playing in the Dark*, 32.

76   *He promoted accessible works*: "Critical Notices," *SLM* 2, no. 10 (Aug. 1836): 596, review of *Introductory Lecture to a Course of Chemistry and Natural Philosophy: Delivered in Hampden Sidney College*, by John W. Draper.

77   *"highly creditable to Cincinnati"*: "Critical Notices," *SLM* 1, no. 13 (Sept. 1835): 778. For a helpful survey of Poe's science writing, see Carroll Dee Laverty, "Science and Pseudo-science in the Writings of Edgar Allan Poe" (PhD diss., Duke University, 1951); see also Madeleine B. Stern, "Poe: 'The Mental Temperament' for Phrenologists," *American Literature* 40, no. 2 (1968): 155–63.

77   *"Pinakidia"*: Pinakidia was the name of tablets stored at the ancient library of Alexandria. Whalen links the series to the need for compact, easily accessible compendia (almanacs, encyclopedias, and so on) of useful information (*Poe and the Masses*, 26); on the political economy of information, see Richard R. John, *Spreading the News: The American Postal System from Franklin to Morse* (Cambridge, Mass.: Harvard University Press, 2009).

78   *intellectual arcana*: Baron Bielfeld, *The Elements of Universal Erudition: Containing an Analytical Abridgment of the Sciences, Polite Arts, and Belles Lettres*, trans. W. Hooper, 3 vols. (London: G. Scott, 1770). On Poe's

erudition, see Burton R. Pollin, ed., *The Collected Writings of Edgar Allan Poe: The Imaginary Voyages*, rev. ed. (New York: Gordian Press, 1994); Shaindy Rudoff, "'Written in Stone': Slavery and Authority in *The Narrative of Arthur Gordon Pym*," ATQ 14, no. 1 (2000): 61–82.

78 *Poe also printed a new tale*: Richard P. Benton, "The Tales: 1831–1835," in *A Companion to Poe Studies*, ed. Eric W. Carlson (Westport, Conn.: Greenwood, 1996), 110–28; Whalen, *Poe and the Masses*, 152.

78 *"We have never perused"*: Notice of the SLM in *Richmond Whig*, ca. Aug. 1835, in TPL, 164.

78 *Poe drew from*: Meredith Neill Posey, "Notes on 'Hans Pfaall,'" *Modern Language Notes* 45, no. 8 (Dec. 1930): 501–7.

78 *"a capital burlesque upon balloonings"*: *Baltimore Republican*, cited on SLM wrapper, May 1835, in Pollin, *Collected Writings*, 1:373.

78 *"but hoax, with these"*: Poe, on p. 580 of the first published version of the story, "Hans Pfaall—A Tale," SLM, 1, no. 2 (June 1835): 565–80.

78 *This jarring instability*: Maurice S. Lee, "Genre, Science, and 'Hans Pfaall,'" in *The Oxford Handbook of Edgar Allan Poe*, ed. J. Gerald Kennedy and Scott Peeples (New York: Oxford University Press, 2018), 338–50; Lee notes shifts between "verisimilitude and rigamarole" on 347.

79 *"is decidedly superior"*: Paulding to White, Dec. 7, 1835, in *The Letters of James Kirke Paulding*, ed. Ralph M. Aderman (Madison: University of Wisconsin Press, 1962), 170–72, printed in Washington's *Daily National Intelligencer*, Dec. 18, 1835, reprinted in SLM 2, no. 2 (Jan. 1836): 138, in TPL, 184.

79 *"He first touches"*: *Winchester Republican*, reprinted in "Opinions of the Press," SLM 1, no. 9 (Aug. 1835), in TPL, 168.

79 *"Mr. Poe possesses an extraordinary faculty"*: From Eastern Virginia, "Letters of Correspondents," reprinted on covering papers of SLM 1, no. 9 (May 1835), in TPL, 156.

79 *"the first genius"*: Philip Pendleton Cooke, *Richmond Compiler*, reprinted in "Opinions of the Press" SLM 1, no. 13 (Sept. 1835), in TPL, 173.

79 *viewed with some suspicion*: On questions about Virginia's youth, see Silverman, *Mournful*, 103, 124.

79 *"small for her age"*: Weiss, *Home Life of Poe*, 85–86; Silverman, *Mournful*, 124.

79 *"My dearest Aunty"*: Poe to Maria Clemm (and Virginia E. Clemm), Aug. 29, 1835, in Ostrom, 1:102–4.

80 *Portrait of a girl (Virginia Poe)*: The portrait of Virginia presented here is a matter of some dispute: in *The Portraits and Daguerreotypes of Edgar Allan Poe* (Charlottesville: University of Virginia Press, 1989), 70, painter Michael Deas denied that it was painted by Thomas Sully and is of Virginia Poe. Cynthia Cirile defends both identifications on her provocative website, "Virginia Clemm Poe: The Myth of Sissy" (www.synastrypress.com/virginia2.html), which she promises to elaborate in a future book. The portrait's owner, Dr. Marion Rundell, has had the painting convincingly authenticated as a Sully. Given its resemblance to a deathbed portrait of Virginia, the timing that places Sully and the Poes together in Richmond, and supporting documentation at the Frick Art Reference Library (see Deas, *Portraits and Daguerreotypes*, 88), the identification of the subject as Virginia Poe strikes me as highly plausible.

80 *constitutional julep*: "A dram of spirituous liquor that has mint steeped in it, taken by Virginians of a morning." John Davis, *Travels of Four Years and a Half in the United States of America* (Bristol: R. Edwards, 1803), 379.

80 *"Mr. Poe was a fine gentleman"*: Woodberry, *Life*, 2:443, in TPL, 168.

81 *"unfortunately rather dissipated"*: White to Lucian Minor, Sept. 8, 1835, in TPL, 167.

81 *"It must be expressly understood"*: White to Poe, Sept. 29, 1835, in TPL, 172.

81 *"My health is better"*: Poe to Kennedy, Jan. 22, 1836, in Ostrom, 1:120.

81 *affidavit declaring*: AHQ, 252; Poe to Kennedy, Jan. 22, 1836, in Ostrom, 1:122.

81 *whatever physical form*: Silverman quotes a secondhand account (Frederick W. Coburn, "Poe as Seen by the Brother of 'Annie,'" *New England Quarterly* 16 [1943]: 471) that Poe slept in a separate room from Virginia and for at least two years did not "assume the position of a husband." Silverman, *Mournful*, 124. According to Marie Bonaparte, "That Poe's marriage was never consummated is accepted by several biographers" (including George Woodberry, Hervey Allen, and Joseph Krutch), *Life and Works*, 78; Bonaparte, grand-niece of Napoleon and evangelist of Freud, attributes much of Poe's writing to "physical impotence" caused by a "mother-fixation"; Kenneth Silverman pursues similar psychoanalytic explanations for Poe's subjects, compulsions, and misfortunes.

81 *Each magazine would review*: See the foundational work of McGill, *Culture of Reprinting*; Lara Langer Cohen, *The Fabrication of American Literature: Fraudulence and Antebellum Print Culture* (Philadelphia: University of Pennsylvania Press, 2012); Clare Pettitt, *Patent Inventions: Intellectual Property and the Victorian Novel* (Oxford: Oxford University Press, 2004).

82 *"scenes of great power"*: Lewis Gaylord Clark, *Knickerbocker*, Nov. 1835, in *TPL*, 176.

82 *"WELL!—here we have it!"*: "Critical Notices," *SLM* 2, no. 1 (Dec. 1835): 54–56.

82 *Fay replied with a comedy sketch*: Theodore S. Fay, *New-York Mirror*, April 9, 1836, in *TPL*, 197.

82 *In an era*: Joanne B. Freeman, *The Field of Blood: Violence in Congress and the Road to Civil War* (New York: Farrar, Straus and Giroux, 2018); Leon Jackson, "'Behold Our Literary Mohawk, Poe': Literary Nationalism and the 'Indianation' of Antebellum American Culture," *ESQ* 48, no. 1–2 (2002): 97–133.

83 *"neglectful of the* totality of effect": "Critical Notices," *SLM* 2, no. 10 (Aug. 1836): 600; Willis described literally setting fire to Poe's poem "Fairyland" in *The American Monthly* (Nov. 1829), in *TPL*, 99.

83 *"the sentiment of Intellectual Happiness"*: Review of Drake and Halleck, *SLM* 2, no. 3 (April 1836): 326–36.

83 *against Coleridge's influential view*: On Poe's aesthetic stances in dialogue with Coleridge, see Barton Levi St. Armand, "'Seemingly Intuitive Leaps': Belief and Unbelief in *Eureka*," *ATQ* 26, no. 1 (1975): 4–15.

83 *"To originate"*: Poe, "Peter Snook," *SLM* 2, no. 2 (Oct. 1836): 716–32.

83 *"does not consist"*: Mary Shelley, *Frankenstein*, ed. D. L. Macdonald and Kathleen Scherf (Peterborough, Ont.: Broadview, 2012); Sean Moreland, "'The Plastic and Prolific Creature': Macranthropic Monstrosity, Good's Lucretius, and Shelley's *Frankenstein*," *Anglistik* 30, no. 3 (2019): 27–44.

84 *endless amusement*: See such phrenological worksheets as *Phrenological Chart by O. Fowler, Practical Phrenologist, Presenting a Synopsis of the Science of Phrenology* (Baltimore: John W. Woods, 1836), New-York Historical Society Library.

84 *The Scottish phrenologist*: George Combe, *The Constitution of Man Considered in Relation to External Objects* (Boston: Carter and Hendee, 1829); Roger Cooter, *The Cultural Meaning of Popular Science: Phrenology and the Organization of Consent in Nineteenth-Century Britain* (New York: Cambridge University Press, 1984).

84 *"We might as well make up our minds"*: Poe, review of *The Manual of Phrenology*, *SLM* 1, no. 8 (Aug. 1835).

85 *"assumed the majesty of a science"*: Poe, review of *Phrenology, and the Moral Influence of Phrenology*, by Mrs. L. Miles, *SLM* 2, no. 3 (March 1836): 286–87.

85 *Phrenology's methods and concepts*: Edward Hungerford, "Poe and Phrenology," *American Literature* 2, no. 3 (1930): 209–31; Stern, "Mental Temperament"; Laverty, "Science and Pseudo-science," 99–121.

85 *"peculiar mania"*: Barnum quoted in Matthew Goodman, *The Sun and the Moon: The Remarkable True Account of Hoaxers, Showmen, Dueling Journalists, and Lunar Man-Bats in Nineteenth-Century New York* (New York: Basic Books, 2008), 294–95.

85 *"hydro-oxygen" lamp*: Iwan Rhys Morus, "'More the Aspect of Magic Than Anything Natural': The Philosophy of Demonstration," in *Science in the Marketplace: Nineteenth-Century Sites and Experiences*, ed. Aileen Fyfe and Bernard Lightman (Chicago: University of Chicago Press, 2007), 336–70; Meegan Kennedy, "'Throes and Struggles . . . Witnessed with Painful Distinctness': The Oxy-hydrogen Microscope, Performing Science, and the Projection of the Moving Image," *Victorian Studies* 62, no. 1 (2019): 85–118; Kentwood D. Wells, "Fleas the Size of Elephants: The Wonders of the Oxyhydrogen Microscope," *The Magic Lantern Gazette*, 2/3, no. 29 (Summer/Fall 2017): 3–34.

86 *Poe publicly accepted*: AHQ, 226.

86 *"Not one person in ten"*: Poe, "Richard Adams Locke," in "The Literati of New York City Part VI," *Godey's Lady's Book*, Oct. 1846, 161.

86 *Like Daniel Defoe's*: "Critical Notices," *SLM* 2, no. 2 (Jan. 1836): 127–29.

88 *"who make no scruple"*: Poe, "Maelzel's Chess-Player," *SLM* 2, no. 5 (April 1836): 318–26.

88 *"train of suggestive reasoning"*: Ibid., 323.

88 *heighten the impression of mechanism*: A similar "epistemology of the hoax" is described in "The Purloined Letter" with an example cribbed from Edmund Burke; see S. L. Varnado, "The Case of the Sublime Purloin; or, Burke's Inquiry as the Source of an Anecdote in 'The Purloined Letter,'" *Poe Newsletter* 1, no. 2 (1968): 27.

89 *"one vast miracle"*: David Brewster, *Letters on Natural Magic, Addressed to Sir Walter Scott* (London: John Murray, 1832), 7.

90 *"succession of unerring steps"*: Poe, "Maelzel's Chess-Player."

91 *"an artificer or artificers"*: William Paley, *Natural Theology; or, Evidences of the Existence and Attributes of the Deity: Collected from the Appearances of Nature* (London: J. Faulder, 1802); Aileen Fyfe, "Publishing and the Classics: Paley's Natural Theology and the Nineteenth-Century Scientific Canon," *Studies in History and Philosophy of Science Part A* 33, no. 4 (2002): 729–51.

91 *"Creator, Governor, and Preserver"*: William Whewell, *Astronomy and General Physics Considered with Reference to Natural Philosophy* (London: William Pickering, 1833), 2.

91 *published the* Bridgewater Treatises: Jonathan R. Topham, "Beyond the 'Common Context': The Production and Reading of the *Bridgewater Treatises*," *Isis* 89, no. 2 (1998): 233–62.

91 *"the Power, Wisdom, and Goodness"*: Jonathan R. Topham, "Biology in the Service of Natural Theology: Paley, Darwin, and the *Bridgewater Treatises*," in *Biology and Ideology from Descartes to Dawkins*, ed. Denis Alexander and Ronald L. Numbers (Chicago: University of Chicago Press, 2010), 88–113.

91 *Whewell found it*: Ronald L. Numbers, *Creation by Natural Law: Laplace's Nebular Hypothesis in American Thought* (Seattle: University of Washington Press, 1977), 20. William Buckland's subsequent treatise on geology likewise endorsed the hypothesis. Richard Yeo, "William Whewell, Natural Theology, and the Philosophy of Science in Mid Nineteenth Century Britain," *Annals of Science* 36, no. 5 (1979): 493–516; Robert M. Young, "Malthus and the Evolutionists: The Common Context of Biological and Social Theory," *Past and Present* 43, no. 1 (1969): 109–45.

91 *"the first cause"*: Whewell, *Astronomy and General Physics*, 189.

92 *"we establish by physical proofs"*: Ibid., 342. Whewell drew firm limits on the reach of mathematics and science: "We may thus, with the greatest propriety, deny to the mechanical philosophers and mathematicians of recent times any authority with regard to their views of the administration of the universe; we have no reason whatever to expect from their speculations any help, when we ascend to the first cause and supreme ruler of the universe." Ibid., 334.

92 *William Buckland explained*: William Buckland, *Geology and Mineralogy Considered with Reference to Natural Theology* (London: W. Pickering, 1836); Charles Coulston Gillispie, *Genesis and Geology: A Study in the Relations of Scientific Thought, Natural Theology, and Social Opinion in Great Britain, 1790–1850* (Cambridge, Mass.: Harvard University Press, 1951).

92 *But even the regular order of nature*: R. M. Young, "Natural Theology, Victorian Periodicals, and the Fragmentation of a Common Context," in *Darwin's Metaphor: Nature's Place in Victorian Culture* (Cambridge, U.K.: Cambridge University Press, 1985), 126–63; J. H. Brooke, "The Natural Theology of the Geologists: Some Theological Strata," in *Images of the Earth: Essays in the History of the Environmental Sciences*, ed. Ludmilla J. Jordanova and Roy Porter, 2nd ed. (Chalfont St. Giles: British Society for the History of Science, 1997), 53–74; John Hedley Brooke, "Natural Theology and the Plurality of Worlds: Observations on the Brewster-Whewell Debate," *Annals of Science* 34, no. 3 (1977): 221–86. Likewise, in human society, the divinely ordained interdependencies among social classes showed *Bridgewater* authors that the existing social order and hierarchy was right and good; see William Kirby, *On the Power, Wisdom, and Goodness of God as Manifested in the Creation of Animals and in Their History, Habits, and Instincts* (London: William Pickering, 1835).

92 *"magnificent bodies scattered"*: Herschel quoted in Goodman, *The Sun and the Moon*, 185; Laura J. Snyder, "'Lord Only of the Ruffians and Fiends'? William Whewell and the Plurality of Worlds Debate," *Studies in History and Philosophy of Science Part A* 38, no. 3 (2007): 584–92.

92 *lunar civilization*: Thomas Dick evoked a "population of intelligent beings far more numerous, and perhaps far more elevated in the scale of intellect, than the inhabitants of our globe"; quoted in Goodman, *The Sun and the Moon*, 193. Natural philosophers including Wilhelm Olbers, who had discovered several large asteroids and comets, also believed it "very probable" that there were rational beings on the moon and other planets. See Goodman, *The Sun and the Moon*, 133, 193; William J. Astore, *Observing God: Thomas Dick, Evangelicalism, and Popular Science in Victorian Britain and America* (Aldershot: Routledge, 2001).

93 *Poe also was fascinated*: The most extensive examination of Poe's relationship to natural theology also offers one of the most thorough studies of his scientific thought, focused on Poe's legal reasoning and his training in rhetoric and logic: Margaret Alterton, *Origins of Poe's Critical Theory* (1925; New York: Russell and Russell, 1965).

93 *Poe also furiously marked up*: Poe's personal copy is held at the Harry Ransom Center at the University of Texas, Austin: Thomas Chalmers, *On the Power, Wisdom, and Goodness of God, as Manifested in the*

Adaptation of External Nature, to the Moral and Intellectual Constitution of Man (Philadelphia: Carey, Blanchard, and Lea, 1833).

93 *"the hardihood of crime"*: Ibid., 97.

94 *"in great confusion"*: White to William Scott, Nov. 24, 1836, in *TPL*, 234.

94 *"Highly as I really think"*: White to Nathaniel Beverley Tucker, Dec. 27, 1836, in *TPL*, 236.

6. Delirious Design

95 *"People want something"*: Henry C. Carey to John Pendleton Kennedy, Nov. 26, 1834, in *TPL*, 142. Carey was later troubled by Poe's facility in sympathizing with madmen in his tales: "Is he not deranged? I should care nothing about aiding him as you propose, but I should like to be sure he was sane." Carey to Kennedy, Oct. 4, 1835, in *TPL*, 175.

95 *"in which a single and connected story"*: Harper & Brothers to Poe, June 19, 1836, Harper and Brothers to Poe (RCL152), Misc. Letters, Edgar Allan Poe Society of Baltimore.

95 *lecturer J. N. Reynolds*: See Aaron Sachs, *The Humboldt Current: Nineteenth-Century Exploration and the Roots of American Environmentalism* (New York: Penguin Books, 2007), on Reynolds and Symmes, and the ongoing confusion about Reynolds's first name: historians often call him Jeremiah Reynolds; his official name appears to be James (382n39). See also Aubrey Starke, "Poe's Friend Reynolds," *American Literature* 11, no. 2 (1939): 152–59; Robert F. Almy, *J. N. Reynolds: A Brief Biography with Particular Reference to Poe and Symmes* (New York: Colophon, 1937).

95 *"warm and rich land"*: Captain John Cleves Symmes, *Circular No. 1* (St. Louis), April 10, 1818.

96 *Reynolds's six-hundred-page account*: J. N. Reynolds, *Voyage of the United States Frigate Potomac, Under the Command of Commodore John Downes, During the Circumnavigation of the Globe in the Years 1831–32–33 and 34* (New York: Harper & Brothers, 1835).

96 *When he spoke on the topic*: J. N. Reynolds, *Address on the Subject of a Surveying and Exploring Expedition to the Pacific Ocean and South Seas: Delivered in the Hall of Representatives on the Evening of April 3, 1836* (New York: Harper & Brothers, 1836); Poe, "South-Sea Expedition," *SLM* 3, no. 1 (Jan. 1837): 68–72.

97 *"an intensity of eager expectation"*: Poe, "South-Sea Expedition," 70.

97 *Matthew Fontaine Maury*: Helen M. Rozwadowski, "Introduction: Reconsidering Matthew Fontaine Maury," *International Journal of Maritime History* 28, no. 2 (2016): 388–93; Charles Lee Lewis, *Matthew Fontaine Maury, the Pathfinder of the Seas* (Annapolis, Md.: United States Naval Institute, 1927); John Grady, *Matthew Fontaine Maury, Father of Oceanography: A Biography, 1806–1873* (Jefferson, N.C.: McFarland, 2015).

97 *"The spirit of literary improvement"*: Poe, *SLM* 2, no. 5, June 1836: 454.

97 *"one of the most courteous"*: William Gowans, *Catalogue of American Books*, no. 28 (1870): 11; AHQ, 267.

98 *"The Monthlies of Gotham"*: Poe in Kevin Hayes, *Poe and the Printed Word* (Cambridge, U.K.: Cambridge University Press, 2000), 53.

98 *published in 1838*: Poe, *The Narrative of Arthur Gordon Pym of Nantucket* (New York: Harper & Brothers, 1838).

98 *One reviewer asked*: "Pym," *Alexander's Weekly Messenger* (Philadelphia), Aug. 22, 1838, in *TPL*, 253.

99 *It drew on Reynolds's*: See Lisa Gitelman, "Arthur Gordon Pym and the Novel Narrative of Edgar Allan Poe," *Nineteenth-Century Literature* 47, no. 3 (1992): 349–61; Johan Wijkmark, "Poe's *Pym* and the Discourse of Antarctic Exploration," *EAPR* 10, no. 3 (2009): 84–116.

99 *"verify some of the most important"*: Poe, *Pym*, 48 (page references to *Pym* refer to the 1999 Penguin Classics edition, ed. Kopley). On the Exploring Expedition, see Nathaniel Philbrick, *Sea of Glory: America's Voyage of Discovery: The U.S. Exploring Expedition, 1838–1842* (London: Penguin, 2004); Sachs, *Humboldt Current*; William Stanton, *The Great United States Exploring Expedition of 1838–42* (Berkeley: University of California Press, 1975).

99 *In that case, its author would be*: On Poe's multiple identities, see Daniel Hoffman, *Poe, Poe, Poe, Poe, Poe, Poe, Poe* (New York: Doubleday, 1972). *Pym's* preface recalls Cervantes's jibes with Don Quixote in that novel's preface and has influenced the twentieth-century "metafictions" of Jorge Luis Borges, Julio Cortázar, and Paul Auster.

100 *"My name is Arthur Gordon Pym"*: Poe, *Pym*, 7.

101 *"materially increased"*: Ibid., 217.

101 *"Note"*: Ibid., 219–21.

101 *"So tempting an opportunity"*: Ibid., 161.

102 *"minute philological scrutiny"*: Ibid., 220. On Poe and philology, see Rudoff, "'Written in Stone.'"

103 *perhaps the very same ship*: In his introduction to the Penguin edition of *Pym*, Kopley suggests that the *Penguin*'s return was heralded by the "spirit of reflection" that penguins reveal by the nests they make in tandem with the albatross; on the "quincunx" form of these nests and their connection to Thomas Browne's *Hydriotaphia, Urn-Burial; or, A Discourse of the Sepulchral Urns, Together with the Quincuncial Lozenge, or Network Plantations of the Ancients, Naturally, Artificially, Mystically Considered* (London: Henry Brome, 1658), see John T. Irwin, "The Quincuncial Network in Poe's *Pym*," in *Poe's "Pym": Critical Explorations*, ed. Richard Kopley (Durham, N.C.: Duke University Press, 1992), 175–87.

103 *Or perhaps Poe meant readers to see*: To note just a few of the book's contrasting interpretations: *Pym* has been seen as a "providential" tale of sin and redemption (Curtis Fukuchi, "Poe's Providential *Narrative of Arthur Gordon Pym*," *ESQ* 27, no. 3 [1981]: 147–56; Richard Kopley, "The 'Very Profound Undercurrent' of *Arthur Gordon Pym*," *Studies in the American Renaissance* [1987]: 143–75), as well as a satire of providential literature (David Vance, "Poe/Defoe—Pym/Crusoe: Providential Indeterminacy in *Arthur Gordon Pym of Nantucket*," *EAPR* 12, no. 2 [2011]: 64–78); the book's ending has been aligned with apocalyptic writing (David Ketterer, *New Worlds for Old: The Apocalyptic Imagination, Science Fiction, and American Literature* [Garden City, N.Y.: Anchor Books, 1974]), and with mythical initiations including the Arthur legend's "reunion with the white goddess" (Carol Peirce and Alexander G. Rose III, "Poe's Reading of Myth: The White Vision of Arthur Gordon Pym," in Kopley, *Poe's "Pym,"* 57–74), as well as myths of a departed race of southern giants (Kent Ljungquist, "Descent of the Titans: The Sublime Riddle of *Arthur Gordon Pym*," *Southern Literary Journal* 10, no. 2 [1978]: 75–92). These "mythical" interpretations resonate with Jungian archetypal transformation (Barton Levi St. Armand, "The Dragon and the Uroboros: Themes of Metamorphosis in Arthur Gordon Pym," *ATQ* 37 [1978]: 57–72) and a Freudian "primal return to an amniotic paradise" (Frederick S. Frank, "The Gothic at Absolute Zero: Poe's *Narrative of Arthur Gordon Pym*," *Extrapolation* 21, no. 1 [1980]: 21–30). The figures carved in Tsalal's chasms might be read as confirmation of Pym's providential mission, a biblical curse, or a racist allegory announcing the damnation of the African race (see Rudoff, "'Written in Stone'"). Alternatively, the Tsalal section has been read as a satire of racial assumptions (John C. Havard, "'Trust to the Shrewdness and Common Sense of the Public': The *Narrative of Arthur Gordon Pym* as a Hoaxical Satire of Racist Epistemologies," in *Deciphering Poe: Subtexts, Contexts, Subversive Meanings*, ed. Alexandra Urakova [Bethlehem Pa.; Lehigh University Press, 2013], 107–20), while Mat Johnson submits *Pym*'s racial imagery to hilarious reconfigurations in his novel, *Pym* (New York: Spiegel & Grau, 2011). According to J. Gerald Kennedy, *Pym* may be read as a "fable of misreading": "As Pym's adventure recurrently demonstrates, interpretive efforts disclose only the 'hideous uncertainty' of all human truth" ("The Invisible Message: The Problem of Truth in *Pym*," in *The Naiad Voice: Essays on Poe's Satiric Hoaxing*, ed. Dennis W. Eddings [Port Washington, N.Y.: Associated Faculty Press, 1983], 124–35).

103 *"the Specter of the Brocken"*: John T. Irwin, *American Hieroglyphics: The Symbol of the Egyptian Hieroglyphics in the American Renaissance* (New Haven, Conn.: Yale University Press, 1980).

103 *"superstitious terrors"*: Poe, *Pym*, 77. Brewster's argument had deep roots in the eighteenth-century Enlightenment: both Hume's *Natural History of Religion* and Adam Smith's "History of Astronomy" argued that the fear caused by surprising natural phenomena was at the origin of the first polytheistic religions, while Burke's *Philosophical Enquiry into the Origin of Our Ideas of the Sublime and Beautiful* pointed out how the manipulation of darkness and shadow could be used by despots to produce fear and awe in their subjects (Edmund Burke, *A Philosophical Enquiry into the Origin of Our Ideas of the Sublime and Beautiful* [London: R. and J. Dodsley, 1757]; Peter Gay, *The Enlightenment: An interpretation. Vol. 1: The Rise of Modern Paganism* (London: Weidenfeld and Nicolson, 1967).

104 *Yet much as in De Quincey's*: Thomas De Quincey, *Confessions of an English Opium-Eater* (London: Printed for Taylor and Hessey, 1822); Frances Wilson, *Guilty Thing: A Life of Thomas De Quincey* (New York: Farrar, Straus and Giroux, 2016).

104 *Poe always took great care*: Leon Jackson analyzes Poe's "obsession with printedness, especially as it manifested itself in his comments on typography and typographical errors," in "'The Italics Are Mine'"; see also his "Poe and Print Culture" and Hayes, *Poe and the Printed Word*, 93–95; McGill, *Culture of Reprinting*.

105 *doublings, inversions, and illusions*: Was the page's composition intended to suggest these images, or are these simply mirages, artifacts of an overeager drive to interpret? If deliberate, do we credit Poe or the typographer? No records of the process of laying out the type of *Pym* have been found, but I have consulted hundreds of title pages from Harper & Brothers published 1835–38. None have the

same forms or visual density as *Pym*'s title page, though a few bear a rough resemblance, such as Lyman Cobb's *North American Reader* (1835) and David Hibbard's *Treatise on Cow Pox* (1835). John Steuart's *Bogotá in 1836–7* (1838), another travel narrative, might be seen to have a comparable above and below visual motif, but with far less text. Richard Kopley brought to my attention another travel account, James O. Pattie, *The Personal Narrative of James O. Pattie, of Kentucky* (Cincinnati: John H. Wood, 1831), whose plot has strong resemblances to Poe's later unfinished novel, *The Journal of Julius Rodman*; the upper half of Pattie's title page similarly clusters the text of a lengthy subtitle into a hull-like or bowl-like shape. It is entirely possible that Poe saw this book and guided *Pym*'s compositor to emulate it. It is also possible that these marks of apparent intention and art are figments of the viewer's imagination—an uncertainty about the presence or absence of intentional design that *Pym* itself develops as a central theme.

105 *symmetry and reversal into* Pym's *structure*: *Pym*'s mirror structure is elaborated with increasing detail in Charles O'Donnell, "From Earth to Ether: Poe's Flight into Space," *PMLA* 77, no. 1 (1962): 85–91; David Ketterer, "Devious Voyage: The Singular Narrative of A. Gordon Pym," *ATQ* 37 (1978): 21–33; Kopley, "'Very Profound Under-current'"; all of these reject the claim of *Pym*'s lack of order expressed in Joseph V. Ridgely and Iola S. Haverstick, "Chartless Voyage: The Many Narratives of Arthur Gordon Pym," *Texas Studies in Literature and Language* 8, no. 1 (1966): 63–80.

106 *the rhetorical figure of chiasmus*: Max Nänny, "Chiasmus in Literature: Ornament or Function?," *Word and Image* 4, no. 1 (1988): 51–59; William E. Engel, *Early Modern Poetics in Melville and Poe: Memory, Melancholy, and the Emblematic Tradition* (Burlington, Vt.: Ashgate, 2012).

106 images of a journey: John T. Irwin explicates "the network of images that Poe groups around the abyss" in *American Hieroglyphics*, 235.

106 *"found these fancies creating their own realities"*: Poe, *Pym*, 206.

106 *warped mirror, a kaleidoscope, a camera obscura, or a magic lantern*: On Poe's optical technologies, see Saltz "'Eyes Which Behold'"; Sharp, "Poe's Chapters on 'Natural Magic'"; William J. Scheick, "An Intrinsic Luminosity: Poe's Use of Platonic and Newtonian Optics," in *American Literature and Science*, ed. Robert J. Scholnick (Lexington: University Press of Kentucky, 1992), 77–93; Barbara Cantalupo, *Poe and the Visual Arts* (University Park: Pennsylvania State University Press, 2014), esp. "Poe's Visual Tricks," 103–22; John Tresch, "Estrangement of Vision: Edgar Allan Poe's Optics," in *Observing Nature—Representing Experience, 1800–1850*, ed. Erna Fiorentini (Berlin: Reimer, 2007), 155–86.

107 *"so lately and so providentially"*: Poe, *Pym*, 123.

107 *any entity, and our judgment of it*: Christopher Herbert, *Victorian Relativity: Radical Thought and Scientific Discovery* (Chicago: University of Chicago Press, 2001). This emphasis on reversibility and "comparative" values is largely in agreement with Dana Nelson's argument that "while on one level *Pym* is a racist text, on another the text provides a reading that counters racist colonial ideology and the racialist, scientific knowledge structure" and "undermines the pretensions of colonial knowledge to disinterested objectivity." Dana D. Nelson, *The Word in Black and White: Reading "Race" in American Literature, 1638–1867* (Oxford: Oxford University Press, 1992), 92, 107.

107 *a definite mystery*: *Pym*'s combination of a rigorous structure, factual details, and passages calling out for multiple interpretations has made it both appealing and frustrating to critics. G. R. Thompson traces Poe's debts to Schlegel's notion of the unfinished "arabesque" novel in "The Arabesque Design of *Arthur Gordon Pym*" (in Kopley, *Poe's "Pym*," 188–213), and summarizes: "Despite the astonishing range of readings . . . there is in *Pym* a coherent and symmetrical structure of events that generates a haunting ambiguity"; *Pym* "exemplifies Poe's method of resonant indeterminateness" ("Edgar Allan Poe and the Writers of the Old South," in *Columbia Literary History of the United States*, ed. Emory Elliott [New York: Columbia University Press, 1988], 274). For an epitome of the delirious pleasures of hyper-interpreting Poe, see David Ketterer, "'Shudder': A Signature Cryptogram in 'The Fall of the House of Usher,'" *Resources for American Literary Study* 25, no. 2 (1999): 192–205.

107 *American reviewers found*: *New York Morning Courier*, July 30, 1838, in Burton R. Pollin, "Pym's Narrative in the American Newspapers: More Uncollected Notices," *PS* 9, no. 1 (1978): 11, 8–9, in *TPL*, 249.

107 *"the very ingenious author"*: *New York Gazette*, July 30, 1838, in Pollin, "American Newspapers," 9, in *TPL*, 249.

107 *Locke wrote to thank*: *New York New Era*, Aug. 1, 1838, in Pollin, "American Newspapers," 10, in *TPL*, 250.

107 *Several other papers noted*: On the reception of *Pym*, and the polarization between those offended by its hoax-like aspects and those impressed with its literary achievement, see Richard Kopley, "Readers Write: Nineteenth-Century Annotations in Copies of the First American Edition of Poe's *The Narrative of Arthur Gordon Pym*," *Nineteenth-Century Literature* 55, no. 3 (2000): 399–408.

107 *"The grave particularity"*: George P. Putnam, "Leaves from a Publisher's Letterbook," *Putnam's Magazine*, Oct. 1869, 471, in *TPL*, 255.

108 *"nautical knowledge"*: *Spectator*, Oct. 27, 1838, quoted in Burton R. Pollin, "Poe's *Narrative of Arthur Gordon Pym* and the Contemporary Reviewers," *Studies in American Fiction* 2 (Spring 1974): 53–54, in *TPL*, 257.

108 *"Literary and Scientific"*: *Gentleman's Magazine*, Nov. 1838, quoted in Pollin, "Contemporary Reviewers," 47–49, in *TPL*, 258. See also Burton R. Pollin, "Poe 'Viewed and Reviewed': An Annotated Checklist of Contemporaneous Notices," *PS* 13, no. 1 (1980): 21.

108 *"very clever extravaganza"*: *Alexander's Weekly Messenger*, Aug. 22, 1838, in *TPL*, 254.

108 *"go on an exploring expedition"*: *Family Magazine*, Sept. 1838, quoted in Pollin, "Contemporary Reviewers," 46, in *TPL*, 254.

108 *"A more impudent attempt"*: William E. Burton in *Burton's Gentleman's Magazine*, Sept. 1838, in *TPL*, 254.

Part III: Philadelphia

109 *"What age can boast"*: This is an excerpt from a mock-epic treatment of American poets published under a pseudonym in 1847: Lavante, *The Poets and Poetry of America: A Satire* (Philadelphia: William S. Young, 1847); in 1887 it was reprinted by Benjamin and Bell of New York with the (shaky) conjecture that its author was Poe.

7. The Athens of America

111 *"Philadelphia is a city"*: Nathaniel P. Willis, "Pencillings by the Way," *New-York Mirror*, Nov. 12, 1831, 58; Nicholas B. Wainwright, "The Age of Nicholas Biddle, 1825–1841," in *Philadelphia: A 300-Year History*, ed. Russell F. Weigley, Nicholas B. Wainwright, and Edwin Wolf II (New York: W. W. Norton, 1982).

111 *"divided into many systems"*: James Russell Lowell, "Our Contributors: Edgar Allan Poe," *Graham's Lady's and Gentleman's Magazine*, Feb. 1845, 49–53. As this chapter shows, it was precisely this dispersion that led reformers to seek a national framework for American science on universal grounds.

111 *it remained preeminent in publishing*: For a comparison of the scientific life of post-Revolutionary Philadelphia, Boston, and New York, see Simon Baatz, "'Squinting at Silliman': Scientific Periodicals in the Early American Republic, 1810–1833," *Isis* 82, no. 2 (1991): 223–44: "In sum, Philadelphia was the center, during the first half of the nineteenth century, of American science and medicine" (225).

112 *Jacksonian appeals to nativist resentment*: Bruce Laurie, *Working People of Philadelphia, 1800–1850* (Philadelphia: Temple University Press, 1980).

112 *foster regular contacts*: Bruce Sinclair, *Philadelphia's Philosopher Mechanics: A History of the Franklin Institute, 1824–1865* (Baltimore: Johns Hopkins University Press, 1974); on comparable (and overlapping, as in the case of Saxton, see below) circuits in London, see Iwan Rhys Morus, *Frankenstein's Children: Electricity, Exhibition, and Experiment in Early-Nineteenth-Century London* (Princeton, N.J.: Princeton University Press, 1998).

113 *"I almost always return from new York"*: Henry to Bache, Oct. 28, 1839, in *The Papers of Joseph Henry*, ed. Marc Rothenberg and Nathan Reingold, 12 vols. (Washington, D.C.: Smithsonian Institution Press, 1972–2012), 4:119–20.

113 *In November 1834*: This visit is condensed from "Henry's Notes on a Trip to Philadelphia, December 5–7, 1834," in *Papers of Joseph Henry*, 2:288–95; Albert E. Moyer, *Joseph Henry: The Rise of an American Scientist* (Washington, D.C.: Smithsonian Institution Press, 1997), 187–90.

114 *"doubt and self-distrust"*: William Dunlap (playwright), diary entry, Oct. 8, 1832, in Thomas Coulson, *Joseph Henry, His Life and Work* (Princeton, N.J.: Princeton University Press, 1950), 94.

115 *"intellectual power"*: Silliman in Moyer, *Joseph Henry*, 130–31.

115 *"stand among the first"*: Torrey in ibid., 130.

115 *"of all the dirty work"*: Henry to Elias Loomis, 1841, *Papers of Joseph Henry*, 5:29; Julia Grummitt, "Joseph Henry and Sam Parker," *Princeton & Slavery*, slavery.princeton.edu.

115 *largely future ministers*: On the importance of theological authors—including the Princeton theologian Charles Hodge—in setting the terms for the reception of science, see Walter H. Conser, *God and the Natural World: Religion and Science in Antebellum America* (Columbia: University of South Carolina Press, 1993).

115 *Henry's visit to Philadelphia*: Moyer, *Joseph Henry*, 187–90; *Papers of Joseph Henry*, 2:288–95.

115 *Carey and Lea's bookstore*: David Kaser, *Messrs. Carey & Lea of Philadelphia: Study in the History of the Booktrade* (Philadelphia: University of Pennsylvania Press, 1957).

116 *show of automata*: In 1835, Maelzel's automata were displayed at the American Museum at Fifth and Chestnut: "These pieces are so constructed, and their entire arrangements so exposed to view, as to force conviction upon the mind of the spectator, that their correct imitation of nature, in their various actions, are the result of mechanism only, without any other agency ... the same Automata so highly spoken of in Brewster's Letters on Magic." McAllister playbill collection, Library Company of Philadelphia.

116 *a 1785 amontillado*: Moyer, *Joseph Henry*, 197.

116 *"composed of an infinite number"*: "Henry's Notes," 288–95.

116 *"I am not altogether pleased"*: Henry in Moyer, *Joseph Henry*, 189.

116 *daily "deviation" of magnetic compasses*: Bache would later publish *Observations at the Magnetic and Meteorological Observatory at Girard College*, 3 vols. (Washington, D.C.: Gales and Seaton, 1840–47); John Cawood, "The Magnetic Crusade: Science and Politics in Early Victorian Britain," *Isis* 70, no. 4 (1979): 493–518; Diane Greco Josefowicz, "Experience, Pedagogy, and the Study of Terrestrial Magnetism," *Perspectives on Science* 13, no. 4 (2005): 452–94; Jenny Bulstrode, "The Eye of the Needle: Magnetic Survey and the Compass of Capital in the Age of Revolution and Reform" (PhD diss., University of Cambridge, 2020).

117 *"kindred spirits ready to discuss"*: Alexander Dallas Bache's Eulogy of James P. Espy, *Annual Report of the Board of Regents of the Smithsonian Institution, 1859* (Washington, D.C.: Thomas Ford, 1860), 109, quoted in Merle M. Odgers, *Alexander Dallas Bache: Scientist and Educator, 1806–1867* (Philadelphia: University of Pennsylvania Press, 1947), 152.

117 *abandoned his wife and children*: Jansen, *Alexander Dallas Bache*, 48–50.

117 *"Experimental Laboratory and Workshop"*: Ibid., 74.

117 *"to extend, not merely to diffuse"*: Bache to Henry, May 28, 1839, in *Papers of Joseph Henry*, 4:224–26.

118 *plans for American science*: Bache's affinities to the Whigs are rightly emphasized in Hugh Richard Slotten, *Patronage, Practice, and the Culture of American Science: Alexander Dallas Bache and the U.S. Coast Survey* (Cambridge, U.K.: Cambridge University Press, 1994), 15–17, though the political dynasty to which Bache belonged included Democrats, and his own eventual success in Washington depended on avoiding party conflict; Jansen, *Alexander Dallas Bache*, 27–66.

118 *"The object of a system"*: Bache to George Mifflin Dallas, Jan. 27, 1833, George Mifflin Dallas Papers, Historical Society of Pennsylvania, quoted in Jansen, *Alexander Dallas Bache*, 98.

118 *Bache led a commission*: Jansen, *Alexander Dallas Bache*, 85.

118 *"first federally sponsored research"*: Ibid., 78.

118 *Bache urged the introduction*: "General Report on the Explosions of Steam Boilers," in Jansen, *Alexander Dallas Bache*, 87. On French steam engine regulations, which served as a model for Bache, see Jean-Baptiste Fressoz, *L'apocalypse joyeuse* (Paris: L'Univers Historique/Seuil, 2012).

118 *wrote up Lewis and Clark's*: Biddle prepared the manuscript for the first edition, before passing it on to James Allen to shepherd it through publication; though only Allen's name appears in print, the first edition is known as the "Biddle-Allen" edition: Meriwether Lewis, William Clark, and Nicholas Biddle, *History of the Expedition Under the Command of Captains Lewis and Clark*, ed. Paul Allen (Philadelphia: Bradford and Inskeep, 1814).

118 *"because he has scalped"*: Biddle in Howe, *What Hath God Wrought*, 391; see Thomas Payne Govan, *Nicholas Biddle: Nationalist and Public Banker, 1786–1844* (Chicago: University of Chicago Press, 1959), 253. As proof of the passions raised by the Bank War, Biddle's brother was killed in a duel defending the reputation of the bank.

119 *"a private misfortune"*: Nicholas Biddle, "The Address," *North American Magazine*, Aug. 1833, 216, in Jansen, *Alexander Dallas Bache*, 133.

120 *"Hurrah for the Yankee experiment!"*: Alfred Mayer, "Henry as a Discoverer," in Moyer, *Joseph Henry*, 213.

120 *Humphry Davy had connected*: "Excerpts from Joseph Henry's European Diary," April 18, 1837, in *Science in Nineteenth-Century America: A Documentary History*, ed. Nathan Reingold (Chicago: University of Chicago Press, 1980), 80.

120 *"understand our prejudices"*: Henry in Moyer, *Joseph Henry*, 219.

120 *"marble statues in many instances"*: Ibid., 214.

120 *The BAAS had been launched in 1831*: Jack Morrell and Arnold Thackray, *Gentlemen of Science: Early Years of the British Association for the Advancement of Science* (Oxford: Clarendon Press, 1981); Laura J. Snyder, *The Philosophical Breakfast Club: Four Remarkable Friends Who Transformed Science and Changed the World* (New York: Broadway Books, 2011).

120 *But after his lecture*: Henry in Moyer, *Joseph Henry*, 225–26; *Papers of Joseph Henry*, 5:136–37, 158, 244–45.

121 *"I am ever more of your opinion"*: Henry to Bache, in Moyer, *Joseph Henry*, 230.

121 *"a great system of public instruction"*: Bache, "Address on Manufactures" (1842), in Jansen, *Alexander Dallas Bache*, 195.

122 *"We are overwhelmed in this country"*: Henry to Wheatstone, Feb. 27, 1846, in *Papers of Joseph Henry*, 6:382–85, in Moyer, *Joseph Henry*, 237.

122 *"The charlatanism of our country"*: Henry to Bache, Aug. 8, 1838, in *Joseph Henry*, 4:97.

122 *"in the way of securing priority"*: Henry to Vaughan, in Moyer, *Joseph Henry*, 193.

122 *"filch with impunity"*: Henry, July 16, 1844, in Moyer, *Joseph Henry*, 145.

122 *"promiscuous assembly of those"*: Henry to M. de la Rive, Nov. 12, 1841, Henry MSS, Smithsonian Institution Archives, in Sally Gregory Kohlstedt, "A Step Toward Scientific Self-Identity in the United States: The Failure of the National Institute, 1844," *Isis* 62, no. 3 (1971): 353.

123 *"real working men in the way of science"*: Henry to Bache, Aug. 9, 1838, in Moyer, *Joseph Henry*, 230.

123 *scientific superstar from Edinburgh*: Combe's vast worldwide following and contacts are examined in James Poskett, *Materials of the Mind: Phrenology, Race, and the Global History of Science, 1815–1920* (Chicago: University of Chicago Press, 2019).

124 *Combe complained*: George Combe, *Notes on the United States of North America, During a Phrenological Visit in 1838–9–40*, 2 vols. (Philadelphia: Carey & Hart, 1841), in Ann Fabian, *The Skull Collectors: Race, Science, and America's Unburied Dead* (Chicago: University of Chicago Press, 2010), 94; Rusert, *Fugitive Science*, 178–79; William R. Stanton, *The Leopard's Spots: Scientific Attitudes Toward Race in America, 1815–1859* (Chicago: University of Chicago Press, 1960); George M. Fredrickson, *The Black Image in the White Mind: The Debate on Afro-American Character and Destiny, 1817–1914* (New York: Harper & Row, 1971).

124 *Combe kept silent*: Cameron A. Grant, "George Combe and American Slavery," *Journal of Negro History* 45, no. 4 (1960): 259–69.

124 *1838 riot in Philadelphia*: Beverly C. Tomek, *Pennsylvania Hall: A "Legal Lynching" in the Shadow of the Liberty Bell* (New York: Oxford University Press, 2014); Samuel Webb, *History of Pennsylvania Hall, Which Was Destroyed by a Mob, on the 17th of May, 1838* (Philadelphia: Samuel Webb, 1838).

124 *"over-anxious" to distance themselves*: Webb, *History of Pennsylvania Hall*, 36.

124 *He cleaned and emptied them*: Stephen Jay Gould, *The Mismeasure of Man*, rev. ed. (New York: W. W. Norton, 1996); Gould's critique of Morton's measurements is tempered by Paul Wolf Mitchell in "The Fault in His Seeds: Lost Notes to the Case of Bias in Samuel George Morton's Cranial Race Science," *PLoS Biology* 16, no. 10 (2018), e2007008, while recognizing the bias intrinsic to Morton's work.

125 *Douglass would eventually*: Frederick Douglass, "The Claims of the Negro, Ethnologically Considered," in *The Frederick Douglass Papers, Series One: Speeches, Debates, and Interviews*, ed. John W. Blassingame and John R. McKivigan, 5 vols. (New Haven, Conn.: Yale University Press, 1979–92), 2:497–525; Rusert, *Fugitive Science*, 126–31.

125 *German naturalist Friedrich Tiedemann*: Mitchell, "Fault in His Seeds."

125 *Polygenesis was directly challenged*: Robert Benjamin Lewis, *Light and Truth, from Ancient and Sacred History* (Portland, Maine: D. C. Colesworthy, 1836); Hosea Easton, *A Treatise on the Intellectual Character, and Civil and Political Condition of the Colored People of the United States* (Boston: Isaac Knapp, 1837); see discussion in Rusert, *Fugitive Science*, chaps. 2–3; Stephen G. Hall, *A Faithful Account of the Race: African American Historical Writing in Nineteenth-Century America* (Chapel Hill: University of North Carolina Press, 2009); Mia Bay, *The White Image in the Black Mind: African-American Ideas About White People, 1830–1925* (New York: Oxford University Press, 2000).

126 *"series of six hundred skulls"*: Agassiz to Henri Milne-Edwards, May 1847, in *Life, Letters, and Works of Louis Agassiz*, ed. Jules Marcou, 2 vols. (New York: Macmillan, 1895), 2:28–29, quoted in Stanton, *Leopard's Spots*, 102n4.

126 *"literally suffering for want of food"*: James Pedder's recollections, 1852, in *A Catalogue of the Books and Manuscripts of Harry Elkins Widener*, comp. A. S. W. Rosenbach, 2 vols. (Philadelphia, 1918), 2:56, in TPL, 248.

126 *practical, commercially oriented publication*: Benjamin Cohen, *Notes from the Ground: Science, Soil, and Society in the American Countryside* (New Haven, Conn.: Yale University Press, 2009); Emily Pawley, *The Nature of the Future: Agriculture, Science, and Capitalism in the Antebellum North* (Chicago: University of Chicago Press, 2020).

127 *"obtain the most unimportant"*: Poe to Paulding, July 19, 1838, in Ostrom, 1:175, and TPL, 248.

127 *"She desires me to thank you"*: Poe to Hiram Haines, April 24, 1840, in Ostrom, 1:215, and TPL, 294.

127 *industrial importance of coal*: Andreas Malm, *Fossil Capital: The Rise of Steam Power and the Roots of Global Warming* (London: Verso, 2016).

127 *Isaac Lea*: Newton Pratt Scudder, *The Published Writings of Isaac Lea, LL.D.* (Washington, D.C.: Government Printing Office, 1885).

127 *e. conchis omnia*: Martin Priestman, *The Poetry of Erasmus Darwin: Enlightened Spaces, Romantic Times* (Burlington, Vt.: Ashgate, 2013), 110–11.

127 *"did not know what it was"*: Lea in Scudder, *Writings of Isaac Lea*; Isaac Lea, "Description of Six New Species of the Genus Unio, Embracing the Anatomy of the Oviduct of One of Them, Together with Some Anatomical Observations on the Genus," *Transactions of the American Philosophical Society* 3, n.s. (1830): 259–73.

127 *a less formal, more inclusive scientific society*: Simon Baatz, "Philadelphia Patronage: The Institutional Structure of Natural History in the New Republic, 1800–1833," *Journal of the Early Republic* 8, no. 2 (1988): 111–38; Thomas Peter Bennett, "The History of the Academy of Natural Sciences of Philadelphia," *Archives of Natural History* 1, no. 1 (1983): 1–14.

128 *"England's prosperity is based"*: Buckland and Lea quoted in Scudder, *Writings of Isaac Lea*, xiii.

128 *in his conchology textbook*: Thomas Wyatt, *A Manual of Conchology, According to the System Laid Down by Lamarck, with the Late Improvements by De Blainville* (New York: Harper & Brothers, 1838).

128 *Thomas Brown's Elements*: Thomas Brown, *Elements of Conchology* (London: Lackington, Allen, 1816).

129 *collective and largely anonymous labor*: Helen Curry et al., eds., *Worlds of Natural History* (Cambridge, U.K.: Cambridge University Press, 2018); Londa Schiebinger, *Secret Cures of Slaves: People, Plants, and Medicine in the Eighteenth-Century Atlantic World* (Stanford, Calif.: Stanford University Press, 2017).

129 *"progressive, even innovative"*: Stephen Jay Gould, "Poe's Greatest Hit," *Natural History* 102, no. 7 (1993): 10–19.

129 *"appear to every person of science"*: Poe, *The Conchologist's First Book: A System of Testaceous Malacology, Arranged Expressly for the Use of Schools* (Philadelphia: Haswell, Barrington, and Haswell, 1839), 3.

129 *"from artificial description to integrative biology"*: Gould, "Poe's Greatest Hit," 16.

130 *"To an upright and well regulated mind"*: Poe, *Conchologist's First Book*, 8.

130 *It also gave him*: Dwight Thomas, however, raises the possibility that Lea "did not play a significant role in Lea & Blanchard's decision" to publish *Tales of the Grotesque and Arabesque*, in "Poe in Philadelphia, 1838–1844: A Documentary Record," 2 vols. (PhD diss., University of Pennsylvania, 1978), 833; on connections between *Pym* and *The Conchologist's First Book*, see J. D. Lilley, "Poe, Movement, and Matter: The Malacological Aesthetics of the *Narrative of Arthur Gordon Pym*," *Arizona Quarterly* 73, no. 4 (2017): 1–31.

8. Methods Grotesque and Arabesque

131 *"two hours a day"*: Burton to Poe, May 11, 1839; AHQ, 78.

131 *"uncalled for severity"*: Burton to Poe, May 30, 1839, in TPL, 262.

131 *One evening he dined*: Recollection of the host's son, Horace Wemyss Smith, quoted in Rosenbach, *Catalogue*, 296, in TPL, 263.

132 *the latest works of science*: S. Augustus Mitchell, *A System of Modern Geography* (Philadelphia: Thomas, Cowperthwait & Co., 1839).

132 *a richly illustrated Synopsis*: Louis-Céran Lemonnier, *A Synopsis of Natural History: Embracing the*

Natural History of Animals, with Human and General Animal Physiology, Botany, Vegetable Physiology and Geology, trans. Thomas Wyatt (Philadelphia: T. Wardle, 1839), reviewed by Poe in *Burton's,* July 1839, 61-62.

132 *natural theology textbook in verse:* Henry Duncan, *Sacred Philosophy of the Seasons, Illustrating the Perfection of God in the Phenomena of the Year,* 2 vols. (1836; Boston: Marsh, Capen, Lyon, and Webb, 1839), reviewed by Poe in *Burton's,* Feb. 1840, 106, and March 1840, 151–52.

132 *"A Chapter on Science and Art":* *Burton's,* March, April, May, and July 1840.

133 *natural philosophy was breaking down:* Simon Schaffer, "Scientific Discoveries and the End of Natural Philosophy," *Social Studies of Science* 16, no. 3 (1986): 387–420; Cahan, *From Natural Philosophy to the Sciences.*

133 *Coleridge noted the change:* Snyder, *Philosophical Breakfast Club,* 2–3; Sydney Ross, "Scientist: The Story of a Word," *Annals of Science* 18, no. 2 (1962): 65–85.

133 *In France, Auguste Comte:* John Tresch, *The Romantic Machine: Utopian Science and Technology After Napoleon* (Chicago: University of Chicago Press, 2012), 253–86.

134 *Immanuel Kant and the idealist philosophers:* Ernst Cassirer, *Kant's Life and Thought* (1918; New Haven, Conn.: Yale University Press, 1981).

134 *updated in John Herschel's:* John F. W. Herschel, *A Preliminary Discourse on the Study of Natural Philosophy* (London: Longman, Rees, 1830); Walter F. Cannon, "John Herschel and the Idea of Science," *Journal of the History of Ideas* 22, no. 2 (1961): 215–39; Erna Fiorentini, "Practices of Refined Observation: The Conciliation of Experience and Judgement in John Herschel's Discourse and in His Drawings," in Fiorentini, *Observing Nature,* 19–42.

134 *Whewell's Philosophy:* William Whewell, *The Philosophy of the Inductive Sciences, Founded upon Their History,* 2 vols. (London: John W. Parker, 1840). Whewell, influenced by Kant as well as natural theology, believed some of the fundamental ideas were innately given. Laura J. Snyder, "William Whewell," in *The Stanford Encyclopedia of Philosophy,* Spring 2019 ed., ed. Edward N. Zalta, plato.stanford .edu/archives/spr2019/entries/whewell/; Michael Ruse, "Darwin's Debt to Philosophy: An Examination of the Influence of the Philosophical Ideas of John F. W. Herschel and William Whewell on the Development of Charles Darwin's Theory of Evolution," *Studies in History and Philosophy of Science* 6, no. 2 (1975): 159–81; William Whewell, review of *A Preliminary Discourse on the Study of Natural Philosophy,* by John Herschel, *Quarterly Review* 90 (1831): 374–407; Snyder, *Philosophical Breakfast Club.*

134 *Bacon's induction was reinforced:* George Daniels's landmark *American Science in the Age of Jackson* somewhat overstates the Baconianism of the era; see the important qualifications in Theodore Dwight Bozeman, "Science and Nineteenth-Century American Culture: A Note on George H. Daniels' Science in the Age of Jackson," *Isis* 63, no. 3 (1972): 397–402. Susan Faye Cannon reframed the era's empiricism within Humboldt's projects of global science in "Humboldtian Science," in *Science in Culture: The Early Victorian Period* (New York: Science History Publications, 1978), 73–110.

135 *roads, canals, railroads, and communications:* Richard R. John, *Network Nation: Inventing American Telecommunications* (Cambridge, Mass.: Harvard University Press, 2015); Howe, *What Hath God Wrought.*

135 *"Science has become":* William Ellery Channing, *The Works of William Ellery Channing,* 6 vols. (Boston: American Unitarian Association, 1903), 6:153; see Kasson, *Civilizing the Machine.*

135 *daily and seasonal changes:* On Bache's involvement with the project to map variations in terrestrial magnetism—jokingly christened the "magnetic crusade" by other American participants—see Jansen, *Alexander Dallas Bache,* 163.

135 *to participate as equal players:* A. Hunter Dupree, "Central Scientific Organisation in the United States Government," *Minerva* 1, no. 4 (1963): 453–69; on the political range of U.S. Humboldtianisms, see Walls, *Passage to Cosmos.* On the dynamics of international competition in science in this period—with England, France, and Prussia seeking to balance cooperation with national progress—see Charles Babbage, *Reflections on the Decline of Science in England, and on Some of Its Causes* (London: B. Fellowes, 1830).

136 *François Arago secured:* John Tresch, "The Daguerreotype's First Frame: François Arago's Moral Economy of Instruments," *Studies in History and Philosophy of Science Part A* 38, no. 2 (2007): 445–76.

136 *"a drawing in which the light":* Alexander Dallas Bache, "The Daguerreotype Explained," *United States Gazette* (Philadelphia), Sept. 25, 1839; see Library Company of Philadelphia, "Catching a Shadow: Daguerreotypes in Philadelphia, 1839–1860," www.librarycompany.org/catchingashadow/.

136 *Cornelius, a skilled metalworker:* William F. Stapp, "Robert Cornelius and the Dawn of Photography," in *Robert Cornelius: Portraits from the Dawn of Photography,* ed. William F. Stapp (Washington, D.C.: Smithsonian Institution Press, 1983), 25–44.

136 *Saxton and Cornelius, working with*: Albert Rung, "Joseph Saxton: Pennsylvania Inventor and Pioneer Photographer," *Pennsylvania History: A Journal of Mid-Atlantic Studies* 7, no. 3 (1940): 153–58; Joseph Henry, "Memoir of Joseph Saxton: 1799–1873," read before the National Academy, Oct. 4, 1874; Notice on Robert Cornelius, *Godey's Lady's Book*, April 1840, 190; J. F. Goddard, "Application of the Daguerreotype to the Taking of Likenesses from the Life," *Chemist; or, Reporter of Chemical Discoveries and Improvements* 2 (May 1841): 142–43; both online at Gary W Ewer, ed., *The Daguerreotype: An Archive of Source Material*, www.daguerreotypearchive.org.

137 *"hanging out their clothes"*: Lewis Gaylord Clark, "The Daguerreotype," *Knickerbocker*, Dec. 1839, 560–61.

137 *"every gradation of light & shade"*: Herschel to William Henry Fox Talbot, May 9, 1839, National Science and Media Museum, Bradford, www.foxtalbot.dmu.ac.uk/letters/transcriptDocnum.php?docnum=3875.

137 *"If you believe everything"*: Cole to William Althorpe Adams, Feb. 26, 1840, photocopy of manuscript letter provided by the New York State Library, Thomas A. Cole Papers, 1821–1863, SC 10635, box 1, folder 4, Adams, Feb. 26, 1840, Ewer Archive.

137 *It was an automatic method*: The notion that a machine—photography in particular—might be a more reliable observer than humans was important for the emerging scientific ideal of objectivity; see Lorraine Daston and Peter Galison, *Objectivity* (Cambridge, Mass.: Zone, 2007).

138 *"The telescope is rather an unfair tell-tale"*: "New Discovery—Engraving, and Burnet's Cartoons," *Blackwood's Edinburgh Magazine*, March 1839, 382–91, Ewer Archive. The author compared this technology (by which nature composes a lasting image of itself) to the idea expressed in Charles Babbage's *Ninth Bridgewater Treatise*, that "every word uttered from the creation of the world has registered itself, and is still speaking, and will speak for ever in vibrations," imagining "the great business of the sun" as "to act registrar likewise, and to give out impressions of our looks" (see further discussion of Babbage below).

138 *Whether the image shows Poe*: Benjamin J. McFarland and Thomas Peter Bennett, "The Image of Edgar Allan Poe: A Daguerreotype Linked to the Academy of Natural Sciences of Philadelphia," *Proceedings of the Academy of Natural Sciences of Philadelphia* (1997): 1–32. To Susan Elizabeth Sweeney, the claim that the man at the right is Poe is "tenuous at best"; "The Horror of Taking a Picture in Poe's 'Tell-Tale Heart,'" *EAPR* 18, no. 2 (2017): 145. "Daguerreotypomanie" was the banner of a famous print by Théodore Maurisset, see Roger Watson and Helen Rappaport, *Capturing the Light: The Birth of Photography, a True Story of Genius and Rivalry* (New York: St. Martin's, 2013), 161.

138 *"most important, and perhaps the most extraordinary"*: Poe, "The Daguerreotype," in *Alexander's Weekly Messenger*, Jan. 15, 1840, 2, and May 6, 1840, 2. Poe added further updates in "A Chapter on Science and Art" in *Burton's*, April 1840, 193: Daguerre's "costly combination of glasses" had been replaced with a single concave "Meniscus" glass; the complicated "dilute nitric acid" had been shown to be unnecessary; a shallower box for developing the plate halved the time of development.

139 *John W. Draper*: Sarah Kate Gillespie, "John William Draper and the Reception of Early Scientific Photography," *History of Photography* 36, no. 3 (2012): 241–54; John William Draper, "On the Process of the Daguerreotype and Its Application to Taking Portraits from the Life," *London and Edinburgh Philosophical Magazine and Journal of Science* 17 (1840): 217–25.

140 *photography became a popular art*: On the central role of Philadelphia's scientists and artisans in the chemical and commercial development of photography, see the detailed reconstructions in Michelle Smiley, "'An American Sun Shines Brighter': Art, Science, and the American Reinvention of Photography" (PhD diss., Bryn Mawr College, 2020).

140 *"a pamphlet of trash"*: Poe, "A Charlatan," *Alexander's Weekly Messenger*, 4, no. 18, April 1829, 2.

141 *"exercises of the analytical"*: Poe, "Enigmatical and Conundrum-ical," *Alexander's Weekly Messenger*, Dec. 18, 1839, 4.

141 *The next month, a reader*: Poe, "Enigmatical," *Alexander's Weekly Messenger*, 4, no. 3, Jan. 15, 1840, 2.

141 *"gaggery, or more delicately"*: Poe, "Our Late Puzzles," *Alexander's Weekly Messenger*, Feb. 12, 1840, p. 2.

142 *"the row increased as the wonder grew"*: Poe, "Our Late Puzzles," *Alexander's Weekly Messenger*, Feb. 12, 1840, 2. On Poe and cryptography, see Shawn James Rosenheim, *The Cryptographic Imagination: Secret Writing from Edgar Poe to the Internet* (Baltimore: Johns Hopkins University Press, 1997).

142 *"The simple truths which science unfolds"*: Poe, "Article on Beet-Root," *Alexander's Weekly Messenger*, Dec. 18, 1839, 2; see also Poe, "Credulity," *Alexander's Weekly Messenger*, May 6, 1840, 2, which suggests that

the very cunning man can be most easily confused and misled simply by telling him the truth: "To act honorably with a scoundrel is so completely to mystify him as to paralyze his utmost exertions."

142 *"is subject to the universal"*: Poe, "Puzzles Again!" *Alexander's Weekly Messenger*, March 25, 1840, 2.

142 *"Peter Pendulum"*: Poe, "The Business Man," *Burton's*, Feb. 1840, 87–89.

142 *"Diddling Considered as One of the Exact Sciences"*: Poe, "Raising the Wind; or, Diddling Considered as One of the Exact Sciences," *Philadelphia Saturday Courier*, Oct. 14, 1843, 1.

143 *"Bugaboo and Kickapoo"*: Poe, "The Man That Was Used Up: A Tale of the Late Bugaboo and Kickapoo Campaign," *Burton's Gentleman's Magazine* 5, no. 2, Aug. 1839, 66–70.

143 *Poe's tale showed*: J. Gerald Kennedy, "'A Mania for Composition': Poe's Annus Mirabilis and the Violence of Nation-Building," *American Literary History* 17, no. 1 (2005): 1–35; Heather Chacón, "Prosthetic Colonialism: Indian Removal, European Imperialism, and International Trade in Poe's 'The Man That Was Used Up,'" *PS* 50, no. 1 (2017): 46–68. On science in the borderlands occupied by Native Americans, former slaves, and slaveholders, see Cameron B. Strang, *Frontiers of Science: Imperialism and Natural Knowledge in the Gulf South Borderlands, 1500–1850* (Chapel Hill: University of North Carolina Press, 2018).

144 *"Grotesque" tales exaggerated literary*: Poe drew upon Walter Scott's analysis of E. T. A. Hoffmann, who "exhibited the fantastic or supernatural grotesque in his compositions, so nearly on the verge of actual insanity, as to be afraid of beings his own fancy created," and was subject to trains of ideas "in which fancy had a large share and reason none at all" ("Novels of Ernest Theodore Hoffmann," in *The Miscellaneous Prose Works of Sir Walter Scott* [Edinburgh: Robert Cadell, 1849], 306). Poe might have also been inspired by a review of Disraeli: "The grotesque is a deviation from Nature, permitted in order that an effect may be produced, which can not be produced by adhering to Nature"; see "The Continuation of Vivian Grey," *New Monthly Magazine*, April 1827, in Lewis A. Lawson, "Poe's Conception of the Grotesque," *Mississippi Quarterly* 19, no. 4 (1966): 200–205.

144 *"Arabesque" tales opened up a dream realm*: See Thompson, "Arabesque Design"; Rachman, "Poe, the Arabesque." David Ketterer sees the arabesque as "the fusion of the senses and the obliteration of reason," an effort "to melt away the rigid pattern . . . imposed by man's reason"; see *The Rationale of Deception in Poe* (Baton Rouge: Louisiana State University Press, 1979), 36–37. For historical context, see Jacob Rama Berman, "Domestic Terror and Poe's Arabesque Interior," *ESC* 31, no. 1 (2005): 128–50.

144 *"succession of richly-coloured"*: Louis Fitzgerald Tasistro, *New-York Mirror* 17, no. 26, Dec. 28, 1839, 215.

145 *The task of a philosophical natural science*: Robert J. Richards, *The Romantic Conception of Life: Science and Philosophy in the Age of Goethe* (Chicago: University of Chicago Press, 2002); Iain Hamilton Grant, *Philosophies of Nature After Schelling* (London: Continuum, 2006).

145 *"something evermore"*: Wordsworth, *The Prelude*, 68, quoted in Jerome McGann, "Rethinking Romanticism," *ELH* 59, no. 3 (1992): 747.

145 *senses, emotions, body, and soul*: Richard Holmes, *The Age of Wonder: How the Romantic Generation Discovered the Beauty and Terror of Science* (New York: Vintage, 2010).

145 *Naturphilosophie fed rigorous research*: Cunningham and Jardine, *Romanticism and the Sciences*; Tresch, *Romantic Machine*; Branka Arsić, "Materialist Vitalism or Pathetic Fallacy: The Case of the House of Usher." *Representations* 140, no. 1 (2017): 121–36.

146 *Emerson and the transcendentalists*: Georges J. Joyaux, "Victor Cousin and American Transcendentalism," *French Review* 29, no. 2 (1955): 117–30.

146 *"wild Pantheism of Fichte"*: Poe, "Morella," *SLM* 1, no. 8 (April 1835): 448–50. In a special issue of *EAPR* 13, no. 2 (2012) on Poe and German idealism, see Sean Moreland, Jonathan Murphy, and Devin Zane Shaw, "'Theory Mad Beyond Redemption': The Post-Kantian Poe," 4–6; Moreland and Shaw, "'As Urged by Schelling,'" 50–80. See also John Limon, *The Place of Fiction in the Time of Science: A Disciplinary History of American Writing* (Cambridge, U.K.: Cambridge University Press, 2009).

146 *"the genius of artists"*: The quotes in this paragraph and the next are found under the heading, "Study of Nature," in "Omniania" in *Burton's Gentleman's Magazine* 6, no. 5, May 1840, 235, unsigned but reliably attributed to Poe; the passage contains a slightly modified citation from Humphry Davy's "Award of the Copley Medal to the Rev. Dr. Buckland," in *The Collected Works of Sir Humphry Davy* (London: Smith, Elder, and Co. Cornhill, 1840), 43–44. Davy's speech highlighted the sublime aesthetic implications of Buckland's geology and natural theology—a stance Poe heartily endorsed.

146 *"those grand monuments of nature"*: Humphry Davy, *Six Discourses Delivered Before the Royal Institution* (London: John Murray, 1827), 55.

146 *geological theory of Cuvier*: Georges Cuvier, *Discours sur les révolutions de la surface du globe: et sur les change-ments qu'elles ont produits dans le règne animal* (Paris: E. d'Ocagne, 1826); Martin J. S. Rudwick, *The Meaning of Fossils: Episodes in the History of Palaeontology* (London: Macdonald, 1972); Martin J. S. Rudwick, *Earth's Deep History: How It Was Discovered and Why It Matters* (Chicago: University of Chicago Press, 2014); David Bates, *Enlightenment Aberrations: Error and Revolution in France* (Ithaca, N.Y.: Cornell University Press, 2002).

146 *"continents broken into islands"*: Poe, "Omniana," pt. 2, *Burton's*, May 1840, 235–36.

147 *"playful effusion"*: John Frost, review in *Alexander's Weekly Messenger*, Dec. 18, 1839, in *TPL*, 282.

147 *"vivid description"*: Anonymous review, *New-York Mirror*, Dec. 28, 1839, in *TPL*, 284.

147 *"too uniformly extravagant"*: Review in *The Era*, Oct. 21, 1838, in "Three More Contemporary Reviews of Pym," *PS*, December 1976, 9, no. 2, 43–44.

147 *a "creole" with African ancestry*: Dayan, "Amorous Bondage," 260–62.

147 *"hideous drama of revivification"*: Poe, "Ligeia," *Tales of the Grotesque and Arabesque* (Philadelphia: Lea and Blanchard, 1840) I:171–92.

147 *This arabesque space*: On the construction of this space (and others), see Henri Justin, *Poe dans le champ du vertige: Des "Contes" à "Eurêka," l'élaboration des figures de l'espace* (Paris: Klincksieck, 1991), and on "arabesque reality," see Ketterer, *Rationale of Deception*, as well as Cantalupo on the "anamorphic" in "Ligeia" in *Poe and the Visual Arts*. Richard Wilbur saw Poe's decors of "flickering candles, wavering torches, and censers full of writhing varicolored flames," along with his recurrent imagery of spirals and whirlpools, as figures of a "hypnagogic state," the descent into dream, and the escape from the material and worldly; Wilbur, "The House of Poe," in *Poe: A Collection of Critical Essays*, ed. Robert Regan (Englewood Cliffs, N.J.: Prentice Hall, 1967), 98–120.

148 *"What was it that so unnerved"*: Poe, "The Fall of the House of Usher," *Tales of the Grotesque and Arabesque*, 75–103.

149 *"Usher" is an intensely self-conscious allegory*: Scott Peeples, "Poe's 'Constructiveness' and 'The Fall of the House of Usher,'" in *The Cambridge Companion to Edgar Allan Poe*, ed. Kevin Hayes (Cambridge, U.K.: Cambridge University Press, 1998), 178–90; Beverley Voloshin, "Explanation in 'The Fall of the House of Usher,'" in *Critical Essays on Poe*, ed. Eric Carlson (Boston: G. K. Hall, 1987), 42–52; Katherine Hayles, *Chaos Bound: Orderly Disorder in Contemporary Literature and Science* (Ithaca, N.Y.: Cornell University Press, 1990); Ketterer, "Shudder."

149 *the alchemists' Great Work*: Barton Levi St. Armand, "Usher Unveiled: Poe and the Metaphysic of Gnosticism," *PS* 5, no. 1 (1972): 1–8; for Wilbur, "When the House of Usher disintegrates or dematerializes at the close of the story, it does so because Roderick Usher has become all soul" ("House of Poe," 110).

149 *"ancestral pile"*: Sean Moreland, "Ancestral Piles: Poe's Gothic Materials," in Kennedy and Peeples, *Oxford Handbook*, 520–41.

149 *repurposing in multiple media*: Jonathan Elmer, "Poe and the Avant Garde," in Kennedy and Peeples, *Oxford Handbook*, 700–717; Scott Peeples, *The Afterlife of Edgar Allan Poe* (Rochester, N.Y.: Camden House, 2004).

149 *"How to Write a Blackwood Article"*: Originally titled "The Psyche Zenobia," *American Museum* 1, no. 3 (Nov. 1838); retitled "How to Write a Blackwood Article," in *Tales of the Grotesque and Arabesque* (1840).

150 *"first white Man"*: Poe, "The Journal of Julius Rodman (Chapter 3)," *Burton's Gentleman's Magazine* 6, no. 3, March 1840, 109–13; quote on promotional paper wrapper.

150 *He took inspiration*: Poe reviewed *Astoria* with admiration for both the author and the subject in the *SLM* 3, no. 1 (Jan. 1837): 59–68.

150 *Poe's first chapters followed Rodman*: On the landscapes at the Philadelphia Academy of the Fine Arts that might have contributed to Rodman's visual aesthetics, see Cantalupo, *Poe and the Visual Arts*, 26–29; on Cole, 44ff.

150 *"the best speculation"*: *Daily Chronicle*, May 30, 1840; *Daily Chronicle*, May 21, 1840; reprinted in *United States Gazette*, in *TPL*, 297.

151 *he detailed his work*: Poe to Burton, June 1, 1840, in *TPL*, 299.

151 *"division of the upper clerks"*: Poe, "The Man of the Crowd," *Burton's Gentleman's Magazine*, 7, no. 6, Dec. 1840, 267–70.

151 *Absolutely alone while entirely surrounded*: On the "kaleidoscopic" visual effects of "The Man of the

Crowd" (and other tales of Poe), see Susan Elizabeth Sweeney, "The Magnifying Glass: Spectacular Distance in Poe's 'Man of the Crowd' and Beyond," *PS* 36, no. 1 (2003): 3–17; Tom Gunning, "From the Kaleidoscope to the X-Ray: Urban Spectatorship, Poe, Benjamin, and Traffic in Souls (1913)," *Wide Angle* 19, no. 4 (1997): 25–61.

9. Dizzy Heights

152 *His one-page prospectus*: Poe, "Prospectus of the Penn Magazine," *Daily Chronicle*, Sept. 11, 1840, 3.

153 *"reigning in his own sphere"*: Willis Gaylord Clark, *Philadelphia Gazette*, June 4, 1840, in *TPL*, 298–99.

153 *"enviable distinction"*: Graham, *Saturday Evening Post*, June 5, 1840, in *TPL*, 299.

153 *"concentrate in Philadelphia"*: Hopkinson to Poe, Jan. 25, 1841, in *TPL*, 318.

153 *"succession of names"*: *Daily Chronicle*, Sept. 14, 1840, in *TPL*, 307.

153 *Falstaff Hotel*: Thomas, *Poe in Philadelphia*, 805–8, in *TPL*, 284.

153 *"I never was in the habit"*: Poe to Joseph E. Snodgrass, April 1, 1841, in Ostrom, 1:263–64.

154 *"some Van Buren men"*: *United States Gazette*, May 21, 1840, in *TPL*, 296; Claude Richard, "Poe and 'Young America,'" *Studies in Bibliography* 21 (1968): 25–58; McGill, *Culture of Reprinting*; David A. Long, "Poe's Political Identity: A Mummy Unswathed." *PS* 23, no. 1 (1990): 1–22.

154 *"I would at once be put"*: Poe to Biddle, Jan. 6, 1841, in Ostrom, 1:254.

154 *four-year subscription*: Ostrom, 1:255.

154 *"a criticism self-sustained"*: Poe, "Prospectus of the Penn."

154 *"Ours has become"*: Abijah M. Ide, Jr. to Poe, Nov. 2, 1843, in *TPL*, 440.

155 *"a severe illness"*: Poe to Lewis J. Cist, Dec. 30, 1840, in Ostrom, 1:251, in *TPL*, 312–13; notice postponing the *Penn* in the *Daily Chronicle*, Dec. 29, 1840, in *TPL*, 312.

155 *"a most advantageous"*: Poe to Wyatt, April 1, 1841, in Ostrom, 1:267, in *TPL*, 316.

155 *"Money is difficult to obtain"*: *Saturday Evening Post*, Feb. 20, 1841, in *TPL*, 318.

155 *"such prospects as are seldom enjoyed"*: George Graham, *Saturday Evening Post*, Feb. 20, 1841, in *TPL*, 318.

156 *he united* Burton's *with* The Casket: *TPL*, 309, 312.

156 *"embellished in a style"*: *Saturday Evening Post*, Dec. 19, 1840, in *TPL*, 312.

156 *"As a stern, just and impartial"*: Graham, *Saturday Evening Post*, Feb. 20, 1841, in *TPL*, 319.

157 *"additional editorial strength"*: Inside front cover of *Graham's*, April 1841, in *TPL*, 320.

157 *One of his first tales*: Poe, "A Descent into the Maelström," *Graham's Magazine*, 18, no. 5, May 1841, 235–41.

157 *a physical impossibility but a mathematical necessity*: Henri Justin, "Le vortex: Raison et vertige," *Les Cahiers du CERLI* 2 (1993): 60–77.

157 *"awe, horror, and admiration"*: Burke, *A Philosophical Enquiry into the Origin of Our Ideas of the Sublime and Beautiful*; Kent Ljungquist, "Poe and the Sublime: His Two Short Sea Tales in the Context of an Aesthetic Tradition," *Criticism* 17, no. 2 (1975): 131–51.

158 *Here detachment brought not horrific*: On the sources for Poe's scientific knowledge in "Maelström," see Aldo Corcella. "A New Poe Source: Thomas Thomson's 'Sketch of the Progress of Physical Science,'" *EAPR* 20, no. 2 (2019): 173–99. Barbara Cantalupo writes, "The narrator's ability to tell his tale depends not on the grace of God but on his power to save himself from demise by using induction and peripheral vision," in *Poe and the Visual Arts*, 120. Limon contrasts the observation in "Crowd" and "Maelström" in *Place of Fiction*, 79. The artists Sandrine Teixido and Aurélien Gamboni have led a multi-year art project in which "Maelström" serves as allegory and instrument for coping with climate change; see Teixido and Gamboni, "A Tale as a Tool: Enquête sur le maelström et le 'devenir-abîme' des mondes," *Techniques et Culture* 75 (2021); ataleasatool.com.

158 *"as a literary people"*: Poe, review of *The Quacks of Helicon*, by Lambert A. Wilmer, *Graham's*, Aug. 1841, 90–93; Cohen, *Fabrication of American Literature*, 24.

158 *Brougham's essays*: Poe, review of *The Critical and Miscellaneous Writings*, by Henry Lord Brougham, *Graham's*, March 1842, 190.

159 *"science of everything"*: Roswell Park, *Pantology; or, A Systematic Survey of Human Knowledge* (Philadelphia: Hogan & Thompson, 1841).

159 *quickly multiplying branches of science*: James A. Secord, *Visions of Science: Books and Readers at the Dawn of the Victorian Age* (Chicago: University of Chicago Press, 2015).

159 *"spoke of me in a species of extravaganza"*: Poe to Frederick W. Thomas, Oct. 27, 1841, in Ostrom, 1:313, and *TPL*, 345.

159   *caused a sensation*: Benjamin to Graham, Oct. 19, 1841, in *TPL*, 344.

159   *"a class of gentlemen"*: Poe, "A Chapter on Autography," pt. 1, *Graham's*, Nov. 1841, 224–34.

159   *"has not altogether succeeded"*: Ibid., 231.

159   *"His good qualities are all"*: Ibid., 229.

159   *"dogmatism, egotism, and other isms"*: Edwin P. Whipple, *Boston Daily Times*, Dec. 1841, in *TPL*, 354.

160   *"congregation of nonsense"*: *Boston Notion*, Dec. 14, 1839, *TPL*, 282.

160   *"insatiable desire . . . to forget ourselves"*: Ralph Waldo Emerson, "Circles" (1841), in *Essays and Lectures* (New York: Library of America, 1983), 401–4, 414; Ralph Waldo Emerson, *Nature* (Boston: James Munroe, 1836).

160   *he shared Nathaniel Hawthorne's conviction*: See Levin, *Power of Blackness*, 34–35; for a contrast between Emerson's and Poe's expansionist cosmologies, see Jennifer Rae Greeson, "Poe's 1848: *Eureka*, the Southern Margin, and the Expanding U[niverse] of S[tars]," in Kennedy and McGann, *Remapping of Antebellum Print Culture*, 123–40; for useful comparison of Emerson, Poe, and Hawthorne, see Van Leer, "Nature's Book," 307–21; on Thoreau's comparable "heterogeneous holism," see Jane Bennett, *Thoreau's Nature: Ethics, Politics, and the Wild* (Lanham: Rowman & Littlefield Publishers, 2002), 72.

160   *"As Americans, we feel proud"*: Poe, review of *Twice-Told Tales*, by Nathaniel Hawthorne, *Graham's*, May 1842, 298–300.

161   *"beginning to acknowledge"*: Poe, "Exordium," *Graham's*, Jan. 1842, 68–69. On this article, and Poe's engagement with post-Kantian criticism, see Paul Hurh, *American Terror: The Feeling of Thinking in Edwards, Poe, and Melville* (Stanford, CA: Stanford University Press, 2015), 75–118.

161   *"frantic spirit of generalization"*: Ibid.

161   *"conglomerate science"*: Poe, "An Appendix of Autographs," *Graham's*, Jan. 1842, 48.

161   *"limit literary criticism"*: Poe, "Exordium," 68.

161   *"art for art's sake"*: John Wilcox, "The Beginnings of l'Art pour l'Art," *Journal of Aesthetics and Art Criticism* 11, no. 4 (1953): 360–77; Peter Galison, "Objectivity Is Romantic," *American Council of Learned Societies Occasional Paper* 47 (1999).

162   *"appertains to no region"*: Poe, "Notes upon English Verse," *Pioneer*, March 1843, 102–12; the later revision was "The Rationale of Verse" in two parts in *SLM* 14, no. 10 (Oct. 1848): 577–85 and 14, no. 11 (Nov. 1848): 673–82; on the wide importance of metrical analysis in poetic criticism at this time, see Jason David Hall, *Nineteenth-Century Verse and Technology: Machines of Meter* (Cham, Switzerland: Springer, 2017).

162   *"resolving thought into its elements"*: Poe, "The Murders in the Rue Morgue," *Graham's* 8, no. 4, April 1841, 166–79.

162   *eccentric seventeenth-century scholar*: This epigraph did not appear in the 1841 version of the tale, whose MS is held at the Philadelphia Free Library; it was added in 1843, while Poe dropped the opening reference to phrenology in 1845. See eapoe.org on successive versions.

163   *"the great nebula in Orion"*: See Simon Schaffer, "The Nebular Hypothesis and the Science of Progress," in *History, Humanity, and Evolution: Essays for John C. Greene*, ed. James R. Moore (Cambridge, U.K.: Cambridge University Press, 1989), 131–64.

164   *emerged from earlier, more primitive forms*: Lawrence Frank, "'The Murders in the Rue Morgue': Edgar Allan Poe's Evolutionary Reverie," *Nineteenth-Century Literature* 50, no. 2 (1995): 168–88.

164   *to identify the culprit*: On the solution, and its much-debated relationship to racist imaginaries, see Elise Lemire, "'The Murders in the Rue Morgue': Amalgamation Discourses and the Race Riots of 1838 in Poe's Philadelphia," in Kennedy and Weissberg, *Romancing the Shadow*, 177–204; Kopley, *Dupin Mysteries*, 27–44. For sources from natural history, see Robert Mitchell, "The Natural History of Poe's Orangutan," *PS/Dark Romanticism* 29, no. 1 (1996): 32–34; Meyers, *Life and Legacy*, 123, on the Masonic Hall of Philadelphia's display of an orange-haired orangutan. Plate 6 of Wyatt's *Synopsis of Natural History*, reviewed by Poe in *Burton's* (July 1839), featured a distinctly humanlike, bearded, and expressive "head of the Orang-Outang."

164   *Dupin working through his "manifold" and "multiform" steps*: For instance, the process of eliminating all other modes of entry than by the window, which led him to discover the loose nail only apparently keeping it shut, and his deduction of the source of the "shrill voice" by ruling out the languages heard by witnesses.

165   *"deep but repulsive interest"*: Horace Greeley, *New-Yorker*, March 27, 1841, in *TPL*, 321.

165 *"human ingenuity cannot concoct a cipher"*: Poe, "A Few Words on Secret Writing," *Graham's*, July 1841, 33–38.

165 *"He can read the hieroglyphics"*: Jesse E. Dow, review of the November *Graham's*, *Washington Index*, Nov. 2, 1841, in *TPL*, 346.

166 *"They who dream by day"*: Poe, "Eleanora: A Fable," *The Gift for 1842* (Philadelphia: Carey and Hart, 1841), 154–62.

166 *"sea of darkness"*: Eric Carlson, *Introduction to Poe: A Thematic Reader* (Glenview, Ill.: Scott, Foresman, 1967), 563.

166 *In early 1840, Peale's Philadelphia Museum*: See advertisement, Feb. 1, 1840, from McAllister Collection, vol. 19, p. 46, Library Company of Philadelphia.

167 *a multimedia artwork*: Poe "The Island of the Fay," *Graham's*, 18, no. 6, June 1841, 253–55.

169 *a threat to the assumption of God's stable*: Whewell, whose language Poe echoes in the introductory paragraphs, discussed Encke's comet in *Astronomy and General Physics*, 151–56; its revolutions were growing shorter and quicker, and eventually would bring it to collapse into the sun; on the stability of planetary orbits—denied by Newton, demonstrated by Laplace—see 129–33. On "Island of the Fay" and sentient matter, see Douglas Anderson, *Pictures of Ascent in the Fiction of Edgar Allan Poe* (New York: Palgrave Macmillan, 2009), 143–177.

169 *Poe's "spirit colloquies"*: The first two were Poe, "The Conversation of Eiros and Charmion," *Burton's*, Dec. 1839, 321–23, and "The Colloquy of Monos and Una," *Graham's*, Aug. 1841, 52–54.

170 *But in a reprint of "A Descent"*: The epigraph was added to the 1845 edition. Joseph Glanvill, a spokesman for the early Royal Society, promoted methods of mechanical natural philosophy and used them to prove the existence of ghosts and demons; see Jonathan I. Israel, *Radical Enlightenment: Philosophy and the Making of Modernity, 1650–1750* (Oxford: Oxford University Press, 2001), 376. Poe also credited Glanvill with the epigraph for "Ligeia": "God is but a great will pervading all things by nature of its intentness. Man doth not yield himself to the angels, nor unto death utterly, save only through the weakness of his feeble will." Poe, "Ligeia," *American Museum*, Sept. 1838, 1, no. 1, Sept. 1838, 25–37. Both quotations, however, appear to be Poe's inventions.

170 *"Sober, quiet, steady Philadelphia"*: Writing under the pseudonym Flib, George Lippard, *Spirit of the Times*, Feb. 7, 1842, 2; Thomas, *Poe in Philadelphia*, 324; see David S. Reynolds, *George Lippard* (Boston: Twayne, 1982).

170 *"the immense amount of torture"*: The penitentiary was inspired both by Quaker insistence on silence as the medium for divine communication and by the utilitarian Jeremy Bentham's panopticon, through which prisoners internalized the supposed gaze from the guard tower. See Charles Dickens, *American Notes: For General Circulation*, 2 vols. (London: Chapman and Hall, 1842), 1:235–38; Joseph Jackson, *Dickens in Philadelphia* (Philadelphia: William J. Campbell, 1912), 1415.

171 *Dickens's own pet bird, Grip*: The stuffed and preserved Grip, purchased by the Poe collector Richard Gimbel, is now on display in the Rare Books Department of the Philadelphia Free Library. It is great.

171 *"Do you know that Godwin"*: Dickens to Poe, March 6, 1842, in *The Letters of Charles Dickens, Volume Three: 1842–1843*, ed. Madeline House, Graham Storey, and Kathleen Tillotson (Oxford: Clarendon Press, 1974), 106–7, in *TPL*, 362.

171 *Dickens promised Poe*: Poe to James R. Lowell, July 2, 1844, in Ostrom, 1:450, in *TPL*, 362.

172 *"would doubtless create a riot"*: *Letters of Charles Dickens*, 3:75n11; Thomas, *Poe in Philadelphia*, 346.

10. The Tide Turns

173 *a gold pocket watch from France*: This was sold in June 2019 at Christie's for $250,000; see https://www.christies.com/en/lot/lot-6210603. It's a beautiful object; one can't help wondering what it sounds like when wrapped in cotton.

173 *"how solicitous"*: Graham, recalling Poe ca. 1842, in "The Late Edgar Allan Poe," *Graham's*, March 1850, 225, in *TPL*, 390.

173 *had begun to chafe*: Poe to Washington Irving, June 21, 1841, in Ostrom, 1:274–75; Poe to Joseph E. Snodgrass, Sept. 19, 1841, in Ostrom, 1:309.

173 *"Such a thing was never heard"*: Poe to Frederick W. Thomas, Oct. 27, 1841, in Ostrom, 1:313. Poe wrote to James Herron, "Now I feel assured of success," on June 30, 1842, in Ostrom, 1:344.

173   "fifty thousand copies": *The Saturday Evening Post* reported that *Graham's* intended "opening the new volume in July next with *fifty thousand copies*," which was to be "the regular and standing edition, being as many as any steel line engraving will yield impressions." *Saturday Evening Post*, March 5, 1842, 2, in *TPL*, 361; Thomas, *Poe in Philadelphia*, 339.

173   "To coin one's brain into silver": Poe to Thomas, July 4, 1841, in Ostrom, 1:292.

174   "It was only on yesterday": Poe to Thomas, Feb. 3, 1842, in Ostrom, 1:324, in *TPL*, 359.

174   "She could not bear": Amanda Bartlett Harris, "Edgar A. Poe," *Hearth and Home*, Jan. 9, 1875, 24, in *TPL*, 358.

174   "not only flatly but discourteously": Poe to Thomas, Feb. 3, 1842, in Ostrom, 1:324–25.

174   "a horrible never-ending oscillation": Poe to George W. Eveleth, Jan. 4, 1848, in Ostrom, 2:641.

174   "love for his wife": Graham, "Late Edgar Allan Poe," 225, in *TPL*, 390.

174   "My reason for resigning": Poe to Thomas, May 25, 1842, in Ostrom, 1:333, in *TPL*, 366–67.

175   "I was continually laboring against myself": Poe to Daniel Bryan, July 6, 1842, in Ostrom, 1:347, in *TPL*, 373.

175   "was quick, it is true": Graham, "Late Edgar Allan Poe," 225, in *TPL*, 390.

175   "3 or 4 hundred": Poe to Bryan, July 6, 1842, in Ostrom, 1:347, in *TPL*, 373.

175   "as one of the best writers": Park Benjamin, *New York New World*, June 4, 1842, in *TPL*, 368.

175   "A thorough analysis of the book": Poe, "Review of New Books," *Graham's*, June 1842, 354–56.

176   "a most outrageous humbug": Poe to Snodgrass, June 4, 1842, in Ostrom, 1:341.

176   *Graham offered him the position*: Graham to Griswold, April 19, 1842, in *TPL*, 364.

176   *thousand-dollar salary*: Graham to Griswold, May 3, 1842, in *TPL*, 365.

176   "We would give more": Dow in the *Index*, June 23, 1842, in *TPL*, 370.

176   "malignant, unjust, and disgraceful": Letter from "FLASH" (a correspondent in Philadelphia), in the *Washington, D.C., Independent*, June 17, 1842, in *TPL*, 370.

176   "a good offer": Poe to Thomas, Sept. 12, 1842, in Ostrom, 1:358, in *TPL*, 379.

176   *collection he was calling* Phantasy Pieces: E. T. A. Hoffmann had given the same title (in German) to a collection of tales: *Fantasiestücke in Callot's Manier*, 4 vols. (Bamberg, Germany: Neues Leseinstitut von C. F. Kunz, 1814–15).

176   *He ran into a young poet*: *TPL*, 371.

176   "He was on a spree": Mary Starr's recollections, quoted in Augustus Van Cleef, "Poe's Mary," *Harper's New Monthly Magazine*, March 1889, 639.

176   "You must have conceived": Poe to J. & H. G. Langley (New York publishers), July 18, 1842, in Ostrom, 1:353.

176   "The Mystery of Marie Rogêt": Poe published the story in three installments in the *Ladies' Companion*, Nov. 1842, Dec. 1842, and Feb. 1843. See Amy Gilman Srebnick, *The Mysterious Death of Mary Rogers: Sex and Culture in Nineteenth-Century New York* (New York: Oxford University Press, 1995); Daniel Stashower, *The Beautiful Cigar Girl: Mary Rogers, Edgar Allan Poe, and the Invention of Murder* (New York: Dutton, 2006). The rumor that Mary Rogers died during an abortion is "driven deeply underground" in Poe's text, according to Laura Saltz: "'(Horrible to Relate!)': Recovering the Body of Marie Rogêt," in Rosenheim and Rachman, *American Face of Edgar Allan Poe*, 237–70. Dana Medoro sees the possibility of terminated pregnancy hovering around all the Dupin tales, in "The Purloined Letter" (as a consequence of the queen's affair), and "The Murders in the Rue Morgue" (whose female victims were "fortune-tellers," a euphemism for abortionists); see Dana Medoro, "So Very Self-Evident: Adultery and Abortion in 'The Purloined Letter,'" *Literature and Medicine* 26, no. 2 (2007): 342–63. "Rogêt," Poe's homonym for the American "Rogers," might also have been a nod to Peter Roget, author of a thesaurus and a *Bridgewater Treatise*—linking Dupin's "ratiocinations" to philological inquiry and natural theology's search for hidden design.

177   "Oh Jupiter!": Poe to Robert Hamilton (editor of the *Ladies' Companion*), Oct. 3, 1842, in Ostrom, 1:365, and *TPL*, 382–3.

177   "roving about the country": John S. Detwiler to E. C. Jellett, n.d., ca. 1842, Joseph Jackson Collection, in *TPL*, 389. Poe was within rambling distance of the recently opened Laurel Hill Cemetery, with its aesthetics of semi-pastoral "death-in-life"; see Aaron Wunsch, "Emporia of Eternity: 'Rural' Cemeteries and Urban Goods in Antebellum Philadelphia," *Nineteenth Century* 28, no. 2 (2008): 14–23; Aaron Sachs, *Arcadian America: The Death and Life of an Environmental Tradition* (New Haven, Conn.: Yale University Press, 2013); on the nineteenth century's death obsession, see Karen Halttunen, *Confidence*

Men and Painted Women: A Study of Middle-Class Culture in America, 1830–1870 (New Haven, Conn.: Yale University Press, 1982), 124–52; Gary Laderman, *The Sacred Remains: American Attitudes Toward Death, 1799–1883* (New Haven, Conn.: Yale University Press, 1996).

177 *"There are epochs"*: Poe to James R. Lowell, July 2, 1844, in Ostrom, 1:448.

177 *"Morning on the Wissahiccon"*: Poe, "Morning on the Wissahiccon," *Opal for 1844* (1843): 249–56. Poe changed the title to "The Elk" in a letter to Lowell, May 28, 1844 (in Ostrom, 1:441), and it has been republished under both titles.

177 *"You stroll to your office"*: Thomas to Poe, May 21, 1841, in *TPL*, 327.

177 *"I wish you joy"*: Poe to Thomas, June 26, 1841, in Ostrom, 1:287. The June 26, 1841, *Baltimore Saturday Visiter* reports on Thomas's appointment: "We are glad to find that this administration is inclined to reward literary as well as political talent"; *TPL*, 332.

178 *"I am a Virginian"*: Poe to F.W. Thomas, June 26, 1841, Ostrom 1:170.

178 *Tyler's son Robert was a poet*: Terence Whalen identified Poe's cryptographic overtures to Tyler (and deciphered one of Poe's unsolved codes) in Whalen, *Poe and the Masses*, 195–224; an independent decryption came from John A. Hodgson, "Decoding Poe? Poe, W. B. Tyler, and Cryptography," *Journal of English and Germanic Philology* 92, no. 4 (1993): 523–34.

178 *from one "W. B. Tyler"*: Poe, "Secret Writing (Addendum III)" *Graham's* 19, no. 6, Dec. 1841, 307.

178 *"new life"*: Poe to Thomas, May 25, 1842, in Ostrom, 1:333.

178 *With a government appointment*: Poe to Thomas H. Chivers, Sept. 27, 1842, in Ostrom, 1:363.

178 *"1124 applicants"*: *Spirit of the Times*, Sept. 10, 1842, in *TPL*, 378.

178 *"treated me most shamefully"*: Poe to Thomas, Nov. 19, 1842, in Ostrom, 370–71, in *TPL*, 383–84.

178 *Thomas commiserated, visiting him*: Thomas, recalling Sept. 17, 1842, quoted in Whitty, "Memoir," xliii–xliv, in *TPL*, 381.

178 *"they have, one and all, declined"*: Dickens to Poe, Nov. 27, 1842, in *Letters of Charles Dickens*, 3:384–85; *TPL*, 388. Dickens's request to his publisher was pro forma: "Pray write me such a reply as I can send to the author of the volumes and to get absolution for my conscience."

179 *"his wearing apparel"*: Poe filed for bankruptcy in the District Court of the United States in Philadelphia on Dec. 19, 1842; it was granted on Jan. 13, 1843. The document was discovered by curators at the Philadelphia branch of the National Archives; see discussion in Ostrom, 1:337–38.

179 *"These events," says the narrator*: Poe, "The Black Cat," *United States Saturday Post* (Philadelphia), Aug. 19, 1843, 1. See Susan Sweeney, "Death, Decay, and the Daguerreotype's Influence on 'The Black Cat,'" *EAPR* 19, no. 2 (2018): 206–32; David Reynolds, "Black Cats and Delirium Tremens: Temperance and the American Renaissance," in *The Serpent in the Cup: Temperance in American Literature*, ed. David Reynolds and Debra Rosenthal (Amherst: University of Massachusetts Press, 1997), 22–59.

180 *"sharp pains"*: Poe, "The Mask of the Red Death: A Fantasy," *Graham's* 20, no. 5, May 1842, 257–59. "Mask" was later changed to "Masque."

180 *"TRUE!—nervous— very, very dreadfully nervous"*: Poe, "The Tell-Tale Heart," *The Pioneer* 1, no. 1, Jan. 1843, 29–31.

181 *the technical core of the daguerreotype*: For further connections to the daguerreotype, including the "torture" of posing, see Sweeney, "Horror of Taking a Picture." On the tale's chiasmic structure, pivoting on the central phrase, "I had directed the ray, as if by instinct, precisely upon the damned spot," see Richard Kopley, *The Threads of "The Scarlet Letter": A Study of Hawthorne's Transformative Art* (Newark: University of Delaware Press, 2003), 105–7.

182 *"a certain coincidence of opinion"*: Poe to Lowell, Feb. 4, 1843, in Ostrom, 1:376–77, and *TPL*, 397.

182 *a secret syndicate, union*: "Suppose, for example, that the élite of our men of letters should combine secretly . . . If we do not defend ourselves by some such coalition, we shall be devoured, without mercy, by the Godeys, the Snowdens, et *id genus omne*." Poe to Lowell, March 30, 1844, in Ostrom, 1:432; see Whalen, *Poe and the Masses*, 38.

183 *"for the sake of his imagination"*: Hawthorne, "The Hall of Fantasy," *Pioneer*, Feb. 1843, in *TPL*, 396.

183 *Poe had favorably reviewed*: Thomas C. Clarke, ed., *American Pocket Library of Useful Knowledge*, 2nd ed. (Philadelphia: Griffith & Simon, 1841); Poe, "Review of New Books," from inside rear paper wrapper of *Graham's*, April 1842.

183 *"I am ugly enough"*: Poe to Thomas, Feb. 25, 1843, in Ostrom, 1:381, and *TPL*, 399.

183 *"now, under the best auspices"*: Poe, "Prospectus for the *Stylus*," *Saturday Museum*, March 4, 1843, 3.

in *TPL*, 398. Clarke and Poe made a legal agreement with the illustrator Felix Darley for up to five drawings per month; *TPL*, 395–96.

183   *"Virginia is nearly recovered"*: Poe to John H. Mackenzie, early 1843, in Ostrom, 1:398, and *TPL*, 401.

183   *"the great object"*: Poe to Thomas, Feb. 25, 1843, in Ostrom, 1:381, and *TPL*, 400

184   *Gathering a few dollars*: Poe to Robert Carter, March 7, 1843, in Ostrom, 1:384, and *TPL*, 403.

184   *"covered all over with the marks"*: Thomas to Robert Tyler, March 8, 1843, in *TPL*, 403.

184   *"somewhat excited"*: Dow to Clarke, March 12, 1843, in *TPL*, 405–6.

185   *"all the Departments"*: Poe to Clarke, March 11, 1843, in Ostrom, 1:386, in *TPL*, 405.

185   *"On the second day"*: Dow to Clarke, March 12, 1843, in *TPL*, 405–6.

185   *"seedy in his appearance"*: John Hill Hewitt describes Poe ca. March 10, 1843, in *TPL*, 404.

185   *"he suffered much from his indiscretion"*: Thomas quoted in Whitty, "Memoir," xlvii, in *TPL*, 405.

185   *"He exposes himself here"*: Dow to Clarke, March 12, 1843, in *TPL*, 405–6.

185   *"a warm bath & supper"*: Poe to Thomas and Dow, March 16, 1843, in Ostrom, 1:388.

186   *"the President, yesterday"*: Thomas to Poe, March 27, 1843, in *TPL*, 408.

186   *Tyler even sent a letter*: Robert Tyler to Blythe, March 31, 1843, in *TPL*, 409. Poe reviewed Robert Tyler's poetry in October that year; in light of the puffs and the criticism his father had received in the press, Poe declared his "absolute impartiality," pointing out flaws while praising Tyler's sense of the beautiful; Poe, review of *Death; or Medorus' Dream*, by Robert Tyler, *Graham's*, Dec. 1843, 319–20.

186   *"ready and anxious"*: *Spirit of the Times*, March 16, 1843, in *TPL*, 407.

186   *"From the celestial visitant"*: Editorial by Poe, *Saturday Museum*, April 1, 1843, in *TPL*, 410.

186   *"a lean-to of three rooms"*: Mayne Reid on Poe, ca. April 1843, in "A Dead Man Defended," *Onward*, April 1869, 306, in *TPL*, 409.

186   *He still planned to launch*: Poe to Lowell, March 27, 1843, in Ostrom, 1:393–94, and *TPL*, 408.

187   *"All three of them happening"*: Recollection from Anne E. C. Clarker, ca. 1843, in *TPL*, 445.

187   *"gentlemanly looking personage"*: T. D. English, "The Doom of the Drinker," in *Saturday Museum*, Dec. 9, 1843, in *TPL*, 443.

187   *"it gives me inexpressible pain"*: Wilmer to John Tomlin, May 20, 1843; AHQ, 401–2, and *TPL*, 412.

187   *"many recent reverses"*: William Poe to Poe, June 15, 1843, in *TPL*, 415.

187   *"a refined and very gentlemanly"*: Darley quoted in Woodberry, *Life*, 2:2–3, and *TPL*, 413.

187   *"some of the pleasantest hours"*: Reid, "Dead Man Defended," 206–7, in *TPL*, 410.

188   *"Very Liberal Offers and No Humbug"*: *Baltimore Sun*, March 31, 1843, in *TPL*, 409.

188   *It was a wish-fulfilling story*: See Terence Whalen, "The Code for Gold: Edgar Allan Poe and Cryptography," *Representations* 46 (1994): 35–57; Rosenheim, *Cryptographic Imagination*; J. Gerald Kennedy, *Strange Nation: Literary Nationalism and Cultural Conflict in the Age of Poe* (New York: Oxford University Press, 2016), 370–74; Marc Shell, *Money, Language, and Thought: Literary and Philosophic Economies from the Medieval to the Modern Era* (Berkeley: University of California Press, 1982), 8–22; Barton Levi St. Armand, "Poe's 'Sober Mystification': The Uses of Alchemy in 'The Gold-Bug,'" *PS* 4, no. 1 (1971): 1–7.

188   *"unique work of a singularly constituted"*: Clarke in *Saturday Museum*, July 8, 1843, in *TPL*, 424.

189   *accusing Poe of plagiarism*: Francis H. Duffee, "The 'Gold Bug'—a Decided Humbug," *Daily Forum*, June 27, 1843, in *TPL*, 419–20.

189   *like many, loathed Griswold*: That summer Lippard had published a serialized satire targeting Griswold and Graham, "The Spermaceti Papers," in *Citizen Soldier*, in *TPL*, 413. Widespread loathing for Griswold is noted in Widmer, *Young America*, 1.

189   *"stamped with the impress"*: George Lippard in *Citizen Soldier*, Nov. 15, 1843, in *TPL*, 440.

189   *"unable to gain admission"*: *United States Gazette*, Jan. 8, 1844, in *TPL*, 441.

189   *"highly intelligent audience"*: Clarke in *Saturday Museum*, Nov. 25, 1843, in *TPL*, 441–42.

189   *"a sound Magazine"*: George Lippard in *Citizen Soldier*, Jan. 10, 1844, in *TPL*, 448.

189   *"probability of your success"*: Lowell to Poe, March 6, 1844, in *TPL*, 454.

190   *"It was Mr. Poe that made Graham's"*: George Lippard in *Citizen Soldier*, Nov. 15, 1843, in *TPL*, 441.

11. The March of Science and Quacks

191   *In Washington on April 1*: The event—along with the institute's history, membership, and topics—is detailed in *Bulletin of the National Institute for the Promotion of Science*, vol. 1 (Washington, D.C.: P. Force, 1841) and analyzed in Kohlstedt, "Step Toward Scientific Self-Identity in the United States," quotation

from 358; George B. Goode, "The Genesis of the United States National Museum," *Smithsonian Annual Report for 1897*, vol. 2 (Washington, D.C.: Smithsonian Institution; 1901).

191 *between musical performances: Bulletin of the National Institute*, 430–34.

192 *"members of Congress, strangers": Daily National Intelligencer*, April 2, 1844, 3; *Bulletin of the National Institute*, Ingersoll on 424.

192 *men of science chose to stay home:* Responses in Kohlstedt, "Step Toward Scientific Self-Identity," 356.

192 *"What do you think":* Henry to Torrey, March 27, 1844, in *Papers of Joseph Henry*, 6:62, in Kohlstedt, "Step Toward Scientific Self-Identity," 354n78.

192 *"found in Washington":* James Smithson, in Heather Ewing, *The Lost World of James Smithson: Science, Revolution, and the Birth of the Smithsonian* (New York: Bloomsbury Publishing USA, 2007), 344.

193 *backed by the* Southern Literary Messenger: The essay appeared in two parts (Jan. and March 1840), apparently written by Beverley Tucker, one of the leading southern defenders of slavery. It could be read as part of the antebellum strategy through which pro-slavery politicians, seeking to extend their grip on government, sponsored scientific ventures and naval investment. The *Southern Literary Messenger* gave a platform to southern naval promoters including Paulding and Maury (and, arguably, Reynolds, whose planned expedition was taken up by Poinsett, see below); see Matthew Karp, *This Vast Southern Empire: Slaveholders at the Helm of American Foreign Policy* (Cambridge, Mass.: Harvard University Press, 2016).

193 *In 1840, Poe echoed this call:* Poe, "Chapter on Science and Art," *Burton's*, March 1840, 149–50.

193 *a national scientific organization:* Washington had a local scientific society, the Columbian Institute, since 1816, but by the 1830s it was nearly defunct.

193 *"the spirit of associations":* Alexis de Tocqueville, "On the Use That the Americans Make of Association in Civil Life," in *Democracy in America*, ed. and trans. Harvey C. Mansfield and Delba Winthrop (1835–40; Chicago: University of Chicago Press, 2000), 489–92. Alongside multiple workers' associations, cooperatives, and syndicates, in 1839 the American Art-Union was formed, and in 1847 the American Medical Association.

193 *"with the Philadelphians":* Torrey to Henry, Nov. 9, 1838, Gray MSS, Gray Herbarium, Harvard University, in Kohlstedt, *Formation of the American Scientific Community*, 50–51.

193 *the extension of U.S.-owned plantations:* Karp, *Vast Southern Empire*; Walter Johnson, *River of Dark Dreams: Slavery and Empire in the Cotton Kingdom* (Cambridge, Mass.: Harvard University Press), 293–302.

193 *Nathaniel Hawthorne's application:* Stanton, *Exploring Expedition*, 306; Philbrick, *Sea of Glory*.

194 *fragility of the institute's conception:* Kohlstedt, "Step Toward Scientific Self-Identity," 344.

194 *"the extent and depth":* The institute's planners declared, "We may not compete with the British Association." "First Circular, Respecting Meetings of Scientific and Literary Men in the United States," *Bulletin of the National Institute*, Oct. 15, 1842, 421.

194 *the Association of American Geologists and Naturalists:* The Association of American Geologists was formed at the instigation of Edward Hitchcock, director of the Massachusetts survey and president of Amherst College, who felt disconnected from fellow geologists; they resolved to meet each year in a different city, starting at the Franklin Institute in 1840 thanks to Henry Darwin Rogers. At first they forbade membership to anyone "not devoted to Geological research with scientific views and objects," but at the 1843 meeting in Boston, with Samuel Morton as president, they added "and Naturalists" to their name; Kohlstedt, *Formation of the American Scientific Community*, 67.

195 *"something more popular":* Bache to Loomis, March 7, 1844, Loomis MSS, BYA, in Kohlstedt, "Step Toward Scientific Self-Identity," 354n78. William Redfield, a steam engineer and meteorologist from New York, worried that the institute's design "might interfere, intentionally or not," with the AAGN's meeting; he chose not to attend (Redfield, Nov. 21, 1842, in Kohlstedt, "Step Toward Scientific Self-Identity," 356n54); for the forty-three invitees who declined the invitation, see "Replies to Circulars," Markoe's journal in Rhees MSS, Huntington Library, and Kohlstedt, "Step Toward Scientific Self-Identity," 358n100.

195 *"without the aid of a band":* Dana to Spencer F. Baird, May 19, 1844, Baird MSS, SIA, in Kohlstedt, "Step Toward Scientific Self-Identity," 360n110.

195 *"have half a hundred persons":* Bache to Lloyd, Nov. 30, 1838, box 2, vol. 1, Bache Papers, in Jansen, *Alexander Dallas Bache*, 203.

195 *"under the control principally":* Bache, Address to Washington Philosophical Society, 1871, in Kohlstedt, "Step Toward Scientific Self-Identity," 362n121.

195 *"If we can have a general attendance"*: Dana to A. A. Gould, April 11, 1844, Gould MSS, Houghton Library, Harvard, in Kohlstedt, "Step Toward Scientific Self-Identity," 359.

195 *"the mass of diluvium"*: Henry to Bache, in *Papers of Joseph Henry*, 6:76, in Jansen, *Alexander Dallas Bache*, 207.

195 *"precipitously moving forward"*: Bache, "The Wants of Science in the United States" (1844), MSS, forensically deciphered by Jansen, in *Alexander Dallas Bache*, 227–28.

196 *"every person has the liberty"*: Henry in Moyer, *Joseph Henry*, 227.

196 *"republican in the proper sense"*: Ibid., 228.

196 He was annoyed at Benjamin Silliman's: Arthur Molella, "At the Edge of Science: Joseph Henry, 'Visionary Theorizers,' and the Smithsonian Institution," *Annals of Science* 41, no. 5 (1984): 445–61.

196 *"in the minds of many"*: Henry to Morse, Feb. 24, 1842, in *Papers of Joseph Henry*, 5:150–51.

196 *"have served, as it were"*: Henry in Moyer, *Joseph Henry*, 231.

196 *"the third and fourth rate men"*: Ibid., 228.

197 artificial climate modification: James Rodger Fleming, *Fixing the Sky: The Checkered History of Weather and Climate Control* (New York: Columbia University Press, 2010), 54–58.

197 Bache expressed reservations: Bache, Henry, Peirce, and Adams, quoted in Peter Moore, *The Weather Experiment: The Pioneers Who Sought to See the Future* (New York: Farrar, Straus and Giroux, 2015), 137–39; on Espy and his longstanding clash with William Redfield, see James Fleming, *Meteorology in America, 1800–1870* (Baltimore: Johns Hopkins University Press, 1990), 23–54.

197 *"The chief portion of Professor"*: Poe, "Marginalia," *Graham's*, Dec. 1846; M183, in *The Collected Writings of Edgar Allan Poe, Vol. 11: The Brevities*, ed. Burton R. Pollin (New York: Gordian Press, 1985), 309.

197 *"ostentatious presentations"*: John Collins Warren of the APS to William Horner, Nov. 20, 1838, APS Archives, in Kohlstedt, *Formation of the American Scientific Community*, 50.

197 the very rogue who had challenged Henry. See page 121 above; Moyer, *Joseph Henry*, 225–27.

197 A seductive lecturer: On Lardner's British career, see Jo N. Hays, "The Rise and Fall of Dionysius Lardner," *Annals of Science* 38, no. 5 (1981): 527–42. On public science in Britain, see Bernard Lightman, *Victorian Popularizers of Science: Designing Nature for New Audiences* (Chicago: University of Chicago Press, 2009); Morus, *Frankenstein's Children*; Hsiang-Fu Huang, "A Shared Arena: The Private Astronomy Lecturing Trade and Its Institutional Counterpart in Britain, 1817–1865," *Notes and Records: The Royal Society Journal of the History of Science* 72, no. 3 (2018): 319–41. Lardner's American career and optical technologies are helpfully traced in Kentwood D. Wells, "Dionysius Lardner: Popular Science Showman of the 1840s," *Magic Lantern Gazette* 29, no. 1 (2017): 3–17; see also Anna Louise Martin, *Villain of Steam: A Life of Dionysius Lardner (1793–1859)* (Carlow, Ireland: Tyndall Scientific, 2015).

197 *"extensive and splendid illustrations"*: *New York Weekly Herald*, Nov. 27, 1841, in Wells, "Lardner," 5.

198 *"destroyed the peace"*: *Gloucester Telegraph*, Nov. 24, 1841, 2, in Wells, "Lardner," 5.

198 *"I see by the papers"*: Henry to Torrey, Dec. 20, 1841, in *Papers of Joseph Henry*, 5:132–33.

198 *"no less a personage"*: Poe, "Three Sundays in a Week," first published as "A Succession of Sundays," *Saturday Evening Post*, 12, no. 1061, Nov. 27, 1841, 1.

199 *"have taken the pills"*: David Meredith Reese, *Humbugs of New York, Being a Remonstrance Against Popular Delusion, Whether in Science, Philosophy, or Religion* (New York: Taylor, 1838), 21; on Reese's ornery character, see Fabian, *Skull Collectors*.

199 *"Both Popery and anti-Popery are impostures"*: Ibid., 210.

199 *"Jefferson, Franklin, Rush, and John Jay"*: Ibid., 144.

199 Beyond their endorsement of a common background: Bache largely kept quiet about the Episcopalianism in which he was raised, while the Presbyterian Henry told students that physical science offers "the most striking proofs of the beneficence, the wisdom, and the power of the Creator" (in Moyer, *Joseph Henry*, 141). Benjamin Peirce was a devout Unitarian who once cried out in a lecture after a mathematical proof, "Gentlemen, there must be a GOD!" (in Edward Hogan, *Of the Human Heart: A Biography of Benjamin Peirce* [Bethlehem, Pa.: Lehigh University Press, 2008], 284).

200 of slavery and the race science: See Fabian, *Skull Collectors*; Stanton, *Leopard's Spots*; Rusert, *Fugitive Science*. On an episode involving Henry from the early Civil War, in which he refused to allow Frederick Douglass to speak at the Smithsonian, see Michael F. Conlin, "The Smithsonian Abolition Lecture Controversy: The Clash of Antislavery Politics with American Science in Wartime Washington," *Civil War History* 46, no. 4 (2000): 301–23.

200 public supporter of slavery: Hogan, *Human Heart*, 203–9. In the 1850s, when Henry and Bache invited

Peirce to lecture in Washington, he insisted on giving two talks on comets and two "on the various powers for mathematics of the different races and nationalities of man"; they steereed him toward discussing "the diversity, 'but not *the inferiority* of the races'" (ibid., 211).

200 *contributor to race science*: Edward Lurie, *Louis Agassiz: A Life in Science* (Chicago: University of Chicago Press, 1960), 258–71; Gould, *Mismeasure of Man*, 74–82.

200 *a "polite" silence toward the era's*: On Bache and Henry sidelining race science in the AAAS, particularly after 1850, see Kohlstedt, *Formation of the American Scientific Community*, 111–14; chapter 15 below.

201 *"intimately connected with the scientific"*: Henry, *Papers of Joseph Henry*, 6:16.

201 *"I shall allow myself no rest"*: Peirce, Cambridge, Nov. 28, 1843, A. D. Bache Papers, series RU 7053, box 5 Coast Survey, Lighthouse Board, Papers, Reports, box 3, 2, Incoming Correspondence, Smithsonian Institution Archives.

201 *"had the support of all the most prominent"*: Henry to James Henry, Jan. 1844, *Papers of Joseph Henry*, 6:15–16; on the natural philosophy textbook, see W. H. C. Bartlett of West Point to Joseph Henry, March 1, 1844, cited in Charles I. Weiner, "Joseph Henry's Lectures on Natural Philosophy: Teaching and Research in Physics, 1832–1847" (PhD diss., Case Institute of Technology, 1965), 56.

201 *the land beneath coastal waters*: For links between Bache's hydrography, Maury's oceanography, imperial expansion, and the Humboldtian global sciences, see Michael Reidy and Helen Rozwadowski, "The Spaces in Between: Science, Ocean, Empire," *Isis* 105, no. 2 (2014): 338–51; for overviews of the survey under Bache, see Slotten, Jansen, and Daniel Kevles, "Practical Pressures and Scientific Payoffs: A Long View of Knowledge and Utility in Federal Research," *Social Research: An International Quarterly* 84, no. 3 (2017): 561–82.

201 *"to the aid of the survey"*: Bache to Elias Loomis, Dec. 13, 1843, both NR II roll S, Record Unit 7470, Reingold Papers, SIA, in Jansen, *Alexander Dallas Bache*, 199.

202 *The survey offered rigorous*: The use of field surveys and expeditions as scientific training ground had been developed by Colby in Ireland, Everest in India, and De la Beche in the British Geological Survey. See Rachel Hewitt, *Map of a Nation: A Biography of the Ordnance Survey* (London: Granta, 2011); Matthew Edney, *Mapping an Empire: The Geographical Construction of British India, 1765–1843* (Chicago: University of Chicago Press, 2009); James Secord, "The Geological Survey of Great Britain as a Research School, 1839–1855," *History of Science* 24, no. 3 (1986): 223–75.

202 *wizard instrumentalist Joseph Saxton*: Joseph Henry, "Memoir of Joseph Saxton: 1799–1873," in *Biographical Memoirs* (Washington, D.C.: National Academy of Science, 1877), 219–316; on Saxton's improvements to barometers, pyrometers, standard lengths, divided angles, and a device to measure tides automatically, see ibid., 308–14.

203 *"national scientific project"*: Slotten, *Patronage*, 119.

203 *"the general scientific agency"*: Dupree, *Science in the Federal Government*, 104.

204 *"Bridgewater Lectures"*: Playbill, Dec. 24, 1843, McAllister Collection, vol. 19, p. 68, Library Company of Philadelphia.

204 *an astronomical two hundred thousand dollars*: Thomas Ollive Mabbott, "Poe and Dr. Lardner," *American Notes and Queries* 3, no. 8 (1943): 117.

205 *"The nose of a mob"*: Poe, "Marginalia," *SLM* 15, no. 6 (June 1849); M226, in Pollin, *Brevities*, 336. In the first installment of "Marginalia," Poe tore into Lardner's unoriginal and "metaphysical" discussion of the apparent size of the sun, a discussion that informs his tale "The Sphinx"; he later made use of Lardner's anecdotes about inventions in "The Thousand-and-Second Tale of Scheherazade"; see Mabbott, "Poe and Dr. Lardner," 115–16; Poe, "Marginalia," *United States Magazine and Democratic Review*, Nov. 1844; M38, in Pollin, *Brevities*, 143–48.

205 *he had been writing for two audiences*: Allen, *Poe and the British Magazine Tradition*. According to J. Gerald Kennedy, Poe was "perpetually investigating the possibility of creating a single literary text capable of satisfying both the popular and the critical taste"; see *A Historical Guide to Edgar Allan Poe*, ed. J. Gerald Kennedy (New York: Oxford University Press, 2001), 67. The tension between the two targeted audiences is central to Elmer, *Reading at the Social Limit*, and Whalen, *Poe and the Masses*.

205 *"Where is the ingenuity"*: Poe to Philip P. Cooke, Aug. 9, 1846, in Ostrom, 2:595. On Poe's self-referential clues, see Louis Renza, "Poe's Secret Autobiography," in *The American Renaissance Reconsidered*, ed. Walter Benn Michaels (Baltimore: Johns Hopkins University Press, 1985), 58–89; see also the ever-expanding decipherments of Poe's allusions by Poe scholars—among the most tireless of them, the late Burton R. Pollin, as in *Discoveries in Poe* (Notre Dame, Ind.: University of Notre Dame Press, 1970).

205　*Poe reviewed a special issue*: Poe, review of *A Brief Account of the Discoveries and Results of the United States Exploring Expedition, Graham's*, Sept. 1843, 164–65.

205　*a Fijian chief named Veidovi*: This grisly episode and others equally disturbing are recounted in Fabian, *Skull Collectors*; see also T. D. Stewart, "The Skull of Vendovi: A Contribution of the Wilkes Expedition to the Physical Anthropology of Fiji," *Archaeology and Physical Anthropology in Oceania* 13, no. 2/3 (1978): 204–14, and Adrienne Kaeppler, "Two Polynesian Repatriation Enigmas at the Smithsonian Institution," *Journal of Museum Ethnography* 17 (2005): 152–62.

205　*Wilkes was court-martialed*: On Wilkes's trial and his clashes with the onboard "scientifics," see D. Graham Burnett, "Hydrographic Discipline among the Navigators," in *The Imperial Map: Cartography and the Mastery of Empire*, ed. James R. Akerman (Chicago: University of Chicago Press, 2009), 185–259; Philbrick, *Sea of Glory*; Stanton, *Exploring Expedition*.

205　*The refurbishment was a hit*: Jane Walsh, "From the Ends of the Earth: The United States Exploring Expedition Collections," in *The United States Exploring Expedition, 1838–1842*, Smithsonian Institution Libraries, Digital Collection 2004, www.sil.si.edu/DigitalCollections/usexex/learn/Walsh-01.htm; Antony Adler, "From the Pacific to the Patent Office: The US Exploring Expedition and the Origins of America's First National Museum," *Journal of the History of Collections* 23, no. 1 (2011): 49–74; Curtis Hinsley, *Savages and Scientists: The Smithsonian Institution and the Development of American Anthropology, 1846–1910* (Washington, D.C.: Smithsonian Institution, 1981).

206　*"gems and gold and iron ores"*: Poe, review of *A Brief Account*, 164–65.

206　*"By a walk through the National Gallery"*: "United States Exploring Expedition," *American Journal of Science and Arts* 44 (1843): 399; see Adler, "From the Pacific to the Patent Office," 69, who attributes the article to G. S. Silliman.

206　*It served as a warrant for future scientific*: See Hinsley, *Savages and Scientists*, 17–20.

206　*"the many able and respectable"*: Poe, review of *A Brief Account*, 165. The *American Journal of Science and Arts* essay similarly concludes "Mr J. N. Reynolds was left behind, yet, though unrewarded for his efforts by the pleasure of accompanying the expedition, and adding to its laurels, his distinguished merits will not be forgotten or disregarded by his countrymen" (408).

207　*led by a "coalition"*: Poe to Lowell, March 30, 1844, in Ostrom, 1:432, in *TPL*, 456.

207　*Congress had granted Samuel Morse*: John, *Network Nation*, chap. 2, on federal largesse toward Morse and incentives for invention provided by patent legislation; patent models and inspections were held in the same building as the collections for the Exploring Expedition.

207　*A girl who grew up near Poe's home*: Lydia Hart Garrigues, ca. 1843, quoted in Mary E. Phillips, *Edgar Allan Poe, the Man* (Chicago: John C. Winston, 1926), 827, in *TPL*, 445.

Part IV: New York City

209　*"I have reached these lands but newly"*: Poe, "Dream-Land," first published in *Graham's*, June 1844; LOA, 79–80.

12. The Market for Novelty

211　*"ASTOUNDING NEWS!"*: Poe, unsigned article appearing in *New York Sun*, April 13, 1844, later anthologized as "The Balloon Hoax"; LOA, 743–55.

211　*Mason's technical improvements*: On the details of the trip, see discussion in Harold Beaver, ed., *The Science Fiction of Edgar Allan Poe* (Harmondsworth, U.K.: Penguin, 1976), 371. On news and woodcuts, see Michael Leja, "News Pictures in the Early Years of Mass Visual Culture in New York: Lithographs and the Penny Press," in *Getting the Picture: The Visual Culture of the News*, ed. Jason Hill and Vanessa Schwartz (London: Bloomsbury, 2015), 146–54.

212　*means of getting news quickly*: Frank M. O'Brien, *The Story of "The Sun": New York, 1833–1918* (New York: George H. Doran, 1918), 146; Hayes, *Poe and the Printed Word*, xi.

212　*they had been taken for a ride*: Naomi Miyazawa, "Edgar Allan Poe and Popular Culture in the Age of Journalism: Balloon Hoaxes, Mesmerism, and Phrenology" (PhD diss., University at Buffalo, State University of New York, 2010); Jeffrey A. Savoye, "Meanderings Here and There in Poe's 'Balloon Hoax,'" *EAPR* 18, no. 2 (2017): 257–62.

212　*"nothing put forth in the Balloon-Story"*: Poe, "Doings of Gotham," *Columbia (Pa.) Spy*, May 25, 1844, 3.

212　*"a far more intense sensation"*: Ibid.

212　*"attempt to hoax"*: *New York Herald*, April 15, 1844, in *TPL*, 460.

212   *"50,000 of the extras"*: *Philadelphia Saturday Courier*, April 20, 1844, in TPL, 461.

212   *"The more intelligent believed"*: Poe's account of the article's reception is challenged in Beaver, *Science Fiction*, and Miyazawa, "Poe and Popular Culture," 3–4.

213   *"coughed none at all"*: Poe to Clemm, April 7, 1844, in Ostrom, 1:437–38, in TPL, 457.

213   *"Where two individuals"*: Poe, "Doings of Gotham," *Columbia (Pa.) Spy*, June 15, 1844, 3.

213   *The Lyceum of Natural History*: Among its founders were the physician Samuel Latham Mitchill and politicians including DeWitt Clinton and Rufus King. John Hendley Barnhart, "The First Hundred Years of the New York Academy of Sciences," *Scientific Monthly* 5, no. 5 (1917): 463–75; and Simon Baatz, *Knowledge, Culture, and Science in the Metropolis: The New York Academy of Sciences, 1817–1970* (New York: New York Academy of Sciences, 1990), 1–256, esp. 9–55.

214   *John W. Draper, continuing his work*: Gillespie, "Reception of Early Scientific Photography"; Kenneth Silverman, *Lightning Man: The Accursed Life of Samuel F. B. Morse* (New York: Knopf, 2003).

214   *Blitz, a celebrated stage magician*: Antonio Blitz, *Fifty Years in the Magic Circle* (Hartford, Conn.: Belknap and Bliss, 1879); James W. Cook, *The Arts of Deception: Playing with Fraud in the Age of Barnum* (Cambridge, Mass.: Harvard University Press, 2001), 179–81.

214   *"the chosen arena of itinerating mountebanks"*: Reese, *Humbugs of New-York*, 17–18.

214   *Ormsby Mitchel, the West Point mathematician*: Robert J. Scholnick, "*Eureka* in Context: Poe, the Newspaper, the Lyceum, and Cosmic Science," in *Poe Writing/Writing Poe*, ed. Richard Kopley and Jana Argersinger (New York: AMS Press, 2013), 31–50.

214   *Daily newspapers had first sprung up*: Michael Schudson, *Discovering the News: A Social History of American Newspapers* (New York: Basic Books, 1978); Frank Luther Mott, *A History of American Magazines, 1741–1850*, vol. 1 (Cambridge, Mass.: Harvard University Press, 1930).

215   *"newsboy system"*: O'Brien, *Story of "The Sun,"* 132.

216   *"eminent physicians and intelligent men"*: Undated handbill (ca. Sept. 1835) for a show in Hingham, Mass., in Cook, *Arts of Deception*, 6n11; see further testimonials in *The Life of Joice Heth, the Nurse of Gen. George Washington, the Father of Our Country, Now Living at the Astonishing Age of 161 Years, and Weighs Only 46 Pounds. Price Six Cents* (New York: printed for the publisher, 1835), New-York Historical Society; Benjamin Reiss, *The Showman and the Slave: Race, Death, and Memory in Barnum's America* (Cambridge, Mass.: Harvard University Press, 2001).

216   *"the great father of caterers"*: Barnum quoted in Cook, *Arts of Deception*, 22.

216   *The onboard death of the captured Fijian chief*: Fabian, *Skull Collectors*, 121–64.

216   *"Dr. Griffin, agent of the Lyceum"*: Barnum in Cook, *Arts of Deception*, 82.

217   *"regarding which there has been"*: Barnum to Kimball, Sept. 4, 1843, draft of a notice later published in several newspapers, in Cook, *Arts of Deception*, 84.

218   *"mob violence"*: The episode is discussed in Kenneth Greenberg, *Honor and Slavery: Lies, Duels, Noses, Masks, Dressing as a Woman, Gifts, Strangers, Humanitarianism, Death, Slave Rebellions, the Proslavery Argument, Baseball, Hunting, and Gambling in the Old South* (Princeton, N.J.: Princeton University Press, 1996), 6.

218   *Peale's sons presented*: Cook, *Arts of Deception*, 102–3.

218   *Barnum's shows, instead, invited viewers*: On Barnum's "operational aesthetic," see Neil Harris, *Humbug: The Art of P. T. Barnum* (Chicago: University of Chicago Press, 1981): "An exhibitor did not have to guarantee truthfulness; all he had to do was possess probability and invite doubt. The public would be more excited by controversy than conclusiveness" (23).

218   *Barnum's exhibitions were among the most important*: John Rickards Betts, "P. T. Barnum and the Popularization of Natural History," *Journal of the History of Ideas* 20, no. 3 (1959): 353–68.

219   *"thronged with strangers"*: Poe, "Doings of Gotham," May 25, 1844.

219   *"far and wide"*: Poe, "Doings of Gotham," *Columbia (Pa.) Spy*, June 1, 1844, 3.

219   *Already he saw preparations*: Poe, "Doings of Gotham," *Columbia (Pa.) Spy*, June 8, 1844, 3; on the military frenzy around the Polk campaign, see Kennedy, "Mania for Composition," and Howe, *What Hath God Wrought*, 683–90.

220   *"the mob-disorder which so lately beset Philadelphia"*: Poe, "Doings of Gotham," June 1, 1844; see Emma Jones Lapsansky, "'Since They Got Those Separate Churches': Afro-Americans and Racism in Jacksonian Philadelphia," *American Quarterly* 32, no. 1 (1980): 54–78; James Brewer Stewart, "The Emergence of Racial Modernity and the Rise of the White North, 1790–1840," *Journal of the Early Republic* 18, no. 2 (1998): 181–217.

220   *launch by the novelist Nathaniel Willis*: Poe praised the *Mirror* but lamented that it kept Willis from

writing novels and poetry, such as his "Unseen Spirits," which Poe called "the best of *all* his poems" in "Doings of Gotham," *Columbia (Pa.) Spy*, July 6, 1844, in *TPL*, 466.

220   *"mentioning that [Poe] was ill"*: Nathaniel Parker Willis, "Letter About Edgar Poe," *New York Home Journal*, Oct. 30, 1858, 2, in *TPL*, 473.

220   *defend a work by John W. Draper*: "Notice of Dr. Draper and the *North American Review*," *Evening Mirror*, Jan. 20, 1845, 2. The authorship of this piece has been disputed (see eapoe.org), but Poe had discussed Draper's work while both were in Virginia; it concerns daguerreotypes, chides Boston's *North American Review*, and details the reception of Draper's book by European scientists in a way strongly reminiscent of the opening of Poe's "Von Kempelen's Experiment."

220   *"monster instrument now nearly"*: Poe, "The Cincinnati Telescope," *Evening Mirror*, Jan. 10, 1845.

220   *"sit at a desk"*: Willis, "Letter." In the Oct. 10, 1844, *Mirror*, Willis announced Poe as a fellow in arms to fight for an international copyright and better pay for authors: "We wish to light beacons for an authors' crusade and we have no leisure to be more than its Peter the Hermit. We solemnly summon Edgar Poe to do the devoir of Coeur de Lion—no man's weapon half so trenchant!," in *TPL*, 473–74.

221   *the clashes Poe himself had stirred up*: On Poe's influence on the vituperative critic Park Benjamin, see Sandra Tomc, "Edgar Allan Poe and His Enemies," in Kennedy and Peeples, *Oxford Handbook*, 559–75.

221   *"A reply from me to Mr. Briggs"*: Willis to Poe, ca. 1845, in Harrison, *Complete Works*, 17:206, cited in Sandra Tomc, "Poe and His Circle," in Hayes, *Cambridge Companion*, 21–41.

221   *calling themselves Young America*: Widmer, *Young America*; Richard, "Poe and 'Young America,'" 25–58; Meredith McGill, "Poe, Literary Nationalism, and Authorial Identity," in Rosenheim and Rachman, *American Face of Edgar Allan Poe*, 271–304.

221   *"an image of rural"*: Duyckinck, 1845, in Perry Miller, *The Raven and the Whale: Poe, Melville, and the New York Literary Scene* (New York: Harcourt, Brace, 1956), 111.

221   *Poe dubbed him an "ass"*: Poe to Frederick W Thomas, Sept. 12, 1842, in Ostrom, 1:359.

222   *Readers had a choice*: Duyckinck, "Literary Situation of 1845," *American Whig Review* (Feb. 1845), in McGill, *Culture of Reprinting*, 286.

222   *"false and lawless state of things"*: Cornelius Mathews, "An Appeal to American Authors and the American Press in Behalf of an International Copyright," *Graham's*, Sept. 1842, 122, in McGill, *Culture of Reprinting*, 287.

222   *first authors to use photography*: Kevin J Hayes, "Poe, the Daguerreotype, and the Autobiographical Act," *Biography* 25, no. 3 (2002): 477–92; Michael Von Cannon, "A Tale of Optics: Poe, Visual Culture, and Antebellum Literary Celebrity," *PS* 47, no. 1 (2014): 36–54, suggests that Poe altered his appearance to suit the daguerreotype. See also Deas, *Portraits and Daguerreotypes*.

223   *"The query is put to us"*: Poe, "Notice of Thingum Bob," *Evening Mirror*, Jan. 14, 1845, 2.

223   *brought to America by P. T. Barnum*: Phineas Taylor Barnum, *The Life of P. T. Barnum, Written by Himself* (New York: Redfield, 1855), 345.

223   *"ingenious pieces of mechanism"*: Poe, "The Swiss Bell-Ringers," *Weekly Mirror*, Oct. 12, 1844. Unlike his exposé of the chess player, which argued that a putative machine was actually controlled by a human, here he argues a set of putative humans are actually machines, coordinated by an electric "conductor." Though this piece is unsigned, the style, topic, and reference to Maelzel make it "characteristically Poe's," according to Mabbott, *Collected Works*, 3:1118.

223   *incomparably productive in 1844*: J. Gerald Kennedy identified this turn in Poe's writing and explores his tales' complex links to the celebration of Manifest Destiny and the "progress" of "Anglo-Saxon civilization," especially around the 1844 election, in "Mania for Composition," and, in extensive comparison with Poe's contemporaries, in *Strange Nation*.

223   *"Through ponderous looms"*: Richard Waterston, *Poem Delivered Before the Mercantile Library Association* (Boston: T. R. Marvin, 1845), in Howe, *What Hath God Wrought*, 569. The poem may be a rewriting of Erasmus Darwin's *Temple of Nature*, canto 4, on cotton and printing: "In level lines the length of woof to spread, / And dart the shuttle through the parting thread. / So Arkwright taught from Cotton-pods to cull / And stretch in lines the vegetable wool."

224   *"a nation of the most powerful"*: Poe, "The Thousand and Second Tale of Scheherazade" *Godey's Lady's Book*, 30, no. 2, Feb. 1845, 61–67.

224   *Poe turned the tables*: Poe, "Some Words with a Mummy," *American Review* 1, no. 4 (April 1845): 363–70.

224 *The American experiment*: Dana Nelson draws out the tale's satire of Morton's theory of polygenesis and racial essentialism, linking the scientists' gathering to wider forms of antebellum white male association in "The Haunting of White Manhood: Poe, Fraternal Ritual, and Polygenesis," *American Literature* 69, no. 3 (1997): 515–46; see also Kennedy, *Strange Nation*.

224 *"'Thou Art the Man'"*: The story's title is a biblical phrase, which Charles Babbage had recalled in his *Ninth Bridgewater Treatise* when imagining a future accusation of a murderous slave trader by "every corporeal atom of his immolated slave": "And Nathan said to David: *Thou art the man*." Charles Babbage, *The Ninth Bridgewater Treatise* (London: J. Murray, 1838), 118–19.

224 *"open, manly, honest"*: Poe, "Thou Art the Man," *Godey's Lady's Book*, 29, no. 5, Nov. 1844, 219–24.

224 *"system of soothing"*: "The System of Doctor Tarr and Professor Fether," *Graham's* 28, no. 5, Nov. 1845, 193–200.

224 *a marketplace crowded with thrills*: Other tales of this time described gimmicky contraptions and delivered surprise endings: "The Premature Burial" (*Dollar Newspaper* [Philadelphia], July 31, 1844), which memorably observed, "The boundaries which divide Life from Death, are at best shadowy and vague," described a bell-ringing mechanism to prevent people from being buried alive; in "The Spectacles," a nearsighted man, too vain to wear glasses, falls in love with a beautiful young woman who turns out to be his seventy-year-old grandmother (*Dollar Newspaper*, March 27, 1844).

225 *"The marvellous increase"*: Poe, "The Angel of the Odd," *Columbian Lady's and Gentleman's Magazine* (New York), Oct. 1844, 158–61; see Ogden, *Credulity*, 129–31.

225 *"the most trifling variation in the facts"*: Poe, "The Mystery of Marie Rogêt," in *Tales* (New York: Wiley and Putnam, 1845), 151–99. On the doctrine of chances in "Marie Rogêt," see Maurice Lee, "Probably Poe," *American Literature* 81, no. 2 (2009): 225–52, where Poe is presented as a proto-pragmatist—in contrast with Lee's earlier "Absolute Poe: His System of Transcendental Racism," *American Literature* 75, no. 4 (2003): 751–81. For the tale's implications for queer studies and Lacan—who, with Derrida, launched a generation of poststructuralist readings of "The Purloined Letter"—see Valerie Rohy, "The Calculus of Probabilities," in Kennedy and Peeples, *Oxford Handbook*, 224–35.

225 *"It is no longer philosophical"*: Poe had already worked out the argument that chance might be foundational to nature's structure in "Doings of Gotham," Letter 6, *Columbia (Pa.) Spy*, June 29, 1844, 3. On transformations in probability from the eighteenth through the nineteenth century as wrought by Laplace, Poisson, Gauss, and others, see Ian Hacking, *The Taming of Chance* (Cambridge, U.K.: Cambridge University Press, 1990)—a narrative culminating in C. S. Peirce's cosmology. On Peirce and chance, see Kenneth Laine Ketner, *His Glassy Essence: An Autobiography of Charles Sanders Peirce* (Nashville: Vanderbilt University Press, 1998); Louis Menand, *The Metaphysical Club: A Story of Ideas in America* (New York: Farrar, Straus and Giroux, 2001). On Poe and Pragmatism, see Umberto Eco and Thomas Sebeok, eds., *The Sign of Three: Dupin, Holmes, Peirce* (Bloomington: Indiana University Press, 1983).

226 *"We see not how"*: Review of *The Temple of Nature*, by Erasmus Darwin, *Literary Miscellany* 1 (1805): 284, in Scholnick, *Poe's "Eureka,"* 57. For more on Erasmus Darwin's cosmology and its influence on Poe, see Karen Weiser, "Poetry in Reason: The Scientific Poems of Edgar Allan Poe and Erasmus Darwin," *PS* 52, no. 1 (2019): 133–48.

226 *chemist John Dalton*: Scholnick, *Poe's "Eureka,"* 48; "The Life and Discoveries of John Dalton," *Living Age*, April 19, 1845.

226 *Reverend John Mason Good*: See Sean Moreland, "Beyond 'De Rerum Naturâ Esqr.': Lucretius, Poe, and John Mason Good," *EAPR* 17, no. 1 (2016): 6–40; on Lucretius's translations, see Scholnick, *Poe's "Eureka,"* 44; see also Daniel Driskell, "Lucretius and 'the City in the Sea,'" *PS* 5, no. 2 (1972): 54–55.

226 *"absolute authority in mercantile"*: Poe, "The Literati of New York City Part II," *Godey's Lady's Book*, June 1846, 266–72.

226 *Dupin had used probabilistic reasoning*: Lee, "Probably Poe."

226 *"what thoughts or sentiments"*: Poe, "The Purloined Letter," in *The Gift for 1845* (Philadelphia: Carey and Hart, 1844), 41–61; Poe mentions something similar when describing the game of whist in "Rue Morgue"—suggesting he cracked that case in part by thinking like the (nonhuman) killer. Even earlier, to outwit Maelzel's chess player, he had to *think like a machine*.

227 *Dupin's empathic method*: Kopley, *Dupin Mysteries*.

227 *Measuring, dividing, calculating*: Paul A. Harris, "Poe-tic Mathematics: Detecting Topology in 'The Purloined Letter,'" *PS* 36, no. 1 (2003): 18–31, sees Poe critiquing algebraic analysis in favor of the descriptive geometry taught at West Point.

227 *The lights in the hall*: Procedures detailed by Practical Magnetizer, *The History and Philosophy of Animal Magnetism, with Practical Instructions for the Exercise of This Power* (Boston: J. N. Bradley, 1843), 10–14.

228 *"the mysticism, or rather the mystic sensuality"*: Nathaniel Hawthorne, *The Blithedale Romance* (Boston: Ticknor, Reed and Fields, 1852), 231; see Taylor Stoehr, "Hawthorne and Mesmerism," *Huntington Library Quarterly* 33, no. 1 (Nov. 1969), 33–60.

228 *"I suppose that as the marketplace"*: Emerson, in *The Journals and Miscellaneous Notebooks of Ralph Waldo Emerson, Volume V: 1835–1838*, ed. Merton M. Sealts Jr. (Cambridge, Mass.: Harvard University Press, 1965), 388, in Stoehr, "Hawthorne and Mesmerism," 36.

229 *"unity and connection between"*: Emerson, "Historic Notes on Life and Letters in New England," in Stoehr, "Hawthorne and Mesmerism," 35.

229 *In the 1820s in Paris*: Bertrand Méheust, *Somnambulisme et médiumnité, 1784–1930: Le défi du magnétisme animal* (Paris: Synthélabo, 1999).

229 *new wave reached London*: Alison Winter, *Mesmerized: Powers of Mind in Victorian Britain* (Chicago: University of Chicago Press, 1998), 163–85. William Morton's related experimental use of medical "ether" as an anesthetic in 1846 was famously greeted with the exclamation "Gentlemen! this is no humbug!" Nathan P. Rice, *Trials of a Public Benefactor, as Illustrated in the Discovery of Etherization* (New York: Pudney and Russell, 1859); thanks to Katja Guenther and Keith Wailoo for tips on antebellum medical history.

229 *It appeared in New York in 1829*: David Schmit, "Re-visioning Antebellum American Psychology: The Dissemination of Mesmerism, 1836–1854," *History of Psychology* 8, no. 4 (2005): 403–34; Practical Magnetizer, *History and Philosophy of Animal Magnetism*, 6–10; Emily Ogden, "Beyond Radical Enchantment: Mesmerizing Laborers in the Americas," *Critical Inquiry* 42, no. 4 (2016): 815–41.

229 *respected researchers*: See *American Journal of Science and Arts* 33, no. 1 (Jan. 1838): 184; John Kearsley Mitchell, "An Essay upon Animal Magnetism, or Vital Induction," in *Five Essays*, ed. Silas Weir Mitchell (Philadelphia: J. B. Lippincott, 1859); Frank Podmore, *Modern Spiritualism: A History and a Criticism* (London: Methuen, 1902), 1:154–78; 234–6.

230 *Mesmerism proposed an empirical and rational approach*: François Azouvi, "Sens et fonction épistémologiques de la critique du magnétisme animal par les Académies," *Revue d'histoire des sciences* 29, no. 2 (1976): 123–42; Tresch, *Romantic Machine*, 31–48.

230 *concede that something was happening*: Thomas Carlyle's wife, Jane, was shocked out of her disbelief when a lower-class magnetizer produced an electric current while shaking her reluctantly proffered hand. Robin Waterfield, *Hidden Depths: The Story of Hypnosis* (New York: Routledge, 2003), 161.

230 *An 1843 manual from Boston*: Practical Magnetizer, *History and Philosophy of Animal Magnetism*, 5. See also Chauncy Hare Townshend, *Facts in Mesmerism or Animal Magnetism: With Reasons for a Dispassionate Inquiry into It* (Boston: Charles C. Little and James Brown, 1841).

230 *"twin brothers by nature"*: Poe, "Phreno-Magnetism," *American Phrenological Journal and Miscellany* 8, no. 4 (1846). See Schmit, "Re-visioning," n36.

230 *"the strange interregnum of the seasons"*: Poe, "A Tale of the Ragged Mountains," *Godey's Lady's Book* 28, no. 4, April 1844, 177–181. On the tale as colonial allegory, see Kennedy, "Mania for Composition"; for its mesmeric and typographical dimensions, see Emily Ogden, *Credulity: A Cultural History of US Mesmerism* (Chicago: University of Chicago Press, 2018); on its chiasmic structure, see Richard Kopley, "Poe's Pym-esque 'A Tale of the Ragged Mountains,'" in Fisher, *Poe and His Times*, 167–77.

231 *the transcendentalist Orestes Brownson*: Poe, "A Chapter on Autography," pt. 1, *Graham's*, Nov. 1841, 224–34.

231 *This weightless, ether-like matter*: Poe's notion of a material fluid nearly indistinguishable from spirit echoed contemporary theories of ether—the invisible substance hypothesized as carrying light, heat, and electromagnetism—and Renaissance views of a fifth element, which Marsilio Ficino described as "a very subtle body; as it were not body and almost soul. Or again, as it were not soul and almost body." *De Triplici Vita* 3.3, in D. P. Walker, *Spiritual and Demonic Magic from Ficino to Campanella* (London: Warburg Institute, 1958), 13.

233 *the anonymous Vestiges*: Robert Chambers, *Vestiges of the Natural History of Creation* (London: John Churchill, 1844), first published anonymously. See James Secord, *Victorian Sensation: The Extraordinary Publication, Reception, and Secret Authorship of "Vestiges of the Natural History of Creation"* (Chicago: University of Chicago Press, 2003); "it made a great sensation; chiefly," said Benjamin Silliman, "because the author cannot be detected" (*American Journal of Science and Arts* 19, no. 1 [April 1845]: 191). On the U.S.

reception of *Vestiges*, see Numbers, *Creation*. On Poe and *Vestiges*, see Scholnick, *Poe's "Eureka,"* and Stamos, *Edgar Allan Poe, "Eureka," and Scientific Imagination*.

233 *The romance (or novel) began*: Citations of *Vestiges* refer to *Vestiges of the Natural History of Creation* (New York: Wiley and Putnam, 1845; 2nd American ed., based on 3rd English ed.), 6–19.

233 *"the formation of bodies in space"*: *Vestiges*, 15.

234 *"difference between mind in the lower animals"*: Ibid., 235.

234 *"LIFE is everywhere ONE"*: This was an oft-quoted phrase from the 1846 follow-up to *Vestiges*, *Explanations: A Sequel to "Vestiges of the Natural History of Creation"* (New York: Wiley & Putnam, 1846), 130–31.

234 *"what but design and intelligence"*: Whewell, *Astronomy and General Physics*, 184.

234 *"that primeval Creative Thought"*: Nichol, *Views of the Architecture of the Heavens*, 22; on the relation between Nichol's *Architecture* and *Vestiges*, see Secord, *Victorian Sensation*, 58–59; Schaffer, "Science of Progress."

234 *"exact account of the phenomena"*: Joseph Henry, "Notes on a Series of Lectures on Geology," Aug. 1841, in Numbers, *Creation*, 26n47.

234 *"the true theory of the universe"*: *Princeton Review*, Jan. 1841, in Numbers, *Creation*, 23n37.

234 *experiments by Andrew Crosse*: James Secord, "Extraordinary Experiment: Electricity and the Creation of Life in Victorian England," in *The Uses of Experiment: Studies in the Natural Sciences*, ed. David Gooding, Trevor Pinch, and Simon Schaffer (Cambridge, U.K.: Cambridge University Press, 1989), 337–83.

235 *"the possible development of higher types"*: *Vestiges*, 272.

235 *"species superior to us in organization"*: "Critical Notices: *Vestiges*," *American Review: A Whig Journal of Politics, Literature, Art, and Science* 1, no. 2 (Feb. 1845): 215.

235 *Vestiges quoted Babbage's book*: *Vestiges*, 155–61.

235 *"the most elaborate attempt"*: Albert A. Dod, "Vestiges of Creation," in *Essays, Theological and Miscellaneous, Reprinted from the "Princeton Review"* (New York: Wiley and Putnam, 1847), 410; originally published as "Vestiges of the Natural History of Creation," *Biblical Repertory and Princeton Review* 17, no. 4 (1845): 505–57.

235 *"swallowed up by the upper classes"*: James Dwight Dana, quoting Silliman, *American Journal of Science and Arts* 49, no. 1 (April 1845): 191.

236 *"undeniably médiocres"*: Poe, "George B. Cheever," in "The Literati of New York City Part II," *Godey's Lady's Book*, June 1846, 267–68.

236 *"elaborate attempt"*: Cheever, in *Vestiges*, xix.

236 *"true character and tendency"*: Bowen, "A Theory of Creation," *North American Review* 60 (April 1845): 427; discussion in Numbers, *Creation*, 31–32.

236 *"blank atheism, cold, cheerless"*: Tayler Lewis, review of *Vestiges of the Natural History of Creation*, *American Review* 1 (May 1845), 525–28, 539–42.

236 *"If the book be true"*: Adam Sedgwick, letter to Charles Lyell, April 9, 1845, John Willis Clark and Thomas McKenny Hughes, *The Life and Letters of the Reverend Adam Sedgwick* (Cambridge: Cambridge University Press, 1890), 2 vols., 2:83–85.

236 *"false conclusions"*: Whelpley, *American Review* 3 (April 1846), in Numbers, *Creation*, 35n19.

236 *"It is admirably calculated"*: Review of *Vestiges*, *Broadway Journal* 1 (Jan. 1845): 52–53, in Scholnick, *Poe's "Eureka,"* 34.

13. A Man of Wonders

238 *"I have just written"*: Joel Benton, "Poe's Opinion of 'The Raven,'" *Forum* 22 (Feb. 1897): 733, in TPL, 495–96.

238 *"ramshackle Nassau Street office"*: Donald G. Mitchell, *American Lands and Letters*, 2 vols. (New York: Scribners, 1898–99), 2:387, in TPL, 484.

239 *February issue*: Poe, "The Raven," *The American Review: A Whig Journal of Politics, Literature, Art and Science* (Feb. 1845); see Thomas and Jackson's reconstruction of the publication sequence and reception in TPL, 496–97.

239 *"Once upon a midnight dreary"*: Poe, "The Raven," LOA, 81–86.

240 *seventeenth-century English poet*: On the influence of Francis Quarles's *Emblems* in the antebellum United States, see Alan Wallach, "The *Voyage of Life* as Popular Art," *Art Bulletin* 59, no. 2 (1977):

234–41; T. O. Beachcroft, "Quarles and the Emblem Habit," *Dublin Review* 28 (Jan.–June 1931): 80–96. On Poe, Quarles, and Renaissance arts of memory, see Engel, *Early Modern Poetics.*

240 *modeled in part on "Lady Geraldine's Courtship":* Richard Kopley and Kevin J. Hayes, "Two Verse Masterworks: 'The Raven' and 'Ulalume,'" in Hayes, *Cambridge Companion,* 191–204.

240 *The literary scholar Betsy Erkkila:* "The Poetics of Whiteness: Poe and the Racial Imaginary," in Kennedy and Weissberg, *Romancing the Shadow,* 41–74. On the interactions of matter/spirit and body/mind in Poe and their relations to both epistemology and antebellum hierarchies, see Dayan, *Fables of Mind;* "Amorous Bondage"; "Poe, Persons, and Property," *American Literary History* 11, no. 3 (1999): 405–25; David Leverenz, "Spanking the Master: Mind-Body Crossings in Poe's Sensationalism," in Kennedy, *Historical Guide to Edgar Allan Poe,* 95–127; Lee, "Absolute Poe."

241 *"the most cherished memory":* Alexander T. Crane, who worked in the office of *The Broadway Journal,* quoted in Mukhtar Ali Isani, "Reminiscences of Poe by an Employee of the Broadway Journal," *PS* 6, no. 2 (1973): 33–34, in *TPL,* 500.

241 *"a poem which would have enriched":* Notice of *American Review,* probably Horace Greeley, in *New York Daily Tribune,* Feb. 3, 1845, in *TPL,* 498.

241 *"may well defy competition":* James Brooks in *Morning Express,* Feb. 5, 1845, in *TPL,* 499.

241 *"You will see in this week's":* Briggs to Lowell, Feb. 6, 1845, in *TPL,* 499–500.

241 *Parodies proved its impact:* "The Owl: A Capital Parody on Mr. Poe's Raven," *Evening Mirror,* Feb. 17, 1845, in *TPL,* 503–4; Snarles, "A Vision," *New World,* April 19, 1845, in *TPL,* 527; C. C. Cooke, "The Gazelle (After the Manner of Poe's 'Raven')," *Evening Mirror,* April 29, 1845, in *TPL,* 528; "The Whippoorwill: A Parody on Mr. Poe's 'Raven,'" *Evening Mirror,* May 30, 1845, in *TPL,* 535; "Turkey," *Boston Jester,* June 1845, in *TPL,* 541.

242 *"I have never seen Poe's 'Raven'":* Lincoln to Andrew Johnston, April 18, 1846, in *The Collected Works of Abraham Lincoln,* ed. Roy P. Basler, 8 vols (New Brunswick, N.J.: Rutgers University Press, 1953–55), 1:377–79, in *TPL,* 635.

242 *"a most favorable impression":* Recollection ca. Feb. 1845 of Charles F. Briggs in "The Personality of Poe," *Independent,* Dec. 13, 1877, 1–2, in *TPL,* 497.

242 *"a pale, thin, and most grave-looking":* Henry T. Tuckerman, "A Memoir of the Author," in John W. Francis, *Old New York* (New York: W. J. Widdleton, 1866), lxxix–lxxx, in *TPL,* 498.

242 *"my Raven is really being talked about":* Smith quoted in J. C. Derby, *Fifty Years Among Authors, Books, and Publishers* (New York: G. W. Carleton, 1884), 547–48, in *TPL,* 497.

243 *"incomparable* Poudres Subtiles": "The Craven: BY POH!" (advertisement), *Evening Mirror,* March 25, 1845, in *TPL,* 521.

243 *"'The Raven' has had a great 'run'":* Poe to Thomas, May 4, 1845, in Ostrom, 1:505, in *TPL,* 530–31.

243 *"more distinct than those of the different dialects":* Lowell, "Our Contributors: Edgar Allan Poe." This was the same problem that Bache, Henry, and Peirce faced in American science—as Poe had recognized in his call for a new national journal—and which drove them, similarly, to seek a national framework for American science.

244 *Poe appeared complete:* The process of creating Poe as an author appearing fully formed is brilliantly analyzed in McGill, "Literary Nationalism," see also Jackson, "Rage for Lions."

244 *"rather wanting in that* nervousness": Joseph Evans Snodgrass, *Baltimore Saturday Visiter,* Jan. 25, 1845, in *TPL,* 494.

244 *"frank, earnest":* Margaret Fuller, *New York Daily Tribune,* Jan. 24, 1845. Even his portrait spoke to her: "The lower part of the face is that of the critic, cold, hard, and self-sufficient; while the upper part, especially the brows, expresses great feeling, and tenderness," in *TPL,* 491–92.

244 *"Whenever his name is mentioned":* Evert A. Duyckinck, *Morning News,* Jan. 25, 1845, in *TPL,* 493–94.

244 *Duyckinck welcomed him:* McGill, *Culture of Reprinting,* 288.

245 *"has attracted far less attention":* Evert A. Duyckinck, *Morning News,* Jan. 18, 1845, in *TPL,* 489–90.

245 *Poe took up the Young Americans' causes:* Poe, "Imitation—Plagiarism," *Evening Mirror,* Feb. 15, 1845, in *TPL,* 502.

245 *he exposed the exploitation:* Poe, "Some Secrets of the Magazine Prison-House," *Broadway Journal,* Feb. 15, 1845, 103–4.

245 *"impudent and flippant critique":* Poe to Mathews, March 15, 1844, in Ostrom, 1:429; also in Richard, "Poe and 'Young America.'"

245 *He also favorably reviewed:* Briggs showed Poe's reviews of Horne and Barrett to Duyckinck for improvements, before he published them in *The Broadway Journal,* in *TPL,* 479–80.

245 *Duyckinck invited him to publish*: Poe to Duyckinck, Feb. 18, 1845, in Ostrom, 1:486; Ezra Greenspan, "Evert Duyckinck and the History of Wiley and Putnam's Library of American Books, 1845–1847," *American Literature* 64, no. 4 (1992): 677–93.

246 *the Society Library*: The Arts Association occupied the library's top floors and displayed Thomas Cole's four-part *Voyage of Life*.

246 *"novel, ingenious, and a capital antidote"*: Evert Duyckinck, *Morning News*, Feb. 28, 1845, in TPL, 407.

246 *"the pernicious influence"*: Evert Duyckinck, *Morning News*, March 1, 1845, in TPL, 509.

246 *"the critical blade of Mr. Poe"*: Willis, *Evening Mirror*, Feb. 27, 1845, in TPL, 507.

246 *"that of a versed and resolute man"*: Duyckinck, *Morning News*, March 8, 1845, in TPL, 509.

246 *"acute and fearless criticism"*: *Daily Tribune*, March 1, 1845, in TPL, 508.

246 *"prefer the dancing dog"*: *Daily Atlas*, March 3, 1845, in TPL, 513.

246 *"would-be-critic"*: *Evening Transcript*, March 5, 1845, in TPL, 513.

246 *lawsuits and duels ensued*: Freeman, *Field of Blood*.

246 *The practice was picked up*: Tomc, "Poe and His Enemies"; Perry Miller, *The Raven and the Whale*; Sidney P. Moss, *Poe's Literary Battles: The Critic in the Context of His Literary Milieu* (Durham, N.C.: Duke University Press, 1963).

247 *Poe suspected that some of the pieces*: Poe, review of *The Waif*, by Longfellow, *Evening Mirror*, Jan. 13–14, 1845, in TPL, 486.

247 *"will serve Longfellow"*: Nathaniel Parker Willis to Charles Sumner, Jan. 16(?), 1845, Charles Sumner Correspondence, 1824–74, HOU HD MS Am 1, box 27: 6821–7072, Houghton Library, Harvard University, quoted in Tomc, "Poe's Enemies," 567.

247 *likely written by Thomas Dunn English*: Review of *Poems*, by Longfellow, *Aristidean*, April 1845, 130–42, in TPL, 529.

247 *In a notice he published*: Poe, "The Magazines," *Broadway Journal*, May 3, 1845, 285.

247 *"Everybody wants to know him"*: Possibly Frances S. Osgood to Sarah Helen Whitman, Jan. 7, 1846, in Whitman, "Life," 13, in TPL, 616.

247 *"Poe had always the bearing"*: Lynch to George W. Eveleth, March 8 and 19, 1854, in TPL, 484. On Poe in the New York salons, see Anne Boyd Rioux, "Lions and Bluestockings," in *Edgar Allan Poe in Context*, ed. Kevin J. Hayes (Cambridge, U.K.: Cambridge University Press, 2013), 129–37.

248 *Among Lynch's visitors*: Poe to Halleck, Jan. 20, 1846, in Ostrom, 1:553, in TPL, 618.

248 *organized much of the intellectual life*: Eliza Richards, *Gender and the Poetics of Reception in Poe's Circle* (Cambridge, U.K.: Cambridge University Press, 2004); Mary Kelley, *Private Woman, Public Stage: Literary Domesticity in Nineteenth-Century America*, rev. ed. (Chapel Hill: University of North Carolina Press, 2002); Cheryl Walker, *The Nightingale's Burden: Women Poets and American Culture Before 1900* (Bloomington: Indiana University Press, 1982).

249 *"not moral obliquity, but want"*: Mary Sargeant Gove, *Lectures to Women on Anatomy and Physiology, with an Appendix on Water Cure* (New York: Harper & Brothers, 1846), 27; Carl J. Guarneri, *The Utopian Alternative: Fourierism in Nineteenth-Century America* (Ithaca, N.Y.: Cornell University Press, 1991); Ann Braude, *Radical Spirits: Spiritualism and Women's Rights in Nineteenth-Century America* (Bloomington: Indiana University Press, 2001).

249 *The regimen was also propounded*: Marie Louise Shew, *Water-Cure for Ladies: A Popular Work on the Health, Diet, and Regimen of Females and Children, and the Prevention and Cure of Diseases* (New York: Wiley and Putnam, 1844), 18. The system had been developed by Vincent Priessnitz of Austria; on the cure as part of alternative women's health care in the nineteenth century, see Susan Cayleff, *Wash and Be Healed: The Water-Cure Movement and Women's Health* (Philadelphia: Temple University Press, 1987).

249 *"entirely subjected by the spirit of Fourier"*: Mary Sargeant Gove Nichols, *Mary Lyndon; or, Revelations of a Life: An Autobiography* (New York: Stringer & Townsend, 1855), 342; Patricia Cline Cohen, "The 'Antimarriage Theory' of Thomas and Mary Gove Nichols: A Radical Critique of Monogamy in the 1850s," *Journal of the Early Republic* 34, no. 1 (2014): 1–20; Jean L. Silver-Isenstadt, *Shameless: The Visionary Life of Mary Gove Nichols* (Baltimore: Johns Hopkins University Press, 2002).

249 *"was so singular"*: Frances S. Osgood, ca. early 1850, quoted in Griswold, "Memoir of the Author," in *The Works of the Late Edgar Allan Poe*, 3 vols. (New York: J. S. Redfield, 1850), 3:xxxvii; TPL, 511–12.

249 *"At my feet little Mrs. Osgood"*: Thomas Dunn English, "Reminiscences of Poe," pt. 3, *New York Independent*, Oct. 29, 1896, 1448, in TPL, 553.

250 *In New York in 1841 she compiled*: Frances Sargent Osgood, *The Poetry of Flowers and Flowers of Poetry: To*

Which Are Added, a Simple Treatise on Botany, with Familiar Examples, and a Copious Floral Dictionary (New York: J. C. Riker, 1841).

251　*"Oh! hasten to my side, I pray!"*: Osgood, in Mary G. De Jong, "Her Fair Fame: The Reputation of Frances Sargent Osgood, Woman Poet," *Studies in the American Renaissance* (1987): 265–83; on gendered forms of expression and sociability, see Julie Ellison, *Delicate Subjects: Romanticism, Gender, and the Ethics of Understanding* (Ithaca, N.Y.: Cornell University Press, 1992).

251　*"a term applied, in despair"*: Poe, "Frances Osgood," in "The Literati of New York City Part V," *Godey's Lady's Book*, Sept. 1846, 126–33.

251　*Poe and Osgood flaunted*: Joanne Dobson, "Sex, Wit, and Sentiment: Frances Osgood and the Poetry of Love," *American Literature* 65, no. 4 (1993): 631–50; Mary De Jong, "'Read Here Thy Name Concealed': Frances Osgood's Poems on Parting with Edgar Allan Poe," *PS* 32, no. 1–2 (1999): 27–36. If you are interested in bizarrely detailed speculation on Osgood and Poe's "affair" (including their offspring), please read John Evangelist Walsh, *Plumes in the Dust: The Love Affair of Edgar Allan Poe and Fanny Osgood* (Chicago: Nelson-Hall, 1980).

252　*"People seem to think"*: An unnamed New York correspondent writing on Jan. 7, 1846, in Sarah Helen Whitman, introductory letter to *The Life and Poems of Edgar Allan Poe* (New York: W. J. Widdleton, 1877), 13.

252　*"think it over fully"*: Orson Squire Fowler, *American Phrenological Journal* (Philadelphia), Sept. 1845, in *TPL*, 566–67. In October, Fowler learned that the story was an invention, and printed a retraction in the *Phrenological Journal*: his mistake came from knowing that "the literary clique to which Poe belongs, Joseph C. Neal included, had given much attention to magnetism," in *TPL*, 572–73.

252　*"clumsily contrived, unnatural"*: Charles A. Dana, review of *Tales*, by Poe, *Harbinger*, July 12, 1845, in *TPL*, 550.

252　*"respectability and influence"*: Poe citing *The Popular Record of Modern Science* in "Marginalia," *Graham's*, March 1848; M200, in Pollin, *Brevities*, 331–33.

252　*"to-do about the tenability"*: Poe, *Broadway Journal*, Sept. 20, 1845.

252　*"absolutely true, although at first"*: Poe, "Marginalia," *Godey's Lady's Book*, Aug. 1845; M130, in Pollin, *Brevities*, 231.

252　*Professor George Bush assured Poe*: T. D. English, review of *Tales*, by Poe, *Aristidean*, Oct. 1845, in *TPL*, 587.

253　*"draw their own conclusions"*: Editorial comment on an excerpt of "Mesmeric Revelation" in *New York New World*, Aug. 3, 1844, in *TPL*, 468; the story first appeared in full in *Columbian Magazine* 2, no. 2, Aug. 1844, 67–70.

253　*abundance of counterfeit reports*: Poe, review of *Human Magnetism, Its Claim to Dispassionate Inquiry, Being an Attempt to Show the Utility of Its Application for the Relief of Human Suffering*, by W. Newnham, *Broadway Journal*, April 5, 1845.

253　*a new mesmeric tale*: Poe, "The Facts in the Case of M. Valdemar," *Whig Journal*, Dec. 1846.

253　*"well known august personage"*: Poe compares Valdemar's appearance (as he had that of Roderick Usher) to that of John Randolph of Roanoke, an eccentric member of the "heroic, Revolutionary generation" who owned slaves but freed them at his death; see Kennedy, *Strange Nation*, 388–94.

253　*concentrated journey*: A link with the concentrated transmissions of telegraphy is established in Adam Frank, "Valdemar's Tongue, Poe's Telegraphy," *ELH* 72, no. 3 (2005): 635–62.

254　*"several good matter-of-fact citizens"*: Horace Greeley, *Daily Tribune*, Dec. 10, 1845, in *TPL*, 603.

254　*"going the round of the newspapers"*: Barrett to Poe, April 1846, in *TPL*, 632.

254　*"Mesmerism in America"*: Poe, "Mesmerism in America: Astounding and Horrifying Narrative," *Sunday Times* (London), Jan. 4, 1846, in *TPL*, 615.

254　*"angry excitement and various rumors"*: "Mesmerism in America," *Popular Record of Modern Science*, Jan. 10, 1846, in *TPL*, 617.

254　*"M. Valdemar's case has been"*: Collyer to Poe, Dec. 16, 1846, printed in *Broadway Journal*, Dec. 27, 1846, in *TPL*, 605.

254　*"to offer one word"*: Poe, *Broadway Journal*, Dec. 20, 1846, in *TPL*, 605.

254　*"poetically imagined"*: Andrew Jackson Davis, *The Magic Staff: An Autobiography* (New York: J. S. Brown, 1857), 317. In April 1846, Poe looked back on the two tales: "Mesmeric Revelation" gave genuine philosophical speculations "a garb of vraisemblance," while "Valdemar" made a "more pronounced effort at verisimilitude for the sake of effect." In the first, he said, "I believe actual truth to be involved," whereas the second was "universally copied and *received as truth*, even in spite of my disclaimer." Poe, *Record*, April 11, 1846, in *TPL*, 631.

255   *"An infinity of error"*: Poe, "Marginalia," *SLM* 15, no. 6 (June 1849); M254, in Pollin, *Brevities*, 393. Poe's cosmological speculations in *Eureka*—positioned outside and against the limits of official science—would bring him up against the rival "cosmopolitics" of Bache and Henry; see chapter 17 below.

255   *"A particular cause"*: Poe "Marginalia," *United States Magazine and Democratic Review*, Nov. 1844; M18, in Pollin, *Brevities*, 127.

255   *"PERVERSENESS"*: Poe, "The Black Cat," *United States Saturday Post* (Philadelphia), Aug. 19, 1843, 1.

256   *"there is no exquisite beauty"*: Poe, "Marginalia," *Graham's*, March 1846; M147, in Pollin, *Brevities*, 254.

256   *preference for the strange*: See also Poe's "Fifty Suggestions (Part I)," *Graham's Magazine*, May 1849, 317–19, 23: "An artist is an artist only by dint of his exquisite sense of Beauty—a sense affording him rapturous enjoyment, but at the same time implying, or involving, an equally exquisite sense of Deformity or disproportion."

256   *"class of fancies, of exquisite delicacy"*: Poe, "Marginalia," *Graham's*, March 1846; M150, in Pollin, *Brevities*, 257–60.

256   *"permanent Impression"*: Babbage, *Ninth Bridgewater Treatise*, 119.

257   *inscribed on the ether*: On the diffuse influence of Babbage's thought experiment about the materiality of thought, see John M. Picker, *Victorian Soundscapes* (New York: Oxford University Press, 2003), chap. 1; for its impact on Poe, see Whalen, *Poe and the Masses*.

257   *"brilliant flowers are the dearest"*: Poe, "The Power of Words," *United States Magazine and Democratic Review*, 26, no. 6, June 1845, 602–04.

257   *"unthought-like thoughts"*: Poe, "To ——" ("To Marie Louise Shew," 1848), in LOA, 88.

14. *The Imp of the Perverse*

258   *"For the last three or four months"*: Poe to Frederick W. Thomas, May 4, 1845, in Ostrom, 1:504.

258   *"He never rests"*: Thomas Dunn English, *Aristidean*, April 1845, in *TPL*, 529.

258   *"within a few years"*: Poe, "Marginalia," *Graham's*, Sept. 1846; M143, in Pollin, *Brevities*, 248.

259   *"heresy of didacticism"*: Poe, "Increase of the Poetical Heresy—Didacticism," *Evening Mirror*, Feb. 3, 1845, in *TPL*, 498.

260   *"without even a difference"*: Poe, review of *Alciphron: A Poem*, by Thomas Moore, *Burton's*, Jan. 1840, 53–56.

260   *"it would create not only ideally"*: Poe, "American Prose Writers, No. 2: N. P. Willis," *Broadway Journal*, Jan. 18, 1845, 37–38.

260   *"the old dogma"*: Poe, "American Poetry," *Aristidean*, Nov. 1845, 373–82.

260   *workers' movements championed*: William Sewell, *Work and Revolution in France: The Language of Labor from the Old Regime to 1848* (Cambridge, U.K.: Cambridge University Press, 1980); Tocqueville, "That Almost All the Americans Follow Industrial Callings," in *Democracy in America*.

261   *"dwells in the mystery"*: Lowell, "Our Contributors: Edgar Allan Poe."

261   *"the spontaneous overflow"*: Wordsworth and Coleridge, *Lyrical Ballads*, 2nd ed. (London: Longman and Rees, 1800), xiv.

261   *"a man who is a scholar"*: James Russell Lowell, *Conversations on Some of the Old Poets* (London: Henry Clarke, 1845), 51.

261   *"If the practice fail"*: Poe, review of *Conversations on Some of the Old Poets*, by Lowell, *Evening Mirror*, Jan. 11, 1845: "To say that a critic could not have written the work which he criticises, is to put forth a flat contradiction in terms."

262   *"We do not believe that traps"*: *Daily Tribune*, Jan. 15, 1845, in *TPL*, 488. Briggs wrote to Lowell on Jan. 17, 1845, "Poe's criticism about your Conversations was extremely laudatory and discriminating; it was the female ass of the *Tribune* [Fuller] that misunderstood him," in *TPL*, 488.

262   *"nothing more than the arranging"*: Poe, "Nature and Art," *Evening Mirror*, Jan. 17, 1845, in *TPL*, 489.

262   *Rather than two opposed terms*: Pushed by Lowell and Fuller, Poe was refining his critique of Coleridge's presentation of the poetic imagination as a kind of miracle akin to the mystery of divine creation. For Poe, the highest human art does resemble the work of divine creation, but simply because God's craftsmanship is the perfected form of the material, compositional, constructive craft of his creatures. This continuity between human and divine construction confirms both Poe's argument

of "the reciprocity of adaptation" (in which each part of nature is both cause and effect within the divine plan) and his eventual pantheism as expressed in *Eureka*. On Poe, Coleridge, and *Eureka*, see St. Armand, "'Seemingly Intuitive Leaps.'"

262 *"mingled rain and hail"*: Alexander T. Crane, *Omaha Sunday World-Herald*, July 13, 1902, in *TPL*, 526.

263 *Each spree was followed*: These oscillations echoed the reversals, awakenings, and doubles within his stories; as for Poe's drug use, like his contemporaries Poe undoubtedly took laudanum as a medicine, but there is no indication that he took opiates habitually. Jeffrey Savoye, "Edgar Allan Poe, Drugs, and Alcohol," Edgar Allan Poe Society of Baltimore, www.eapoe.org/geninfo/poealchl.htm.

263 *The amount of alcohol consumed*: "Per capita alcohol consumption tripled in the first third of the nineteenth century, so that by the 1830s the typical drinker was consuming nearly a half pint of distilled alcohol every day." Amanda Claybaugh, "Temperance," in *American History Through Literature, 1820–1870*, ed. Janet Gabler-Hover and Robert Sattlemeyer (Detroit: Charles Scribner's Sons, 2006), 1152–58; Mark Lender and James Martin, *Drinking in America: A History* (New York: Simon & Schuster, 1987).

263 *"attacked with a terrible paroxysm"*: Chivers in *Chivers' Life of Poe*, ed. Richard Beale Davis (New York: E. P. Dutton, 1952), 39–52, in *TPL*, 538.

263 *"Life is too short"*: Lynch to Poe, June 27, 1845, in *TPL*, 542.

263 *"a little tipsy"*: Lowell to G. W. Woodberry, March 12, 1884; AHQ, 461.

263 *"wrong impression of my darling Eddie"*: Clemm to Lowell, March 9, 1850; AHQ, 461–62.

263 *"not half the noble-looking person"*: Chivers, *Life*, 39–52; *TPL*, 538.

263 *"Here comes Poe with his Raven"*: James Russell Lowell, *A Fable for Critics* (New York: George P. Putnam, 1848), 78.

264 *"a bigotry the most obstinately blind"*: Poe, review of *A Fable for Critics*, by Lowell, *SLM* 15, no. 3 (March 1849): 189–91. In this 1849 review, Poe objected not to Lowell's views but their intensity: Lowell's "fanaticism about slavery is a mere local outbreak of the same innate wrong-headedness which, if he owned slaves, would manifest itself in atrocious ill-treatment of them, with murder of any abolitionist who should endeavor to set them free." According to Whalen, Poe saw "fanaticism" either for or against slavery interfering with the fragile political "neutrality" of the national sphere: "admission to the national literary market meant turning his back on the momentous political and social struggles of the day." Whalen, *Poe and the Masses*, 138.

264 *"haul down Poe's name"*: Briggs to Lowell, June 27, 1845, in *TPL*, 542.

264 *"to give up the B. Journal"*: Poe to Duyckinck, June 26, 1845, in Ostrom, 1:512, in *TPL*, 542.

264 *"remained in a state of intoxication"*: T. D. English, "A Card: Mr. English's Reply to Mr. Poe," *Evening Mirror*, June 23, 1846, in *TPL*, 540.

264 *"in bed pretending to be sick"*: Chivers, *Life*, 61, in *TPL*, 545.

264 Osgood's poem *"So Let It Be"*: In *Broadway Journal*, April 5, 1845, in *TPL*, 523–24.

264 *Poe eventually received an apology*: Edward J. Thomas to Poe, July 5, 1845, in *TPL*, 547.

265 *issue due on July 5 did not appear*: in *TPL*, 547.

265 *"like a giant refreshed"*: *Evening Mirror*, July 12, 1845, in *TPL*, 550.

265 *"There is Poe with coolness"*: note by Evert Duyckinck, quoted in James B. Reece, *Poe and the New York Literati*, PhD thesis, Duke University, 1954, 95, in *TPL*, 559

265 *"sole editor"*: AHQ, 751–52, and *TPL*, 551.

265 *Calvinist sense of predestined damnation*: William Mentzel Forrest, *Biblical Allusions in Poe* (New York: Macmillan, 1928); Kenneth Alan Hovey, "Poe's Materialist Metaphysics of Man," in Carlson, *Companion to Poe Studies*, 347–66.

265 *impulse to set his life on fire*: On philosophical dimensions of Poe's perverseness, see Stanley Cavell, "Being Odd, Getting Even," in Rosenheim and Rachman, *American Face of Edgar Allan Poe*, 3–36. Baudelaire reckoned Poe's eventual death was "almost a suicide—a suicide prepared for a long time," in "Edgar Poe, sa vie et ses oeuvres," *Revue de Paris*, no. 6 (March 1852): 138–56.

266 *"has been, and is still"*: Poe to Neilson Poe, Aug. 8, 1845, in Ostrom, 1:515, in *TPL*, 559.

266 *Boston Lyceum*: Announced in Boston's *Evening Transcript*, Sept. 30, 1845, in *TPL*, 572.

266 *"prose-writings had been eagerly read"*: Thomas Wentworth Higginson, "Short Studies of American Authors, II: Poe," *Boston Literary World*, March 15, 1879, 89, in *TPL*, 577–78.

266 *"poetry enough for one night"*: *Daily Evening Traveller*, Oct. 17, 1845, in *TPL*, 578.

266 *"felt that we had been under"*: Thomas Wentworth Higginson, "Short Studies of American Authors, II: Poe," *Literary World*, March 15, 1879, in *TPL*, 578.

266 *"elegant and classic production"*: Joseph T. Buckingham in *Boston Courier*, Oct. 18, 1845, in *TPL*, 579.

266 *"A poem delivered before"*: Cornelia Wells Walter, *Evening Transcript*, Oct. 18, 1845, in *TPL*, 579.

266 *"We have been quizzing"*: *Broadway Journal*, Oct. 25, 1845, in *TPL*, 581.

267 *"his best poems"*: Walter, *Evening Transcript*, Oct. 30, 1845, in *TPL*, 584.

267 *"The Bostonians are very well"*: Poe, "Boston and the Bostonians," *Broadway Journal*, Nov. 1, 1845, in Harrison, *Complete Works*, 13:9–13.

267 *"perfectly damnable"*: Henry Norman Hudson to Duyckinck, Nov. 24, 1845, in *TPL*, 594; coverage in national press included *United States Journal* (Washington, D.C.), Oct. 22, 1845, in *TPL*, 584.

267 *"very kindly and human"*: Walt Whitman, "Broadway Sights," in *Specimen Days & Collect* (Philadelphia: Rees Welsh, 1882), 17, in *TPL*, 597.

267 *"I have had a tough time"*: Poe to Chivers, Nov. 15, 1845, in Ostrom, 1:535, in *TPL*, 590–91.

267 *"every body is at us"*: Poe, "Editorial Miscellany," *Broadway Journal*, Dec. 6, 1845, in *TPL*, 603.

267 *"one of his fits of drunkenness"*: Thomas Dunn English, "Reminiscences of Poe," *Independent* 48, no. 2498, Oct. 15, 1896, 1381, in *TPL*, 606.

268 *"close the publication"*: English, Ibid.

268 *"To trust in friends is but so so"*: Walter, *Evening Transcript*, Jan. 2, 1846, in *TPL*, 614.

268 *"jealousies and heart-burnings"*: T. D. English, "Reminiscences of Poe," pt. 3, *New York Independent*, Oct. 29, 1896, 1448.

268 *"dealt him some smart raps"*: English prissily continued, "As I happened to have a heavy seal ring on my little finger, I unintentionally cut him very severely, and broke the stone in the ring, an intaglio cut by Lovatt, which I valued highly" (ibid.).

268 *"It is our intense sense of self"*: Poe, "A Chapter of Suggestions," *Opal* (1845), in Pollin, *Brevities*, 465–47.

269 *"little cottage at the top of a hill"*: The salubrious setting is described by health reformer Mary Gove Nichols, "Reminiscences of Edgar Poe," *Sixpenny Magazine* (London), Feb. 1, 1863, 471, in *TPL*, 644.

269 *"Our charming friend Mrs. Osgood"*: Hewitt to Poe, April 15, 1846, in *TPL*, 634.

269 *"A heavy shower had come up"*: Richard Henry Stoddard, *Recollections, Personal and Literary* (New York: A. S. Barnes, 1903), 151, in *TPL*, 585.

270 *"We should be inclined"*: *Broadway Journal*, April 12, 1845; other 1845 commentary on *Vestiges* in *The Broadway Journal*: Jan. 25, May 15, Aug. 30, Dec. 6.

270 *Irish astronomer William Parsons*: Simon Schaffer, "On Astronomical Drawing," in *Picturing Science, Producing Art*, ed. Peter Galison and Caroline Jones (New York: Routledge, 1998), 441–74; Michael Hoskin, "Rosse, Robinson, and the Resolution of the Nebulae," *Journal for the History of Astronomy* 21, no. 4 (1990): 331–44; Omar Nasim, *Observing by Hand: Sketching the Nebulae in the Nineteenth Century* (Chicago: University of Chicago Press, 2014).

270 *"there can be little"*: Rosse, March 19, 1846, in Schaffer, "On Astronomical Drawing," 46. Among those who doubted Rosse's claim of the nebula's resolvability was John Herschel: though Herschel was hostile to the nebular hypothesis, on theological grounds and as a transgression of the limits of astronomy (which should avoid "first and last things"), he defended his father's program of mapping and examining nebulae; Schaffer, "Science of Progress," 137–38.

271 *"nebulae comparatively near"*: *Explanations*, 6.

271 *"that nearly all the scientific men"*: Ibid., 124–26.

272 *"unwise and ungenerous"*: Brewster, *North British Review* 4 (Feb. 1846): 487–504. Brewster's review of *Vestiges* a year earlier (*North British Review* 3 [Aug. 1845]: 470–515) "was later included in many of the American editions of *Vestiges*" (Robert Chambers, *Vestiges of the Natural History of Creation and Other Evolutionary Writings*, ed. James Secord [Chicago: University of Chicago Press, 1994], 228); Secord includes a remarkable bibliography of published replies to *Vestiges* and *Explanations*.

272 *"the professional establishment"*: Asa Gray, *North American Review* 63 (1846): 506.

272 *"the production of an intelligent"*: Henry, May 6, 1846, in *Papers of Joseph Henry*, 6:281. On the reception of *Vestiges* among Henry's theological colleagues in Princeton, see Bradley J. Gundlach, *Process and Providence: The Evolution Question at Princeton, 1845–1929* (Grand Rapids, Mich.: Wm. B. Eerdmans, 2013).

272 *"magnificent cosmogony"*: J. H. Allen, *Christian Examiner* 40 (May 1846): 334–35, 344. One continuing supporter of *Vestiges* among established American men of science was the geologist Henry Darwin Rogers, who told his brother in 1845 it "singularly accords with views sketched by me at times in my lectures" (in Numbers, *Creation*, 34n17).

272 *"We have only to decide whether"*: Francis Bowen, *North American Review* 60 (April 1845): 426–78.
272 *"subterranean approaches to the citadel"*: William H. Allen, *Methodist Quarterly Review* 28 (April 1846): 295.
273 *"have more of effect"*: John S. Dwight, review of *The Raven*, by Poe, *Harbinger* (West Roxbury, Mass.), Dec. 6, 1845, in TPL, 602.
273 *"by means of the mechanism of verse"*: T. D. English, review of *The Raven*, by Poe, *Aristidean*, Nov. 1845, in TPL, 599.
273 *"more remarkable for its mechanical"*: Lawrence Labree, review of *The Raven*, by Poe, *New York Illustrated Magazine*, Dec. 6, 1845, in TPL, 602.
274 *"has what he thinks a taste"*: Poe to Philip P. Cooke, Aug. 9, 1846, in Ostrom, 1:595, and TPL, 661.
274 *"peep behind the scenes"*: Poe, "Philosophy of Composition," *Graham's*, April 1846, 163–67.
274 *"prefer having it understood"*: A timeworn cliché from Philip Sidney's 1594 *Defence of Poesy* (repr., Cambridge, Mass.: Hilliard, 1831).
275 *a stance similar to that of Maelzel*: See chapter 5 above; Paul Grimstad, "Antebellum AI: 'Maelzel's Chess-Player' and Poe's Reverse Constraints," *Poetics Today* 31, no. 1 (2010): 107–25.
275 *"air of method"*: Poe once boasted that the "hair-splitting" of his Dupin tales was "all done for effect": "People think them more ingenious than they are—on account of their method and *air of* method." Poe to Philip P. Cooke, Aug. 9, 1846, in Ostrom, 1:595, in TPL, 661.
275 *"much less inspired"*: Baudelaire, *Oeuvres complètes*, ed. Claude Pichois, 2 vols. (Paris: Gallimard, 2005), 2:343–342.
275 *"Machine for Composing Hexameter Latin Verses"*: "The Eureka," *Illustrated London News*, July 19, 1845; John Clark, in *The General History and Description of a Machine for Composing Hexameter Latin Verses* (Bridgewater: Frederick Wood, 1848); Jason David Hall, "Popular Prosody: Spectacle and the Politics of Victorian Versification," *Nineteenth-Century Literature* 62, no. 2 (2007): 222–49, thanks to Thomas Vranken for this discovery.
277 *"It is my design"*: Poe, "Philosophy of Composition," 163, italics added.
277 *a creator who was able to plot*: This was the view Babbage laid out in his *Ninth Bridgewater Treatise* (written without the approval of the *Bridgewater* authors and against their basic assumptions), using his own calculating engine as an illustration; see chapter 12 above.
277 *diagnosed him with heart disease*: John Carl Miller, *Building Poe Biography* (Baton Rouge: Louisiana State University Press), 99, in TPL, 732.
278 *"high genius"*: Poe on Margaret Fuller, "The Literati of New York City," *Godey's Lady's Book*, June 1846, 72–78; subsequent quotations: on Briggs and Hiram Fuller, *Godey's*, May 1846; on Clark, *Godey's*, Sept. 1846; and on English, *Godey's*, June 1846.
278 *"gray, watery, and always dull"*: Briggs, *Evening Mirror*, May 26, 1846, in TPL, 634.
278 *"an assassin in morals"*: T. D. English, "Mr. English's Reply to Mr. Poe," *Morning Telegraph*, June 23, 1846, in TPL, 648.
278 *"in a condition of sad"*: Hiram Fuller, *Evening Mirror*, July 20, 1846, in TPL, 656.
279 *a printed reply to these attacks*: He first sent it to Godey, who cravenly paid to have it published elsewhere; "Mr. Poe's Reply to Mr. English and Others," *Spirit of the Times*, July 10, 1846, in TPL, 652.
279 *filed a declaration of grievances*: Sidney P. Moss, *Poe's Major Crisis: His Libel Suit and New York's Literary World* (Durham, N.C.: Duke University Press, 1970), 77, in TPL, 656.
279 *"Mr. Edgar A. Poe, the poet"*: *St. Louis Daily Reveille*, April 12, 1846, in TPL, 633–34.
279 *Even his old friend*: Snodgrass in the *Baltimore Saturday Visiter*, April 18, 1846, in TPL, 635.
279 *literary satire "1844"*: T. D. English, "1844," pt. 15, *Weekly Mirror*, Oct. 31, 1846, in TPL, 668.
279 *"Epitaph on a Modern 'Critic'"*: "Epitaph on a Modern 'Critic,'" *Knickerbocker*, Nov. 1846, in TPL, 669.
279 *"You are now perhaps"*: Simms to Poe, July 30, 1846, in TPL, 660.

Part V: To the Plutonian Shore

281 *"And this was the reason"*: Poe, "Annabel Lee" (1849), in LOA, 102–3.

15. A Spectacle for Angels

283 *"monster of hideous conformation"*: Poe, "The Sphinx," *Arthur's Ladies' Magazine*, Jan. 1846, 15–16.
284 *Another cholera epidemic*: In 1847 an outbreak killed fifteen thousand in Mecca; by 1848 the disease

would strike Berlin, Paris, and London; and in 1849 it ravaged the United States. Rosenberg, *Cholera Years*, 101.

284 *"very ill"*: Nichols, *Mary Lyndon*, 342, in TPL, 669.

284 *"scant and poverty-stricken"*: Nichols, *Reminiscences*, 12–13, in TPL, 669.

284 *Shew, a kindhearted nurse*: For Shew's amazing recollections of Poe, see Miller, *Building Poe Biography*, 88–145. Poe wrote her a beautiful late poem, "To ——," in LOA, 88.

284 *"His friends ought not to wait"*: Hiram Fuller, *Evening Mirror*, Dec. 16, 1846, in TPL, 673.

284 *"dangerously ill"*: The *Morning Express*, Dec. 15, 1846, in TPL, 672.

284 *"brain fever"*: Walt Whitman, *Brooklyn Daily Eagle*, Dec. 18, 1846, in TPL, 673.

285 *"one of the most original men"*: Nathaniel Willis, "Hospital for Disabled Labourers with the Brain," *New York Home Journal*, Dec. 26, 1846, in TPL, 674.

285 *"an asylum for those who have been ruined"*: Hiram Fuller, *Evening Mirror*, Dec. 26, 1846, in TPL, 675.

285 *"An Invocation for Suffering Genius"*: John Edward Reilly, "Ermina's Gales: The Poems Jane Locke Devoted to Poe," in *Papers on Poe: Essays in Honor of John Ward Ostrom*, ed. Richard P. Veler (Springfield, Ohio: Chantry Music Press, 1972), 206–20, in TPL, 674.

285 *"Poe and his wife are both down"*: Cornelia Wells Walter, *Bostonian*, Dec. 26, 1846, in TPL, 675–76.

285 *"reformation of habits"*: Walter, *Evening Transcript*, Dec. 31, 1846, in TPL, 677–78.

285 *"of what is true"*: Poe to Willis, Dec. 30, 1846, published in the *Home Journal*, Jan. 9, 1847, in Ostrom, 1:612.

285 *an advocate of philosophies*: Richard Fusco, "Poe and the Perfectibility of Man," *PS/Dark Romanticism* 19, no. 1 (1986): 1–6; in a review of an essay on human perfectibility in *Burton's* (July 1839, 58), Poe disparaged the "eloquent madness of Turgot, Price, Priestly [*sic*], Condorcet, and de Stael"—an identical list (with the exception of de Staël) to that of the philosophers Ellison prefers.

286 *"the august purposes for which the Deity"*: Poe, "The Domain of Arnheim," *Columbian Magazine*, March 1847, 125.

286 *beings not limited by bodies*: Ibid. On Poe's angels as an "amalgam of Neoplatonic, Gnostic, and Islamic attributes" in pursuit of "cosmic wonder," see William E. Engel, "Fantastic Places, Angelic Spaces," in Phillips, *Poe and Place*, 169–92.

286 *vague terms that debase divine creation*: This passage recalls the prefect's inspection of the minister's rooms in "The Purloined Letter," which lowered his search to the level of an ordinary thinker.

286 *"the Almighty design"*: Poe, "Domain of Arnheim," 127.

287 *Poe's tale began with an exposition*: The 1842 version of the tale, "The Landscape Garden," ends with the explanation of Ellison's theories; the 1847 version describes the result, the Domain of Arnheim, the artwork that applies them.

287 *"merely pastoral care"*: Poe, "Domain of Arnheim," 127–29.

289 *Poe wove nature and art together*: Lorraine Daston and Katharine Park, *Wonders and the Order of Nature, 1150–1750* (New York: Zone Books, 1998); M. H. Abrams, *Natural Supernaturalism: Tradition and Revolution in Romantic Literature* (New York: W. W. Norton, 1971); Pamela Smith, *The Body of the Artisan: Art and Experience in the Scientific Revolution* (Chicago: University of Chicago Press, 2018); Ann-Deborah Lévy-Bertherat, *L'artifice romantique: De Byron à Baudelaire* (Paris: Klincksieck, 1994).

289 *"second nature"*: Leo Marx, *The Machine in the Garden: Technology and the Pastoral Ideal in America* (New York: Oxford University Press, 1964); William Cronon, *Nature's Metropolis: Chicago and the Great West* (New York: W. W. Norton, 2009).

289 *"Landor's Cottage"*: Poe, *Flag of Our Union*, June 9, 1849, 2.

289 *an ode to the power of vision*: Laura Saltz's "'Eyes Which Behold'" emphasizes the tale's appeal to the materiality of vision, referring to Brewster's stereoscope; though this technology became widespread only after Poe's death, his awareness of the material process of dioramas, dissolving views, and other optical devices dependent on ocular physiology confirms her reading; for another aesthetic genealogy, see Catherine Rainwater, "Poe's Landscape Tales and the 'Picturesque' Tradition," *Southern Literary Journal* 16, no. 2 (1984): 30–43.

289 *Thomas Cole's celebrated series*: On Poe and Cole, see Cantalupo, *Poe and the Visual Arts*; Wallach, "*Voyage of Life* as Popular Art"; Joy S. Kasson, "*The Voyage of Life*: Thomas Cole and Romantic Disillusionment," *American Quarterly* 27, no. 1 (1975): 42–56; on Humboldtian and panoramic aspects of U.S. landscape painting, see Jennifer Raab, *Frederic Church: The Art and Science of Detail* (New Haven, Conn.: Yale University Press, 2015).

290 *from beyond the veil*: E. W. Pitcher puts "Arnheim" into dialogue with the protracted phenomenol-
     ogy of death that Poe detailed in "The Colloquy of Monos and Una," suggesting its journey symbolizes
     the passage from life, to death, to a prolonged period of "torpor" before release or rebirth in a divine
     paradise—with "Arnheim" an anagram of "NEAR HIM." E. W. Pitcher, "The Arnheim Trilogy:
     Cosmic Landscapes in the Shadow of Poe's *Eureka*," *Canadian Review of American Studies* 6, no. 1 (1975):
     27–35.
290 *eldritch powers into an alien penitentiary*: Sean Moreland, ed., *The Lovecraftian Poe: Essays on Influence,
     Reception, Interpretation, and Transformation* (Lanham, Md.: Rowman & Littlefield, 2017).
291 *echoes Milton's in* Paradise Lost: Dayan, *Fables of Mind*, 104.
291 *"mystical philosophy of nature"*: *Philadelphia Saturday Courier*, July 25, 1846, in *TPL*, 659.
291 *"My poor Virginia still lives"*: Poe to Shew, Jan. 29, 1847, in Ostrom, 2:617–18.
291 *"She called me to her bedside"*: Shew to J. H. Ingram, March 28, 1875, in *TPL*, 684.
291 *carried the notice*: Obituary in *Daily Tribune* and *New York Herald*, Feb. 1, 1847, in *TPL*, 685.
291 *"of the sweet emotion made up of respect"*: Poe to Locke, March 10, 1847, in Ostrom, 2:624, and *TPL*, 684.
292 *"Deep in earth my love is lying"*: The MS is held by the New York Public Library, Rare Books Room;
     see Victor H. Palsits, "Two Manuscripts in the New York Public Library—Part I: The Manuscript of
     Poe's 'Eulalie,'" *Bulletin of the New York Public Library* 18 (Dec. 1914): 1462.

16. The Plots of God

293 *"brain fever, brought on"*: Shew to J. H. Ingram, Jan. 23, 1875, in *TPL*, 694.
293 *an eerie complex poem*: Poe, "Ulalume," in *LOA*, 89, in *TPL*, 714; Allen, *Israfel*, 2:735.
293 *"grave and melancholy"*: Mary Bronson, reminiscence from 1860, quoted in Carroll D. Laverty, "Poe in
     1847," *American Literature* 20, no. 2 (1948): 165–66, in *TPL*, 700.
294 *"I stand amid the roar"*: Poe, "A Dream Within a Dream," *Flag of Our Union*, March 31, 1849, and in
     LOA, 97–98.
295 *"understands the mere Literature"*: Henry, quoted in diary of John R. Buhler, March 3, 1846, in *Papers
     of Joseph Henry*, 6:282.
296 *"advance the taste for the sublime"*: Review of *Cosmos*, by Humboldt, *Broadway Journal*, July 12, 1845,
     14–15, in Scholnick, *Poe's "Eureka,"* 75–76.
296 *"very much spoken of"*: Agassiz, in Lurie, *Agassiz*, 128; Scholnick, "*Eureka* in Context," 38–39.
296 *Bache warmly welcomed Agassiz*: Bruce, *Launching of American Science*, 43–63. Peirce, Henry, Agassiz,
     and Bache, the PHAB four of the emerging scientific establishment, had got the band together.
296 *"large number of new members"*: *Boston Daily Journal*, Sept. 25, 1847, clippings in Coll 305 A + B I-II
     Assn. Amer. Geologists & Naturalists Papers, 1840–47, Academy of Natural Sciences of Philadelphia.
296 *Agassiz gave three rousing lectures*: Kohlstedt, *Formation of the American Scientific Community*, 76; Lurie,
     Agassiz; Bruce, *Launching of American Science*.
297 *precursor to the National Weather Service*: John Fuller, *Thor's Legions: Weather Support to the U.S. Air Force
     and Army, 1937–1987* (Boston: American Meteorological Society, 2015), 2.
297 *"Science is triumphant"*: Bache to Peirce, Dec. 3, 1846, Peirce Papers, Houghton Library, Harvard
     University, in Molella, "At the Edge of Science," 451.
297 *"Its magnitude and exact place"*: "The Planet Neptune," in *American Review: A Whig Journal of Politics,
     Literature, Art, and Science* 6, no. 2 (Aug. 1847): 145.
297 *"happy accident"*: On Peirce's role in the shifting debates over the Neptune discovery, see John G.
     Hubbell and Robert W. Smith, "Neptune in America: Negotiating a Discovery," *Journal for the History
     of Astronomy* 23, no. 4 (1992): 270; the episode is narrated in Hogan, "Hurrah for Young America," in
     Of the Human Heart, 15–25.
297 *"As one zealous"*: Gray to Peirce, March 26, 1848, in R. C. Archibald, "The Writings of Peirce,"
     American Mathematical Monthly 32 (1925): 22; Hubbell and Smith, "Neptune in America," 284. On
     Mitchel's popular science, see Russell McCormmach, "Ormsby MacKnight Mitchel's 'Sidereal Mes-
     senger,' 1846–1848," *Proceedings of the American Philosophical Society* 110, no. 1 (1966): 35–47; Scholnick,
     "*Eureka* in Context." Neptune's contested discovery prompted jokes about the relationship between
     popular entertainment and the authority of established science: New York's *Yankee Doodle* printed a
     letter from a "Professor of Astronomy and Celestial Trigonometry" at Columbia University, asking
     for information about a "new planet" seen illuminating the southern sky above New York; the editor

identified it as "Planet Barnum"—"unquestionably a most potential planet," which had "presided over the birth of a great many wonderful and curious creatures," including Joice Heth, the Feejee Mermaid, and Tom Thumb. The American Museum's limelight transformed the night sky, just as Neptune's vagaries shined a spotlight on rifts within official science. "The New Planet," *Yankee Doodle*, Jan. 1, 1847, 153.

298 *"resolved into bright points of light"*: Bond to President Everett, Sept. 22, 1847, in *American Journal of Science and Arts* 54 (1847): 426–27; see Bessie Zaban Jones and Lyle Gifford Boyd, *The Harvard College Observatory* (Cambridge, Mass.: Belknap Press, 1971), 67–68.

298 *"totally incapable of advancing Astronomy"*: Peirce, in *Boston Daily Journal*, Sept. 22, 1847. He would later embrace the theory again. His son Charles subsequently developed his own progressive cosmology in which even the laws of nature evolve: C. S. Peirce, "The Doctrine of Necessity Examined," *Monist* 2, no. 3 (1892): 321–37; C. S. Peirce, "Evolutionary Love," *Monist* 3, no. 2 (1893): 176–200.

298 *"It is time that we should have"*: Henry to Elias Loomis, Dec. 28, 1848, Loomis MSS, Beinecke Library, Yale University, in Kohlstedt, *Formation of the American Scientific Community*, 76.

298 *"My health is better—best"*: Poe to George W. Eveleth, Jan. 4, 1848, in Ostrom, 2:640.

298 *"endeavour to re-establish"*: Poe to Chapin, Jan. 17, 1848, in Ostrom, 2:644.

298 *Notices of Poe's comeback*: Announcement in *John-Donkey*, Feb. 3, 1848, in *TPL*, 719.

298 *"listened with enchained attention"*: *Morning Express*, Feb. 4, 1848, in *TPL*, 721.

298 *"Mr. Poe fastened the attention"*: *Weekly Universe*, Feb. 6, 1848, in *TPL*, 722. For a firm historical contextualization of *Eureka*, see Scholnick, "*Eureka* in Context"; setting *Eureka* within the age's growing opposition between literature and science, see Laura Saltz, "Making Sense of *Eureka*," in Kennedy and Peeples, *Oxford Handbook*, 424–44, and Limon, *Place of Fiction*.

298 "Physical, Metaphysical and Mathematical": Poe, *Eureka*, 7; this and subsequent citations are from the critical edition of *Eureka*, ed. Stuart Levine and Susan F. Levine (Urbana: University of Illinois Press, 2004). Poe would print his lecture—with some modifications—as *Eureka*; on the relation between the lecture and its printed form, see Ronald W. Nelson, "Apparatus for a Definitive Edition of Poe's *Eureka*," *Studies in the American Renaissance* (1978): 161–205. For sharp and revealing commentary, see Barbara Cantalupo, "*Eureka*: Poe's 'Novel Universe,'" in Carlson, *Companion to Poe Studies*, 323–44; Levine and Levine, *Eureka*; Beaver, *Science Fiction*.

299 "He who from the top": Poe, *Eureka*, 7–8. On paradoxical perspectives throughout Poe's work, see Henri Justin, "An Impossible Aesthetics or an Aesthetics of the Impossible?," in Kopley and Argersinger, *Poe Writing/Writing Poe*, 127–42.

299 "no such things as axioms": Situating Poe in mid-nineteenth-century debates on method, see Susan Welsh, "The Value of Analogical Evidence: Poe's *Eureka* in the Context of a Scientific Debate," *Modern Language Studies* 21, no. 4 (1991): 3–15.

300 *the method of Johannes Kepler*: A popular biography of Kepler was published in 1830: John Elliot Drinkwater Bethune, *The Life of Galileo Galilei, with Illustrations of the Advancement of Experimental Philosophy; Life of Kepler* (London, 1830).

300 *"deductions or inductions"*: Poe, *Eureka*, 15–16.

301 *"radiated spherically"*: Ibid., 24.

301 "No other principles exist": Ibid., 28. These are "the *sole* properties" by which matter is manifested to mind—a claim echoing the schemes of Kant in his early, widely read *Metaphysical Principles of Natural Philosophy* and of Roger Boscovich, who argued that material atoms were points of positive and negative energy. Thomas Holden, *The Architecture of Matter: Galileo to Kant* (Oxford: Oxford University Press, 2004), 236–72.

301 *forming swirling clouds of diffused matter*: The exposition that follows (from *Eureka*, 56) on the forming of solar systems agrees with *Vestiges*, which began with the nebulae already spread out across the universe—although Poe adds discussion of the newly discovered Neptune as the most remote planet in our system.

302 *"mere assemblages of more or fewer differences"*: Poe, *Eureka*, 27–28. Crucially, for Poe, more heterogeneity means more life, awareness, intelligence.

302 *"Matter, finally, expelling"*: Ibid., 102–3.

302 *culmination of being*: G. R. Thompson, "Unity, Death, and Nothingness: Poe's 'Romantic Skepticism,'" *PMLA* 85, no. 2 (1970): 297–300.

303 *"the great law of periodicity"*: Poe, *Eureka*, 103.

304 *"And now—this Heart Divine"*: Ibid., 103–4.

304 *"The utter impossibility"*: Ibid., 104.

304 *"the regathering"*: Ibid., 104.

304 *"apart and independently,* in the bosom": Ibid., 102.

305 *"In the meantime bear in mind"*: Ibid., 106. John Limon ("How to Place Poe's Arthur Gordon Pym
     in Science-Dominated Intellectual History and How to Extract It Again," *North Dakota Quarterly* 51
     [1983]: 31–47) suggests a debt to the *Naturphilosophie* of Lorenz Oken: "There is no dead matter; it is
     alive through the eternal that is in it . . . Everything is God, that is there, and without God there is
     absolutely nothing"; Alfred Tulk's translation of Oken appeared in London in 1847 (*Elements of Physi-
     ophilosophy* [London: Ray Society, 1847], 38). Limon notes another popular treatment of *Naturphiloso-
     phie*: J. B. Stallo, *General Principles of the Philosophy of Nature: With an Outline of Some of Its Recent Developments
     Among the Germans, Embracing the Philosophical Systems of Schelling and Hegel, and Oken's System of Nature* (Boston:
     William Crosby and H. P. Nichols, 1848). See also Courtney Fugate, "The German Cosmological
     Tradition and Poe's *Eureka*," *EAPR* 13, no. 2 (2012): 109–34.

305 *weirdly fractal symmetries*: On the metaphysical import of figures of self-reference and recursion
     in *Eureka*, see Ruth M. Harrison, "Poe Möbius: An Exploration of Poe's Fractal Universe," *PS* 36, no.
     1 (2003): 32–44.

305 *"captivating and energetic"*: "Spes Credula," *Morning Express*, Feb. 9, 1848, in *TPL*, 722.

305 *"rhapsody of the most intense brilliancy"*: Field, *Memories of Many Men*, 224; *TPL*, 720. Barbara Canta-
     lupo links the "involution of idea" in *Eureka*'s argument and form with the rhetoric and effect of a
     mesmeric trance in "Of or Pertaining to a Higher Power: Involution in *Eureka*," *ATQ* 4, no. 2 (1990):
     81–95.

305 *"we should need the tongue of an archangel"*: Poe, *Eureka*, 85.

306 *"chain of* graduated impression": Ibid., 78.

306 *The long dashes and italics*: On *Eureka*'s punctuation, see Dayan, *Fables of Mind*, 55–56; Cantalupo,
     "Novel Universe," 338.

306 *Poe spun the reader's perceptions*: See Cantalupo, "Of or Pertaining to a Higher Power"; on *Eureka*'s
     "kinetic design," see Jennie Chu Seo-Young, "Hypnotic Ratiocination," *EAPR* 6, no. 1 (2005): 5–19.

306 *debates over the planet's discovery*: Loomis, *American Journal of Science and Arts* 5 (Jan. 1848): 135–36, 153;
     Anton Pannekoek, "The Discovery of Neptune," *Centaurus* 3, no. 1 (1953): 126–37.

307 *"evolution, ceaseless and irresistible"*: Nichol, *Views of Astronomy*, 32; Frederick W. Conner, "Poe & John
     Nichol: Notes on a Source of *Eureka*," in *All These to Teach: Essays in Honor of C. A. Robertson*, ed. Robert A.
     Bryan et al. (Gainesville: University of Florida Press, 1965); on Nichol and nebulae, see Thomas De
     Quincey, "System of the Heavens as Revealed by Lord Rosse's Telescopes," *Eclectic Magazine of Foreign
     Literature*, Oct. 1846, 9, 2.

307 *"ocular evidence"*: Poe, *Eureka*, 67. Poe recalls a letter "purporting to be from Dr. Nichol" that two
     years earlier "went the rounds of our newspapers" suggesting that Nichol was considering aban-
     doning the hypothesis, but it remained prominent in his 1848 New York lectures, *Views of Astronomy*,
     39–42.

307 *"unencumbered* Consistency": Poe, *Eureka*, 15. Poe admired Whewell's *Bridgewater Treatise* above the
     rest of the series; *Eureka*'s notion of "consistency" recalls Whewell's confirmation criteria: prediction,
     consilience, and coherence, in *Philosophy of the Inductive Sciences, Founded upon Their History*, 2 vols. (London:
     John W. Parker, 1840); Laura Snyder, "Renovating the *Novum Organum*: Bacon, Whewell, and Induc-
     tion," *Studies in History and Philosophy of Science Part A* 30, no. 4 (1999): 531–57; Henry Cowles, "The Age
     of Methods: William Whewell, Charles Peirce, and Scientific Kinds," *Isis* 107, no. 4 (2016): 722–37.

307 *"In Divine constructions"*: Poe, *Eureka*, 89.

308 *"Attraction and Repulsion"*: Ibid., 54.

308 *gravity, is simply a consequence*: In an appendix to the *Ninth Bridgewater Treatise*, Babbage recalled that
     philosophers since Newton "have occasionally speculated on the existence of some more compre-
     hensive law, of which gravity itself is a consequence" (179).

308 *"In fictitious literature"*: Poe, *Eureka*, 89; Barbara Cantalupo, "Preludes to *Eureka*: Poe's 'Absolute
     Reciprocity of Adaptation' in 'Shadow' and 'The Power of Words,'" *PS/Dark Romanticism* 29, no. 1
     (1996): 17–21.

309 "ultra-tropical vegetation": Poe, *Eureka*, 64–65. New planets forming inside the orbit of Mercury

might bring "a new modification of the terrestrial surface," thus producing "a race both materially and spiritually superior to Man" (ibid., 65).

309 *If the universe is infinite*: Edward Robert Harrison, *Darkness at Night: A Riddle of the Universe* (Cambridge, Mass.: Harvard University Press, 1987); P. S. Wesson, "Olbers' Paradox in Astronomy: History, Nature, and Resolution," *Science Progress* (1989): 133–46; Stanley Jaki, *The Paradox of Olbers' Paradox* (New York: Herder and Herder, 1972).

309 *"Were the succession of stars"*: Poe, *Eureka*, 75. The Levines' edition of *Eureka* discusses Nichol's use of Olbers, 150n181; Stamos compellingly argues that "Poe's solution to Olbers's paradox is the key to understanding the science of Poe's cosmology in *Eureka*" (*Edgar Allan Poe, "Eureka," and Scientific Imagination*, 217).

309 *"the existence of Evil becomes intelligible"*: Poe, *Eureka*, 104.

310 *recently translated Hindu and Buddhist*: For possible Hindu and Buddhist influences, see Carol Maddison, "Poe's Eureka," *Texas Studies in Literature and Language* 2, no. 3 (1960): 366–67; Alan Hodder, "Asian Influences," in Myerson, Petrulionis, and Walls, *Oxford Handbook of Transcendentalism*, 27–37; David Schmit, "The Mesmerists Inquire About 'Oriental Mind Powers': West Meets East in the Search for the Universal Trance," *Journal of the History of the Behavioral Sciences* 46, no. 1 (2010): 1–26.

310 *"Star after star"*: Erasmus Darwin, *The Botanic Garden: A Poem in Two Parts Containing the Economy of Vegetation* (London: Jones, 1825), 62, cited and discussed in Scholnick, *Poe's "Eureka,"* 61–71.

310 *John Mason Good*: Sean Moreland, "Some Power Unseen," unpublished MS.

311 *"It is not that the Deity"*: Poe, "Mystery of Marie Rogêt," 198.

311 *It interrupted "progress"*: Despite the message of consolation, Poe's cyclical vision was a universe apart from the view of upward progress offered by Condorcet, Bentham, Priestley, Comte, and many nineteenth-century philosophers of history.

311 *intensely theological text*. *Eureka's* theological message was central for contemporary readers, though de-emphasized today; see exceptionally Marilynne Robinson, "On Edgar Allan Poe," *New York Review of Books*, Feb. 5, 2015, 4–6; and Harry Lee Poe, *Evermore: Edgar Allan Poe and the Mystery of the Universe* (Waco, Tex.: Baylor University Press, 2012). *Eureka's* background included not just romanticism and radical science but the debates about atheism, deism, and voluntarism that focused around *Vestiges*; its pantheism resonated with discussions about Goethe, Spinoza, Schelling, and Hegel, and their commentators Coleridge, Carlyle, and Emerson. See Tracey Matysik, "Spinozist Monism: Perspectives from Within and Without the Monist Movement," in *Monism*, ed. T. H. Weir (New York: Palgrave Macmillan, 2012), 107–34; Pierre Macherey, "Un chapitre de l'histoire du panthéisme: La religion Saint-Simonienne et la réhabilitation de la matière," in *Philosophie de la nature*, ed. Olivier Bloch (Paris: Publication de la Sorbonne, 2000), 357–66; on the tensions and public transformations of public religion, see John Lardas Modern, *Secularism in Antebellum America* (Chicago: University of Chicago Press, 2011).

311 *"within the Life divine"*: Marie Louise Shew's friend the theological student John Henry Hopkins Jr. recalled, "I did all I could to persuade him to omit the bold declaration of Pantheism at the close, which was not necessary to the completeness or beauty of the lecture. But I soon found that that was the dearest part of the whole to him; and we got into quite a discussion on the subject of Pantheism. For some time his tone and manner were very quiet, though slowly changing as we went on; until at last, a look of scornful pride worthy of Milton's Satan flashed over his pale, delicate face & broad brow, and a strange thrill nerved and dilated for an instant his slight figure, as he exclaimed, 'My whole nature utterly revolts at the idea that there is any Being in the Universe superior to myself!' I knew then that there was no use in further argument." Hopkins to Shew, Feb. 9, 1873, in *TPL*, 731.

311 *separation, judgment, and opposition*: On psychological and historical dimensions of Poe's schemas of separation and union, see J. James Livingston, "Subjectivity and Slavery in Poe's Autobiography of Ambitious Love," *Psychohistory Review* 21, no. 2 (1993): 175–96.

312 *a parallel with the "United States"*: W. C. Harris, "Edgar Allan Poe's Eureka and the Poetics of Constitution," *American Literary History* 12, no. 1/2 (2000): 1–40; Greeson, "Poe's 1848"; Susan Manning, "'The Plots of God Are Perfect': Poe's Eureka and American Creative Nihilism," *Journal of American Studies* 23, no. 2 (1989): 235–51; Matthew Taylor, *Universes Without Us: Posthuman Cosmologies in American Literature* (Minneapolis: University of Minnesota Press, 2013).

312 *"revolutionize the world"*: Poe to George W. Eveleth, Feb. 29, 1848, in Ostrom, 2:650.

312 *"A poet who has achieved"*: Poe, "Fifty Suggestions (Part I)," *Graham's*, May 1849, 317–19.

312 *"startling work, properly located"*: Freeman Hunt, *Hunt's Merchants' Magazine*, Aug. 1848, in *TPL*, 747.

313 *"new and startling"*: *Brooklyn Daily Eagle*, July 31, 1848.

313 *"tendency to elucidate"*: Nathaniel Willis, "Mr. Poe's Eureka," *New York Home Journal*, Aug. 12, 1848, 3, in *TPL*, 746–51.

313 *"throws aside all previous systems"*: *Daily Tribune*, Aug. 3, 1848, in *TPL*, 751.

313 *"He calls his work a poem"*: *New Church Repository*, Aug. 1848, in *TPL*, 748.

313 *"the system of Pantheism"*: John H. Hopkins, review of *Eureka*, by Poe, *Literary World*, July 29, 1848, in *TPL*, 745–46.

313 *"A new porter"*: T. D. English, *John-Donkey*, Aug. 12, 1848, in *TPL*, 752.

313 *leaping analogies were banned*: Welsh, "Value of Analogical Evidence."

314 *"Mr. Poe is already"*: *Boston Journal*, Feb. 12, 1848, in *TPL*, 724.

314 *But many of the learned astronomers*: Ormsby Mitchel may seem an exception to this professionalizing turn—an institutionally established, professional astronomer who also gave popular lectures (which Walt Whitman disdained in his poem "When I Heard the Learn'd Astronomer")—yet Mitchel's 1847 lectures were offered precisely in order to earn funds to rebuild his institution, the Cincinnati Observatory, destroyed by a fire. On Whitman and science, see Robert Scholnick, "'The Password Primeval': Whitman's Use of Science in 'Song of Myself,'" *Studies in the American Renaissance* (1986): 385–425.

314 *Membership was open*: Sally Kohlstedt, "Creating a Forum for Science: AAAS in the Nineteenth Century," in *The Establishment of Science in America: 150 Years of the American Association for the Advancement of Science*, ed. Sally Kohlstedt, Michael Mark Sokal, and Bruce V. Lewenstein (New Brunswick, N.J.: Rutgers University Press, 1999), 7–49.

314 *for Matthew Maury*: Maury's oceanography had already earned recognition from the BAAS, Arago, and Humboldt, and a favorable review from Poe.

314 *Maury competed against Bache's Coast Survey*: On clashes between Maury and Bache, see Steven J. Dick, *Sky and Ocean Joined: The US Naval Observatory, 1830–2000* (Cambridge, U.K.: Cambridge University Press, 2003); Walls, *Passage to Cosmos*, 142–45; and Grady, *Matthew Fontaine Maury*, 96–98, 112–17.

315 *shadowy brotherhood*: Dana Nelson, *National Manhood: Capitalist Citizenship and the Imagined Fraternity of White Men* (Durham, N.C.: Duke University Press, 1998), which examines nineteenth-century scientific sociability in part through Christopher Newfield's compelling concept of "corporate individualism," in *The Emerson Effect: Individualism and Submission in America* (Chicago: University of Chicago Press, 1996). The term "Lazzaroni" usefully identifies the clique Bache had formed by the early 1850s, though the coherence and endurance of the group can be overestimated; see Mark Beach, "Was There a Scientific Lazzaroni?," in *Nineteenth-Century American Science: A Reappraisal*, ed. Nathan Reingold and George Daniels (Evanston, Ill.: Northwestern University Press, 1972), 115–32.

315 *a standard, methodical regularity*: Kohlstedt, *Formation of the American Scientific Community*, 96.

315 *"expose the wiles of the pretender"*: Draft copy of AAAS address, n.d., for outgoing Presidential Address, 1851, quoted in Kohlstedt, *Formation of the American Scientific Community*, 109; see Jansen, *Alexander Dallas Bache*, 238–47.

316 *Despite early attempts*: Nott and Peter Browne lectured in Charleston, drawing on Morton's skulls, Gliddon's mummies, and hair samples; Agassiz provided pious support. When publishing the meeting's *Proceedings*, Henry decided to use the Smithsonian's presses to publish work on algae rather than Browne's on hair; Bache said of Browne's lecture and Agassiz's response, "The remarks at the close of the meeting were altogether of too popular a cast to require their printing" (Bache to Lewis Gibbes, May 30, 1850, Lewis Gibbes Papers, Library of Congress, quoted in Stanton, *Leopard's Spots*, 154); Kohlstedt (*Formation of the American Scientific Community*, 112–13) describes Bache steering Agassiz away from race science to avoid public excitement and controversy; Lurie, *Agassiz*, 260, 257–71.

316 *"Huge smoking cities"*: Poe, "The Colloquy of Monos and Una," *Graham's*, Aug. 1841, 52–54.

317 *"It is no use to reason"*: Poe to Clemm, July 7, 1849, in Ostrom, 2:820.

17. Falling Star

318 *"South and West"*: Poe to Nathaniel Willis, Jan. 22, 1848, Ostrom, 2:359.

318 *"like distant thunder"*: Sarah Helen Whitman, *New York Home Journal*, March 18, 1848, in John E. Reilly, "Mrs. Whitman's Poems to Poe," in "Poe in Imaginative Literature" (PhD diss., University of Virginia, 1965), 187, eapoe.org.

318 *"a deeply rooted prejudice"*: Lynch to Whitman, March 10, 1848, in TPL, 728.

318 *"They fill my soul with Beauty"*: Poe, "To Helen," LOA, 95–7.

318 *"Your beautiful invocation has reached"*: Osgood to Whitman, March 26, 1848, quoted in Caroline Ticknor, *Poe's Helen* (New York: C. Scribner's Sons, 1916), 48, in TPL, 729–30.

319 *"A man calling himself Poe"*: John R. Thompson quoted by Thomas Dimmock, "Notes on Poe," *Century Magazine*, June 1895, 316, in TPL, 749.

319 *drunkenly challenged an editor*: Poe mentions having challenged John Moncure Daniel the previous year in a letter to Maria Clemm, ca. Aug. 28–29, 1849, Ostrom 2:830. Daniel himself later wrote: "Thousands have seen him [Poe] drunk in the streets of this city . . . in a state approaching mania. Whenever he tasted alcohol he seldom stopt drinking it so long as he was able . . . His taste for drink was a simple disease—no source of pleasure nor of excitement." *Richmond Semi-Weekly Examiner*, Oct. 19, 1849, in TPL, 750.

319 *"unmistakably a gentleman"*: Thompson, in Dimmock, "Notes," 316, *Richmond Semi-Weekly Examiner*, Oct. 19, 1849, in TPL, 750.

319 *"the existence of spiritual influences"*: Poe to Whitman, Oct. 1, 1848, in Ostrom, 2:696.

319 *"My brain reeled"*: Poe to Whitman, Oct. 1, 1848, Ostrom, 2:382–91.

319 *"Had I youth and health and beauty"*: Poe, quoting Whitman's letter back to her, Oct. 1, 1848, in Ostrom, 2:696.

320 *"And if not, Helen"*: Poe to Whitman, Oct. 1, 1848, in Ostrom, 2:700.

320 *who would call herself "Annie"*: A well-informed blogger, "Undine," at worldofpoe.blogspot.com, presents a healthy skepticism toward Nancy Richmond's letters—and toward other accounts, including Whitman's, that have contributed to some of the more melodramatic aspects of the Poe myth. A definitive retelling of these episodes would be most welcome, as Undine's overall image of Poe—emphasizing philosophical depth, idealism, and principled complexity—is one with which the present book is in agreement.

320 *"a state of wild & delirious"*: Whitman to J. H. Ingram, Oct. 25, 1875, in TPL, 766.

320 *"My mother did say more than once"*: Whitman to Ingram, March 20, 1874, in TPL, 767.

321 *"burden of responsibility"*: Whitman to Mary Hewitt, Sept. 25 or 27, 1850; AHQ, 584.

321 *"No person could be long"*: Whitman to Ingram, March 20, 1874, in TPL, 767.

321 *"seemed a portent or an omen"*: Whitman to Ingram, March 16, 1874, in TPL, 768.

321 *"Hast thou not stooped"*: Whitman to Hewitt, Sept. 27–28, 1850; Whitman, "To Arcturus," TPL, 768. This poem was collected in her *Hours of Life, and Other Poems* (Providence: G. H. Whitney, 1853).

321 *"You say that all depends"*: Poe to Whitman, Nov. 24, 1848, in Ostrom, 2:730.

322 *"Keep up heart"*: Poe to Whitman, Dec. 16, 1848, in Ostrom, 2:741.

322 *"That marriage may never"*: Poe as quoted by Mary Hewitt, letter to Sarah Whitman, Oct. 2, 1850, in TPL, 778.

322 *"a communication was handed"*: Whitman to Ingram, July 21, 1874, in TPL, 767; "Poe and Whitman," https://providenceathenaeum.org/about/history/poe-whitman/.

322 *"While he was endeavouring"*: Mrs. Whitman recalls her last day with Poe, Dec. 23, 1848, in a letter to Hewitt, Sept. 27–28, 1850, in TPL, 780.

322 *"For Annie"*: LOA, 98–100.

323 *under the heading* ODD POEM: *New York Home Journal*, April 28, 1849. In the spring of 1848, Poe also wrote "The Bells," with the aid of Marie Louise Shew. The poem's merger of an unusual, repetitive rhythm with a music that underlined and at times stretched its words' conventional meaning replayed the strengths of "The Raven"—the sonic "tintinnabulation" of the words pointing to a reality beyond them; on the deliberate, musical oddness of Poe's poetics, see Jerome McGann, *The Poetry of Edgar Allan Poe: Alien Angel* (Cambridge, Mass.: Harvard University Press, 2014), 169–82.

323 *"They are a heartless"*: Poe to Annie Richmond, Jan. 21 (?), 1849, in Ostrom 2:419.

323 *"'Over the Mountains'"*: Poe, "Eldorado," LOA, 101.

323 *"pays well as times go"*: Poe to Willis, April 20, 1849, in Ostrom, 2:790–91.

324 *"mystical and cabalistical"*: Poe, "X-ing a Paragrab," *Flag of Our Union*, May 12, 1849, 2.

324 *reworking the letter Poe quoted*: Poe, "Mellonta Tauta," *Godey's Lady's Book*, Feb. 1849, 133–38.

324 *"I think that human exertion"*: Poe to Lowell, July 2, 1844, in Ostrom, 1:449.

324 *"very minute and elaborate paper"*: Poe, "Von Kempelen and His Discovery," *Flag of Our Union*, April 14, 1849.

324 *"a check to the gold-fever"*: Poe to Duyckinck, March 8, 1849, in Ostrom, 2:786.

324 *The most horrifying of Poe's late tales*: Poe, "Hop-Frog; or, The Eight Chained Ourang-Outangs," *Flag of Our Union*, March 17, 1849.

325 *"I never knew what it was"*. Poe to F. W. Thomas, May 4, 1845, in Ostrom, 1:504.

325 *Poe identified with a slave in revolt*: In Dayan's reading, "What Mabbott regards as 'a terrible exposition of the darkness of a human soul' is Poe's envisioned revenge for the national sin of slavery"; "Amorous Bondage," 258. On Poe's probing of the national conscience in the light of slavery and imperial warfare, see Kadir Djelal, "Edgar Allan Poe: America's Conscience and Epistemic Anxiety," in *Poe Alive in the Century of Anxiety*, ed. Luisa Juárez (Madrid: Instituto Franklin, 2010), 17–30.

325 *After years of deprivation*: For related commentary on "Hop-Frog," see Levin, *Power of Blackness*, 122; Louis D. Rubin Jr., *The Edge of the Swamp: A Study in the Literature and Society of the Old South* (Baton Rouge: Louisiana State University Press, 1989), 183–89. Paul Gilmore associates "Hop-Frog" with Barnum's "What Is It?" exhibition, featuring the acrobat Harvey Leach as the taxonomically unsettling "wild-man of the prairies," in *The Genuine Article: Race, Mass Culture, and American Literary Manhood* (Durham, N.C.: Duke University Press, 2001), 106.

325 *Virginia's death (and perhaps other losses)*: Poe, "Annabel Lee," in LOA, 102–3. For Silverman, "in Poe's promise of never-ending remembrance of her girlish beauty, and his pleasure in joining her in early death, Annabel Lee represents all of the women he loved and lost" (*Mournful*, 401–2).

326 *"If Eddy gets to Richmond"*: Clemm to a friend, July 9, 1849; AHQ, 619.

326 *"cholera bulletins"*: Recollection of George Lippard, quoted in T. C. Duncan Eaves, "Poe's Last Visit to Philadelphia," *American Literature* 26, no. 1 (1954): 46–47, in *TPL*, 817.

326 *"brightly radiant"*: John Sartain in Silverman, *Mournful*, 416–7.

326 *"I have been so ill"*: Poe to Clemm, July 7, 1849, in Ostrom, 2:820–21.

327 *Patterson was well pleased*: Poe to Patterson, July 19, 1849, in Ostrom, 2:828; Aug. 7, 1849, in Ostrom, 2:829–30; Patterson to Poe, Aug. 21, 1849, www.eapoe.org/misc/letters/t4908210.htm.

327 *to accommodate her three children*: Silverman, *Mournful*, 97.

327 *by joining the Richmond chapter*: AHQ, 624. *Temperance Banner* announces Poe's initiation, Aug. 31, 1849, in *TPL*, 830.

327 *"refined, high-bred, and chivalrous"*: Susan A. T. Weiss, "The Last Days of Edgar A. Poe," *Scribner's*, March 1878, 709.

328 *"I went down and was amazed"*: Royster to Clemm, Sept. 22, 1849; AHQ, 634.

328 *"quite a full, and very fashionable audience"*: Ibid.

328 *"We were never more delighted"*: *Richmond Whig*, Aug. 21, 1849, in *TPL*, 826.

328 *"The Poetic Principle"*: Poe, *Sartain's Union Magazine*, Oct. 1850, 231–39.

328 *"the principle of Unity"*: Poe to Charles Fenno Hoffman, Sept. 20, 1848, Ostrom 2: 380, a letter explaining and defending *Eureka*.

329 *"the desire of the moth for the star"*: Percy Shelley, "To ——" ("One word is too often profaned"), in *Posthumous Poems of Percy Bysshe Shelley* (London: John and Henry L. Hunt, 1824), 200. Shelley's poem invokes a devotional, otherworldly form of love and was addressed to Jane Williams; her husband, Edward Williams, drowned with Shelley in 1822 when their boat sank off the Italian coast.

329 *"his eyes dark and restless"*: Edward V. Valentine, sharing reminiscences of his brother William W. Valentine in letter to J. H. Ingram, Sept. 28, 1874, in *TPL*, 841.

330 *"unusually silent and preoccupied"*: Weiss, "Last Days of Poe," 712.

330 *"I never was received"*: Poe to Clemm, ca. Aug. 28–29, 1849, in Ostrom, 2:830–31.

331 *"We were standing on the portico"*: Weiss, "Last Days of Poe," 713–14; AHQ, 636.

331 *"He was very sad"*: Royster to Clemm, Oct. 11, 1849, in *TPL*, 843.

331 *He might have traveled to Philadelphia*: AHQ, 637; Matthew Pearl, in *The Poe Shadow: A Novel* (New York: Random House, 2006), argues against this Philadelphia trip, citing a record for a letter sent to Poe from Maria Clemm, received poste restante, which Poe never claimed.

331 *"the candidates saw that every voter"*: William Hand Browne to J. H. Ingram, Jan. 13, 1909, in *TPL*, 844.

331 *"There is a gentleman"*: Joseph W. Walker to Snodgrass, Oct. 3, 1849, in *TPL*, 844.

331 *"instantly recognized the face"*: Joseph E. Snodgrass, "Death and Burial of Edgar A. Poe," *Life Illustrated* (New York), May 17, 1856, 24, in *TPL*, 844–45.

332 *"but his answers were incoherent"*: Moran to Clemm, Nov. 15, 1849, in *TPL*, 846.

332 *crying out "Reynolds!"*: The attending physician, John Moran, described Poe's cries in a letter to Maria Clemm; issues surrounding this report, and other facts pertaining to Poe's death, are critically scrutinized by William T. Bandy in "Dr. Moran and the Poe-Reynolds Myth," in Benjamin Franklin Fisher, ed., *Myths and Reality: The Mysterious Mr. Poe* (Baltimore: The Edgar Allan Poe Society, 1987), 26–36.

332 *"utterly without ostentation"*: Reverend Clemm to E. R. Reynolds, Feb. 20, 1889, in *TPL*, 848.

332 *"neat mahogany coffin"*: Henry Herring, in Woodberry, *Life*, 2:448, and *TPL*, 848.

332 *"the burial ceremony"*: George P. Clark, "Two Unnoticed Recollections of Poe's Funeral," *Poe Newsletter* 3, no. 1 (June 1970): 1–2, in *TPL*, 848.

332 *"Wrote, hastily, two or three"*: Griswold, cited in James A. Harrison, *Complete Works of E. A. Poe* (New York: Thomas Crowell & Co., 1902) 1:359.

333 *"Edgar Allan Poe is dead"*: R. W. Griswold, writing as "Ludwig," *Daily Tribune*, Oct. 9, 1849, reprinted in *Weekly Tribune*, Oct. 20, 1849; *Richmond Enquirer*, Oct. 13, 1849; *Saturday Evening Post*, Oct. 20, 1849, in *TPL*, 849.

333 *"was a most eccentric genius"*: *New York Herald*, Oct. 9, 1849, in *TPL*, 850.

333 *"will be read with profound regret"*: *Richmond Whig*, Oct. 9, 1849, in *TPL*, 851.

333 *"He had talents"*: Cornelia Wells Walter, *Evening Transcript*, Oct. 9, 1849, in *TPL*, 851.

334 *"Poor Poe! He was an original"*: Kennedy, diary entry, Oct. 10, 1849, in *TPL*, 852.

334 *"the slanderous and malicious miscreant"*: Wilmer, *Merlin*, Baltimore, 1827, 25; AHQ 654–55.

334 *"melancholy death"*: Longfellow, quoted in *SLM* 15, no. 11 (Nov. 1849); AHQ 655.

334 *"a quiet, patient, industrious"*: N. P. Willis, *Home Journal*, Oct. 13, 1849, in E. C. Stedman and G. E. Woodberry, eds. *The Works of Edgar Allan Poe* (1895)—Vol. X: Poems (Chicago: Stone and Kimball, 1895), 10:242–3.

334 *"I assure you, My dear Madam"*: Neilson Poe to Clemm, Oct. 11, 1849, Original Autograph MS, Enoch Pratt Library; AHQ 643.

334 *"Annie my Eddy is dead"*: Clemm to Richmond, Oct. 9, 1849, in *TPL*, 850.

334 *"Oh my mother, my darling"*: Richmond to Clemm, Oct. 10, 1849, in *TPL*, 854.

335 *"employed in setting up the copy"*: Griswold, Oct. 31, 1849, Original Autograph MS, Harvard University Library; AHQ 660.

335 *"Memoir of the Author"*: Griswold, in *Works of the Late Edgar Allan Poe*, 3:vii–xxxix.

335 *"scarcely any virtue"*: Griswold, "Memoir of the Author"; AHQ 673.

335 *contacted Poe directly*: See, for example, the "Spiritual trance-speaker," Lizzie Doten, whose *Poems of the Inner Life* (Boston: William White, 1863) included six new poems first recited "under the direct influence" of Poe, including "Resurrexi," "The Streets of Baltimore," and "Farewell to Earth."

335 *Even the cause of his death*: Death by brain tumor was argued by Leon Neyfakh, "Poe's Mysterious Death: The Plot Thickens!," *New York Observer*, Oct. 16, 2007, in agreement with Pearl in *Poe Shadow*; the Baltimore cardiologist Michael Benitez's observations were reported in "Poe's Death Is Rewritten as Case of Rabies, Not Telltale Alcohol," *New York Times*, Sept. 15, 1996; John Evangelist Walsh (author of the Osgood-Poe potboiler *Plumes in the Dust*) speculates he was murdered, in *Midnight Dreary: The Mysterious Death of Edgar Allan Poe* (New York: Macmillan, 2000); see Natasha Geiling, "The (Still) Mysterious Death of Edgar Allan Poe," *Smithsonian Magazine*, Oct. 7, 2014, www .smithsonianmag.com/history/still-mysterious-death-edgar-allan-poe-180952936/; Scott Peeples, *The Afterlife of Edgar Allan Poe* (Woodbridge: Boydell & Brewer, 2004), 155–64.

Conclusion: From a Lighthouse

337 *"potent chemistry"*: *North American Review* 83 (Oct. 1856): 427–55; AHQ 685; for incisive reflections on recurrent moral critiques (and misreadings) of Poe, see Paul Lewis, "From Emerson to Edmundson: The Case Against Poe," *EAPR*, 11, no. 2 (2010): 73–84.

337 *struggled to incorporate him*: F. O. Matthiessen, *American Renaissance: Art and Expression in the Age of Emerson and Whitman* (Oxford: Oxford University Press, 1941); F. O. Matthiessen, "Poe," *Sewanee Review* 54, no. 2 (1946): 175–205. Matthiessen gave Margaret Fuller even less attention, and none to Frederick Douglass. On successive canons of American literature, see Kermit Vanderbilt, *American Literature and the Academy: The Roots, Growth, and Maturity of a Profession* (Philadelphia: University of Pennsylvania Press, 1986), 501; Gerald Graff, *Beyond the Culture Wars: How Teaching the Conflicts Can Revitalize American Education* (New York: W. W. Norton, 1993); on Poe's eventual canonization, see Jerome McGann, "Literary History and Editorial Method: Poe and Antebellum America," *New Literary History* 40, no. 4 (2009): 825–42.

338 *"almost more a scientist"*: D. H. Lawrence, *Studies in Classic American Literature* (New York: Thomas Seltzer, 1923), 93.

338 *Dostoevsky, Conrad, and Kafka*: Lois Davis Vines, *Poe Abroad: Influence, Reputation, Affinities* (Iowa City: University of Iowa Press, 2002), Kafka on 204; Martin Ray, *Joseph Conrad: Memories and Impressions* (Boston: Brill Rodopi, 2007), 8.

339 *"on bended knees"*: Roberto Bolaño. "Advice on the Art of Writing Short Stories," in *Between Parentheses: Essays, Articles, and Speeches, 1998–2003* (New York: New Directions, 2011), 350–70; Emron Esplin, "Poe and His Global Advocates," in Kennedy and Peeples, *Oxford Handbook*, 597–617.

339 *same time as Charles Darwin*: Alfred Russel Wallace, *Edgar Allan Poe: A Series of Seventeen Letters Concerning Poe's Scientific Erudition in Eureka and His Authorship of Leonainie* (New York: privately printed, 1904).

339 *Peirce tested his own theories*: Jonathan Elmer, "Peirce, Poe, and Protoplasm," *PS* 52, no. 1 (2019): 29–49; Eco and Sebeok, *Sign of Three*.

339 *Eddington was a Quaker*: Matthew Stanley, *Practical Mystic: Religion, Science, and A. S. Eddington* (Chicago: University of Chicago Press, 2007).

340 *"the work of a man"*: Eddington to Quinn, Sept. 29, 1940; AHQ, 555–56.

340 *"a very beautiful achievement"*: Einstein to Gimbel, 1933, Gimbel Collection, Free Library of Philadelphia. Einstein wrote, *"Eine sehr schöne Leistung eines ungewöhnlich selbständigen Geistes."*

340 *"very witty and remarkable"*: Einstein to Quinn, Aug. 6, 1940, Quinn Papers, University of Pennsylvania; see René van Slooten, "Edgar Allan Poe—Cosmologist?," *Scientific American Blog*, Feb. 1, 2017, blogs.scientificamerican.com/guest-blog/edgar-allan-poe-cosmologist/.

340 *Einstein was years into his battle*: To Georges Lemaître, Einstein wrote, "Your calculations are correct, but your physics is atrocious," quoted in Andre Deprit, "Monsignor Georges Lemaître," in *The Big Bang and Georges Lemaître*, ed. A. Barger (Dordrecht: Reidel, 1984), 370.

341 *"to imagine an evolving Universe"*: Paolo Molaro and Alberto Cappi, "Edgar Allan Poe: The First Man to Conceive a Newtonian Evolving Universe," arXiv preprint, arXiv:1506.05218, 2015; Alberto Cappi, "Edgar Allan Poe's Physical Cosmology," *Quarterly Journal of the Royal Astronomical Society* 35 (1994): 177–92.

341 *solving "Olbers's paradox"*: Harrison, *Darkness at Night*.

341 *Friedmann published in 1924*: "Über die Möglichkeit einer Welt mit konstanter negativer Krümmung des Raumes," *Zeitschrift für Physik* 21, no. 1 (1924): 326–32. Friedmann died in 1925 at age thirty-seven, a few months after a balloon trip to a record-setting altitude. Tom Siegfried, *Strange Matters: Undiscovered Ideas at the Frontiers of Space and Time* (Washington, D.C.: Joseph Henry Press, 2002), 118–22; on Friedmann and Poe, see Cappi, "Poe's Physical Cosmology."

341 *"hypothesis of the primeval atom"*: Georges Lemaître, "The Beginning of the World from the Point of View of Quantum Theory," *Nature* 127, no. 3210 (1931): 706.

342 *"The essential characteristic of Einstein's"*: Paul Valéry in "Au sujet d'*Eureka*," in Cantalupo, "*Eureka*: Poe's 'Novel Universe,'" 329. See Jimena Canales, *The Physicist and the Philosopher: Einstein, Bergson, and the Debate that Changed Our Understanding of Time* (Princeton: Princeton University Press, 2016).

342 *theories of a multiverse*: See Cappi, "Poe's Physical Cosmology," 188n50; Stamos, *Edgar Allan Poe, "Eureka," and Scientific Imagination*, 219, notes the similarity to Penrose's cyclical universe (see Roger Penrose, *Cycles of Time: An Extraordinary New View of the Universe* [New York: Random House, 2010]) and connections to string theory (see van Slooten, "Edgar Poe—Cosmologist?"); Stamos, a philosopher of science, presents Poe as an inspiring theorist of aesthetics and imagination in science. See also Mary-Jane Rubinstein, *Worlds Without End: The Many Lives of the Multiverse* (New York: Columbia University Press, 2014), and John Gribbin, *In Search of the Multiverse: Parallel Worlds, Hidden Dimensions, and the Quest for the Frontiers of Reality* (New York: Wiley, 2010); David Kaiser, *Quantum Legacies: Dispatches from an Uncertain World* (Chicago: University of Chicago Press, 2020).

342 *cryptology, information theory*: James Gleick, *The Information: A History, a Theory, a Flood* (New York: Vintage, 2011), 120, 171, 217, 375, on Poe and codes, Babbage, and information metaphysics.

343 *"It is somewhat remarkable"*: Douglass, "Claims of the Negro, Ethnologically Considered," 2:507. On the deep entanglement of post-war molecular biology and eugenics, see Daniel Kevles, *In the Name of Eugenics: Genetics and the Uses of Human Heredity* (Cambridge, Mass.: Harvard University Press, 1995).

344 *the White House in the election of 2016*: Beyond the similarity between Barnum's *Art of Money Getting* and Donald Trump's *Art of the Deal*, in a 2016 interview Trump embraced the comparison, saying, "We

need P. T. Barnum, a little bit, because we have to build up the image of our country." *Meet the Press*, NBC, Dec. 30, 2016, www.msnbc.com/transcripts/mtp-daily/2016-12-30.

345 *an underlying "philosophy of composition"*: In this interpretation, Poe's philosophy holds just as firmly to truth as it does to the inevitability of contradiction and the permanence of doubt; it acknowledges the effective realities constituted by scientists, mystics, and artists by presenting a non-reductive notion of unitary reality (a multiplicity of modes of being, embraced within an ineffable "Spirit Divine"— which equates to "nothingness," as G. R. Thompson emphasized in "Unity, Death, and Nothingness"). Poe unfolds the participatory "panentheism" of Spinoza and the perspectivism of Leibniz over time—as did Hegel and Schelling—but without linear progress. The process philosophies of Bergson and Whitehead offer useful comparison, because they acknowledge the power of the empirical and mathematical sciences while embedding them within a vital, though elusive, totality, as do Peirce's evolutionary cosmology and James's "pluriverse." In other words, the multiplicity Daniel Hoffman discerned in *Poe, Poe, Poe, Poe, Poe, Poe, Poe* is not limited to Poe's character and styles; it also applies to his pluralist view of reality. Further, as Poe would recognize, his view of nature and knowledge, like any other truth, is bound up with the specific occasions, organs, and media of its realization and expression. For this reason, his consistently inconsistent "philosophy" needs to be approached through the story of his life and times.

346 *To hold back the destruction*: Variations on this recurrent existential/epistemological/psychological pattern in Poe are explored in, for example, J. Gerald Kennedy, *Poe, Death, and the Life of Writing* (New Haven, Conn.: Yale University Press, 1987); Justin, *Poe dans le champ du vertige*; Robert Shulman, "Poe and the Powers of the Mind," *ELH* 37, no. 2 (1970): 245–62.

346 *In a little-known manuscript*: Poe, "The Light-House," a name given by Woodberry in 1909, was published for the first time in 1942 and first included in a collection in 1951; three of its four MS sheets (in brown writing on light blue paper) are held at the Houghton Library at Harvard; the first is in the Henry W. and Albert A. Berg Collection of English and American Literature, the New York Public Library. Joyce Carol Oates reimagined the story in "Poe Posthumous: Or, The Light-House" in *Wild Nights!: New Stories* (New York: Harper Collins, 2009); its traces may lurk in Junji Ito's spiral-obsessed Manga, *Uzumaki* (San Francisco: Viz, 2013), and Jeff VanderMeer's *Annihilation* (New York: Farrar, Straus and Giroux, 2014). Late in the writing of this book, Robert Eggers's *The Lighthouse* was released, a visually stunning film based on Poe's fragment.

Acknowledgments

My enthusiastic thanks go to the staff of the Van Pelt Library and Kislak Center at the University of Pennsylvania; the New York Public Library's Rare Books Division, Berg Collection, and Pforzheimer Collection; the Harry Ransom Center of the University of Texas, Austin; the Library Company of Philadelphia; the Historical Society of Philadelphia; the Archives of the Academy of Natural Sciences, Philadelphia; the Free Library of Philadelphia and its Rare Books Department; the New York Historical Society Library; the Huntington Library; the Peabody Library of Johns Hopkins University; the Special Collections of the United States Military Academy at West Point; the Smithsonian Institution Archives; the Poe Museum, Richmond; and the Poe Houses and Museums in Philadelphia, Baltimore, and New York.

Welcome support for research was provided by the National Endowment for the Humanities (NEH grant FA-252132-17); the Cullman Center for Scholars and Writers at the New York Public Library; the Institute for Advanced Study at Princeton; the Huntington Library; and the Max Planck Institute for History of Science, Berlin.

Simon Schaffer has given indispensable advice on this book's nebulously shifting forms since its inception; the eminent Richard Kopley generously provided priceless commentary and suggestions; Clare Carlisle talked things over and cleared away inessentials; Sean Moreland suggested successors to "The Light-House"; Jonathan Elmer sharpened my view of Poe's media; Michelle Smiley lit up early photography; James Delbourgo helped sort the eighteenth century; Sally Gregory Kohlstedt shared her knowledge of American scientists and institutions; Richard John provided an antebellum crash course; talking with and reading Jim Secord has been endlessly beneficial;

Britt Rusert opened up African American science; Axel Jansen illuminated Bache and state-building; Josh Nall rightly insisted on lyceums; David Kaiser rectified certain stray conjectures; René van Slooten shared his discoveries in Lemaître and Einstein; and J. Gerald Kennedy shared his keen eye and vast knowledge. Thanks also to Helen McKenna-Uff and Joanne Schillizzi of Philadelphia's Poe House, to Susan Glassman at the Wagner Free Science Institute, to Jean Strouse, Marie d'Origny, Ken Alder, and James Chandler, for manifold and multiform support.

I'm happy to be able to thank friends and colleagues from the Warburg Institute, the University of Pennsylvania, and various scholarly constellations. For chats and sometimes accidental encouragement, I also thank Esther Allen, Robby Aronowitz, Tamara Barnett-Herrin, Nicolas Barreyre, Etienne Benson, Carin Berkowitz, Charles Bernstein, Bob Brain, Jeremy Brooker, Graham Burnett, David Ciepley, Brian Connolly, Ruth Schwartz Cowan, Lorraine Daston, Debbie Davis, Emily Dolan, Marwa Elshakry, Emron Esplin, Erna Fiorentini, François Furstenberg, Rivka Galchen, Peter Galison, Bernard Geoghegan, Cathy Gere, Andrea Goulet, Michael Hagner, Kevin Hayes, Edward Jones-Imhotep, Matthew Jones, Dan Kevles, Ava Kofman, Rob Kohler, Elaine Lafay, Bruno Latour, Michael Leja, Rebecca Lemov, Rhodri Lewis, Bernard Lightman, Dana Medoro, Iwan Morus, Projit Mukharji, Fred Nocella, Emily Ogden, Joe O'Neill, Katherine Pandora, Christelle Rabier, Karen Russell, Robert Scholnick, Otto Sibum, Laura Stark, Anne H. Stevens, Dimitri Topitzes, Jean-Christophe Valtat, Thomas Vranken, Liliane Weissberg, Simon Werrett, Aaron Wunsch, and Jason Zuzga—and Aaron Davis for lending me *The Science Fiction of Edgar Allan Poe* in 1994.

My editor, Alex Star, has made working on this book a great pleasure and an ongoing education; I offer sincere thanks to him, Ian Van Wye, Brianna Panzica, Gretchen Achilles, Tanya Heinrich, Susan VanHecke, Ingrid Sterner, Armando Veve, and everyone at Farrar, Straus and Giroux. Alex Jacobs of Elyse Cheney Agency has profoundly shaped the project over many years; Adam Eaglin's continued support has been crucial. Claire Sabel put in decisive hours on research assistance and endnotes. I also thank my students—especially the survivors of "Poe's Experiments."

Warm thanks go to my mother, for permanently fusing Baltimore with rainbow-colored astronomy, and to all my family; to Joseph for welcome diversions; and to Clare most of all.

Index

Page numbers in *italics* refer to illustrations.

Illustration Credits

204 McAllister Collection, Library Company of Philadelphia

206 *United States Magazine*, October 1865, 289

209 New York *Sun*, April 13, 1844, Free Library of Philadelphia

215 National Portrait Gallery

215 Playbills and Programs from New York City Theaters (TCS 65), Harvard Theatre Collection, Houghton Library, Harvard University

217 New York: William Van Norden and Frank Leslie, 1850, Library of Congress GV1821 B25

228 By "A Practical Magnetizer" (Boston: J. N. Bradley, 1843), National Library of Medicine

244 From "Our Contributors: Edgar Allan Poe," by Lowell (*Graham's Lady's and Gentleman's Magazine*, February 1845)

250 Frontispiece to Caroline May, ed., *American Female Poets* (Lindsay & Blakiston, 1848)

251 Frances S. Osgood, *The Poetry of Flowers and Flowers of Poetry* (New York: Riker, 1841)

261 Massachusetts Digital Commonwealth, Watertown Free Public Library

262 Museum of Fine Arts, Boston

271 Frontispiece to John Pringle Nichol, *The Stellar Universe: Views of Its Arrangements, Motions, and Evolutions* (Edinburgh: John Johnstone, 1848); Linda Hall Library of Science, Engineering, and Technology

276 *The Illustrated London News*, July 19, 1845

281 Science Museum / Science and Society Picture Library

290 National Gallery of Art, 1971.16.2

303 Diagram by author

321 John Hay Library, Brown University

321 American Antiquarian Society

327 In *Edgar Allan Poe, the Man* by Mary E. Phillips (Cambridge, Mass.: John C. Winston, 1926)

349 Science Museum / Science and Society Picture Library